Physiology of Cold Adaptation in Birds

NATO ASI Series

Advanced Science Institutes Series

A series presenting the results of activities sponsored by the NATO Science Committee, which aims at the dissemination of advanced scientific and technological knowledge, with a view to strengthening links between scientific communities.

The series is published by an international board of publishers in conjunction with the NATO Scientific Affairs Division

A	Life Sciences	Plenum Publishing Corporation
B	Physics	New York and London
C	Mathematical and Physical Sciences	Kluwer Academic Publishers Dordrecht, Boston, and London
D	Behavioral and Social Sciences	
E	Applied Sciences	
F	Computer and Systems Sciences	Springer-Verlag
G	Ecological Sciences	Berlin, Heidelberg, New York, London,
H	Cell Biology	Paris, and Tokyo

Recent Volumes in this Series

Volume 169—Evolutionary Tinkering in Gene Expression
   edited by Marianne Grunberg-Manago, Brian F.C. Clark, and Hans G. Zachau

Volume 170—*ras* Oncogenes
   edited by Demetrios Spandidos

Volume 171—Dietary ω3 and ω6 Fatty Acids: Biological Effects and Nutritional Essentiality
   edited by Claudio Galli and Artemis P. Simopoulos

Volume 172—Recent Trends in Regeneration Research
   edited by V. Kiortsis, S. Koussoulakos, and H. Wallace

Volume 173—Physiology of Cold Adaptation in Birds
   edited by Claus Bech and Randi Eidsmo Reinertsen

Volume 174—Cell and Molecular Biology of Artemia Development
   edited by Alden H. Warner, Thomas H. MacRae, and Joseph C. Bagshaw

Volume 175—Vascular Endothelium: Receptors and Transduction Mechanisms
   edited by John D. Catravas, C. Norman Gillis, and Una S. Ryan

Series A: Life Sciences

Physiology of Cold Adaptation in Birds

Edited by
Claus Bech and
Randi Eidsmo Reinertsen
University of Trondheim
Trondheim, Norway

Plenum Press
New York and London
Published in cooperation with NATO Scientific Affairs Division

Proceedings of a NATO Advanced Research Workshop on
Physiology of Cold Adaptation in Birds,
held June 6–10, 1988,
in Loen, Norway

Library of Congress Cataloging in Publication Data

NATO Advanced Research Workshop on Physiology of Cold Adaptation in Birds
(1988: Loen, Norway)
  Physiology of cold adaptation in birds / edited by Claus Bech and Randi Eidsmo Reinertsen.
    p.   cm.—(NATO ASI series. Series A, Life sciences; vol. 173)
  "Proceedings of a NATO Advanced Research Workshop on Physiology of Cold Adaptation in Birds, held June 6–10, 1988, in Loen, Norway"—T.p. verso.
  Bibliography: p.
  Includes index.
  ISBN 0-306-43237-4
  1. Birds—Physiology—Congresses. 2. Cold adaptation—Congresses. I. Bech, Claus, 1951-   II. Reinertsen, Randi Eidsmo. III. North Atlantic Treaty Organization. Scientific Affairs Division. IV. Title. V. Series: NATO ASI series. Series A, Life sciences; v. 173.
QL698.N38   1988
598.2'542—dc20

89-16102
CIP

© 1989 Plenum Press, New York
A Division of Plenum Publishing Corporation
233 Spring Street, New York, N.Y. 10013

All rights reserved

No part of this book may be reproduced, stored in a retrieval system, or transmitted in any form or by any means, electronic, mechanical, photocopying, microfilming, recording, or otherwise, without written permission from the Publisher

Printed in the United States of America

PREFACE

The papers in this volume were presented at the NATO Advanced Workshop: Physiology of Cold Adaptation in Birds, held 10.-15. June, 1988, at Hotel Alexandra, Loen, Norway. The workshop was generously supported by the NATO Scientific Affairs Division, Brussel, Belgia (grant ARW 585/87). We are also grateful to the College of Science, University of Trondheim, Norway, for supporting the participation of students.

Most recent symposia and meetings dealing with thermoregulation in vertebrates have emphasized on mammalian studies, and this book seems to be the first one ever published which solely deals with thermoregulation in birds. Birds and mammals maintain their internal body temperature constant and independent of fluctuating ambient temperatures. The present knowledge of mammalian and avian thermoregulation is extensive, but major questions are still unanswered. Furthermore, birds differ from mammals in many fundamental physiological aspects. Thus, even though both mammals and birds regulate their body temperature within strictly defined limits, the mechanisms utilized by birds are often different from those used by mammals. Only by knowing the mechanisms underlying thermoregulation in both groups that regulate their body temperature, can we fully understand the basic mechanisms of thermoregulation in homeotherms.

The main themes for the workshop were central nervous control mechanisms regulating body temperature, the nature of effector mechanisms, including shivering and non-shivering thermogenesis, and also factors initiating changes in the effector mechanisms. We greatly appreciate, through the NATO Scientific Affair's Division, to have been able to organize such a workshop, and it is our hope that the publication of this book will stimulate further research in the field of avian thermoregulation.

We greatly acknowledge and appreciate the enthusiastic responses of Professor William R. Dawson and Professor Eckhart Simon in participating in the organizing committee of the workshop. Thanks are also given to all individual participants who made it an outstanding event. We are especially indepted to Professor John O. Krog, who, through his speech, reminded us about the great impact of Norwegian scientists on the field of comparative physiology.

Lastly, we would like to thank the publisher, Plenum Press. Thanks are especially given to Janie Curtis and Gregory Safford, who provided valuable advice and assistance in publication of this book.

                                                Claus Bech
                                                Randi Eidsmo Reinertsen

CONTENTS

CENTRAL MECHANISMS OF THERMOREGULATION

Nervous control of cold defence in birds 1
   E. Simon

The shivering response in Common Eider ducks 17
   J.B. Mercer

On the thermosensitivity of the spinal
    cord in pigeons 27
   R.E. Reinertsen and C. Bech

MECHANISMS OF HEAT PRODUCTION

Shivering and nonshivering thermogenesis
    in birds: A mammalian view 37
   E. Connolly, J. Nedergaard and B. Cannon

Muscular nonshivering thermogenesis in
    cold-acclimated ducklings 49
   H. Barré, C. Duchamp, J.-L. Rouanet,
   A. Dittmar and G. Delhomme

Nonshivering thermogenesis in winter-acclimatized
    King Penguin chicks 59
   C. Duchamp, H. Barré, D. Delage, G. Berne,
   P. Brebion and J.-L. Rouanet

Apparent non-muscular thermogenesis in
    cold-exposed phasianid birds 69
   E. Hohtola, R. Imppola and R. Hissa

Shivering in aerobic and anaerobic muscles
    in bantams (Gallus domesticus) 77
   A. Aulie and Ø. Tøien

METABOLIC ADAPTATIONS

Metabolic acclimatization to cold
    and season in birds 83
   W.R. Dawson and R.L. Marsh

Enzyme activities in muscles of seasonally
  acclimatized house finches 95
  C. Carey, R.L. Marsh, A. Bekoff,
  R.M. Johnston and A.M. Olin

Energy substrates and metabolic
  acclimatization in small birds 105
  R.L. Marsh and W.R. Dawson

Thermogenic capacity of greenfinches and
  siskins in winter and summer 115
  S. Saarela, B. Klapper and G. Heldmaier

Heat increment of feeding in the Kestrel,
  Falco tinnunculus, and its natural
  seasonal variation. 123
  D. Masman, S. Daan and M. Dietz

Thermoregulation and energetics of
  arctic seabirds 137
  G.W. Gabrielsen and F. Mehlum

Thermoregulatory adaptations to cold in
  winter-acclimatized Long-
  tailed ducks (Clangula hyemalis) 147
  B.M. Jenssen and M. Ekker

RESPIRATION AND CIRCULATION

Respiration and gas exchange in birds 153
  J. Piiper and P. Scheid

Gas exchange during cold exposure in
  Pekin ducks (Anas platyrhynchos) 163
  C. Bech and H. Johannesen

Respiratory responses of the mallard
  to external and internal cooling 173
  H. Johannesen and C. Bech

The respiratory pattern and expiratory
  gas concentrations in torpid
  hummingbirds Colibri coruscans 179
  M. Berger and K. Johansen

Energy metabolism and patterns of ventilation
  in euthermic and torpid hummingbirds 187
  T.L. Bucher and M.A. Chappell

Respiration by birds at high altitude
  and in flight 197
  M.H. Bernstein

Body and brain temperatures in pigeons
    at simulated high altitudes 207
  M.H. Bernstein

Circulatory adaptations to cold in birds 211
  U. Midtgård

Control of cardiorespiration during shivering
    thermogenesis in the pigeon 223
  W. Rautenberg

PHYSIOLOGY OF HYPOMETABOLISM

Sleep, hypometabolism, and torpor in birds 231
  H.C. Heller

Adaptive capacity of the pigeon's daily
    body temperature rhythm 247
  R. Graf, H.C. Heller, S. Krishna,
    W. Rautenberg and B. Misse

Thermal and feeding reactions of pigeons
    during food scarcity and cold 255
  M.E. Rashotte, D. Henderson
    and D.L. Phillips

Metabolism and body temperature during
    circadian sleep and torpor in the
    fed and fasting pigeon 265
  N.H. Phillips and R.J. Berger

Autonomic and behavioral temperature regulation
    as a part of the response complex to
    food scarcity in the pigeon 275
  J. Ostheim and W. Rautenberg

Body mass, food habits, and the use of
    torpor in birds 283
  B.K. McNab

BREEDING AND INCUBATION

Energy saving during breeding and molt in birds 293
  J.-P. Robin, Y. Handrich,
    Y. Cherel and Y. Le Maho

Effect of clutch size on efficiency of
    heat transfer to cold eggs in
    incubating Bantam hens 305
  Ø. Tøien

Emu winter incubation: Thermal, water,
    and energy relations 315
  W.A. Buttemer and T.J. Dawson

Energy saving in incubating Eiders 325
  G.W. Gabrielsen

ADAPTATIONS TO COLD IN CHICKS

Adaptations to cold in bird chicks 329
  R.E. Ricklefs

Energy partitioning in Arctic tern chicks
    (Sterna paradisaea) and possible metabolic
    adaptations in high latitude chicks 339
  M. Klaassen, C. Bech,
  D. Masman and G. Slagsvold

Energetics of avian growth: the causal link
    with BMR and metabolic scope 349
  R.H. Drent and M. Klaassen

Strategies for homeothermy in Eider
    ducklings (Somateria mollissima) 361
  J.B. Steen, H. Grav, B. Borch-Iohnsen
  and G.W. Gabrielsen

Body temperatures under natural conditions
    of two species of alcid chicks 371
  D. Vongraven, F.J. Aarvik and C. Bech

Participants 375

Index 379

NERVOUS CONTROL OF COLD DEFENCE IN BIRDS

Eckhart Simon

Max-Planck-Institut für physiologische und klinische Forschung
W.G.-Kerckhoff-Institut
D-6350 Bad Nauheim, FRG

INTRODUCTION

What is different in cold defence between mammals and birds? Actually nothing which might be important, if one compares their capacities for cold defence. As exemplified according to Scholander

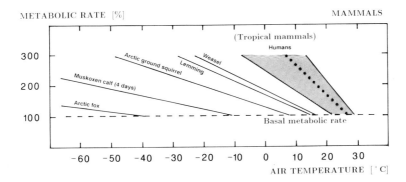

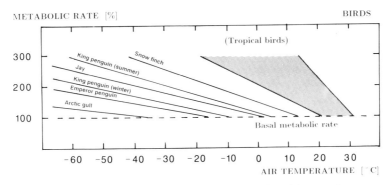

Fig. 1. Metabolic cold defence as a function of ambient air temperature of homeotherms adapted to polar climates (Scholander, 1950, modified).

(1950) in Fig. 1, which contains in addition a few more recent data, the range of lower critical temperatures and the capacity in activating metabolic cold defence are comparable in mammals and birds inhabiting polar regions.

Nervous control of thermoregulatory activities in birds also has long been assumed to be analogous to that in mammals. For this reason corresponding studies on birds have been scarce in comparison to the wealth of information gained from studies on mammals about generation and central processing of temperature signals and about the modes of nervous and hormonal control of thermoregulatory effectors. Especially with regard to central control of body temperature homeostasis, the pre-eminent role attributed in mammals to the rostral brain stem as the dominant site not only of signal integration but also of central thermoreception was tacitly accepted also for birds. While experimental evidence for thermoreception and, to some extent, thermoregulatory integration in the extrahypothalamic central nervous system (Simon, 1974) had disclosed the high degree of redundancy in the organization of the central thermostat, the paradoxical effects of hypothalamic stimulation in birds made it necessary to reconsider the established concepts of homeothermic temperature regulation. Rather than stressing the comparative aspects of the interesting differences found between mammals and birds, it is the aim of this review to demonstrate, how exemplary and seemingly fundamental deviations of avian from mammalian control of cold defence, have guided experimental strategies to elucidate the common properties of central nervous function behind the discrepant observations.

GENERAL PROPERTIES OF NERVOUS CONTROL OF COLD DEFENCE

Table 1. Comparison of effectors and strategies in cold defence and of modes of temperature signal generation and processing in birds and mammals.

	MAMMALS	BIRDS
1. *Thermoeffectors*:		
postural and instrumental behavior	+	+
control of thermal insulation by piloerection	+	+
vasomotor control of thermal conductance	+	+
metabolic heat production by shivering	+	+
nonshivering thermogenesis (*muscle tissue*)	(+)	(+)
nonshivering thermogenesis (*brown adipose tissue*)	+	-
2. *Strategies*:		
euthermic cold defence	+	+
facultative hypothermia	+	+
hibernation	+	-
3. *Generation and processing of thermosensory inputs*:		
variability of input-effector coupling	+	+
multiple-input system (thermosensors in: skin, lower brainstem, spinal cord, deep body tissues)	+	+
hypothalamic thermosensory input	+	?

If one considers the availability of effector mechanisms for cold defence listed in Table 1, congruence between mammals and birds prevails. General absence in birds of thermogenetic brown adipose tissue, which is relevant in only a minority of mammals, appears as the only significant difference. While this does not preclude non-shivering thermogenesis in birds (Saarela and Vakkuri, 1982; Barre et al, 1985), the presence of brown adipose tissue seems to be essential for the ability to hibernate, since only some mammals equipped with this tissue, but no birds, utilize this strategy of cold defence.

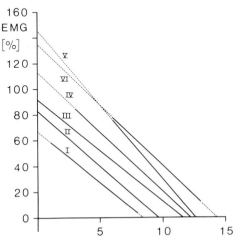

Fig. 2. Spontaneously established relationship between instrumental cold defence and thermoregulatory shivering of pigeons at 6 external cold loads between +12°C (I) and -11°C (VI) air temperature (Schmidt and Simon, 1979).

In the organization of thermoregulatory feedback, birds and mammals seem to share the basic properties of biological systems controlling homeostatic parameters. *Variability of input-effector coupling*, indeed, has been most strikingly exemplified in birds by simultaneous, opposing effects which hypothalamic temperature had on autonomic and behavioral heat defence of pigeons (Schmidt and Simon, 1982). While major differences in input-effector coupling may have a structural basis (Satinoff, 1978), functional variations of coupling to the degree of no activation of one and full activation of another effector is demonstrated in Fig. 2 by the spontaneous compensatory replacement of thermoregulatory shivering by instrumental request for external heat, and *vice versa*, in cold exposed pigeons given the choice between these two modes of cold defence (Schmidt and Simon, 1979).

Multiplicity of temperature signal generation is also shared by mammals and birds, with the same order of magnitude found for the response of metabolic cold defence to the combined thermal input from body core and shell (Simon et al., 1986). Inter-class differences further seem to be not more pronounced than interspecies differences in the contributions of regional thermosensory inputs from the skin and spinal cord. However, a major difference has been disclosed in the hypothalamic contributions: they are generally weak or absent in birds, in contrast to the pronounced thermosensory contribution found in all mammals. While this difference does not invalidate the multiple-input character of the avian thermoregulatory system, its incompatibility with established concepts of hypothalamic thermoregulation has questioned the tacitly assumed general congruence of central thermoregulatory functions in birds and mammals.

EVALUTION OF THERMAL PROPERTIES OF THE AVIAN HYPOTHALAMUS

The challenge to concepts of hypothalamic temperature regulation provoked by weak hypothalamic thermosensitivity in birds (Rautenberg et al., 1972) has been enforced by the observation of inappropriate effects of hypothalamic thermal stimulation in birds (Simon-Oppermann et al., 1978; Schmidt and Simon, 1982). Fig. 3 shows how the metabolic response to lowering core temperature of a duck is suppressed when hypothalamic temperature is lowered to a level which would induce maximum stimulation of cold defence in a mammal. At normal hypothalamic temperature, heat extraction from the body immediately elicits cold defence, with heat generation rising by several watts per kg body weight per 1°C decrease in core temperature. A low hypothalamic temperature, however, permits pronounced hypothermia to develop.

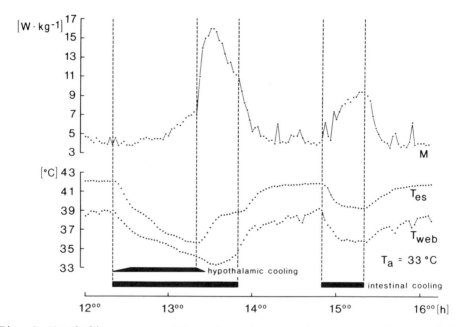

Fig. 3. Metabolic response (M) to lowering core temperature (measured in esophagus; T_{es}) by heat extraction with an intestinal cooling thermode (lower black bars) with (first period; upper black bar) and without (second period) selective hypothalamic cooling from 41.5 to 31.2°C with chronically implanted thermodes in warm ambient conditions. Note the suppression of cold defence by hypothalamic cooling and its increase to maximum values, in combination with a further drop in foot temperature (T_{web}) at rising T_{es}, during rewarming of the hypothalamus (Inomoto and Simon, 1981).

While shivering is depressed and panting facilitated by hypothalamic cooling in birds, both effects are combined with a reduction in sensitivity with which either effector responds to changes of extrahypothalamic body temperature. The impression of a non-specific, local temperature dependence of intrahypothalamic signal transmission has received support from the evaluation of comparable temperature effects on hypothalamic control of non-thermal homeostatic parameters. Effector activity controlling body fluid osmolality was shown to be depressed by hypothalamic cooling. This applies not only to the control of renal water output by antidiuretic hormone (ADH) but also to extrarenal electrolyte output in birds possessing supraorbital salt secreting glands (Hammel et al., 1977; Hori et al., 1986). Fig. 4 shows that hypothalamic cooling reduces osmotic sensitivity of ADH plasma concentration without a change in osmotic threshold.

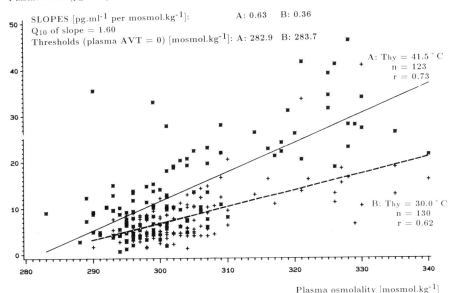

Fig. 4. Relationship between plasma osmolality and plasma concentration of arginine vasotocin (AVT), the avian antidiuretic hormone (ADH), in ducks receiving osmotic loads, when hypothalamic temperature (T_{hy}) was selectively clamped at either a normal (A; asterisks, upper solid regression line) or a hypothermic level (B, +, lower interrupted line) with chronically implanted thermodes. The slopes are significantly different ($P < 0.001$) (Nolte, 1988).

Accordingly, the temperature effect is sufficiently explained by assuming a positive temperature coefficient - corresponding to an apparent Q_{10} of 1.60 - of the overall gain of signal-to-effector transmission in hypothalamic control of ADH release.

From the data of Table 2 positive temperature coefficients may be derived in both mammals and birds for the influence of hypothalamic temperature on the sensitivity of metabolic cold response to changes in body temperature. The hypothalamic Q_{10} values are of the same order of magnitude as the Q_{10} evaluated in Fig. 4 for hypothalamic control of ADH release. Temperature dependence of intrahypothalamic signal processing is obviously similar in systems controlling thermal and non-thermal homeostatic parameters. Further, birds and mammals do not seem to differ in this respect as indicated by the similarity of hypothalamic Q_{10} values evaluated for the sensitivity with which their metabolic heat defence systems react to extrahypothalamic thermal stimulation. This similarity is, however, combined with opposing effects on the threshold temperatures of cold defence activation.

<u>Table 2.</u> Relationship found between core temperature and metabolic heat production of conscious ducks and rabbits as a function of hypothalamic temperature (T_{hy}). By means of water perfused thermodes, the hypothalamus was thermally clamped at either its normal or at a hypothermic level. Proceeding from a steady state condition, T_{core} was reduced by heat extraction with an intestinal thermode and allowed to re-increase. The relationship is described by presenting the core temperatures at the threshold of metabolic activation and at an elevated metabolic rate of 8 $W \cdot kg^{-1}$ (Simon, 1981).

DUCKS:

	T_{core} at M = 3.8 $W \cdot kg^{-1}$	T_{core} at M = 8.0 $W \cdot kg^{-1}$
T_{hy} = 41.5°C	40.78±0.12°C (n=12)	40.07±0.14°C (n=12)
T_{hy} = 31.1°C	39.53±0.15°C (n=12)	38.27±0.20°C (n=12)

local hypothalamic Q_{10} = 1.70 of relationship between T_{core} and M

RABBITS:

	T_{core} at M = 3.0 $W \cdot kg^{-1}$	T_{core} at M = 8.0 $W \cdot kg^{-1}$
T_{hy} = 39.0°C	39.47±0.08°C (n=30)	37.51±0.10°C (n=30)
T_{hy} = 33.8°C	41.39±0.10°C (n=30)	38.62±0.15°C (n=30)

local hypothalamic Q_{10} = 1.90 of relationship between T_{core} and M

Values are means ± standard errors; n = number of experiments

At the neuronal level, single unit recording with simultaneous local thermal stimulation in the hypothalamus of conscious ducks revealed a close correlation between the control discharge rates of warm responsive neurons at normal hypothalamic temperature and their local temperature coefficients (Eissel and Simon, 1980). As indicated by Fig. 5, this interrelationship became more close when samples of neurons were grouped according to their Q_{10} values, which were calculated from the steady state discharge rates of the neurons at normal and lowered hypothalamic temperature. Obviously the positive local temperature coefficient exhibited by a neuron was, to a significant degree, a function of its control discharge rate. About half of these neurons changed their activity with changing extrahypothalamic temperature and were thereby identified as (thermoreactive) interneurons. The relationship shown in Fig. 5 suggests that the synaptic drive of these neurons at normal temperature was reduced by a certain fraction per °C with falling temperature, hence local thermoresponsiveness of these neurons would be attributable to a Q_{10} > 1 of intrahypothalamic synaptic transmission.

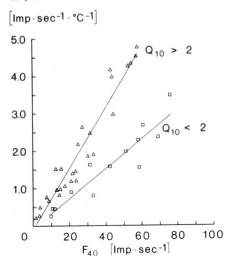

Fig. 5. Relationship between temperature coefficients and control discharge rates at a local temperature of 40°C (F_{40}) of warm responsive hypothalamic neurons recorded in the anterior hypothalamus of conscious ducks with a chronically implanted device permitting insertion of microelectrodes and local thermal stimulation with water perfused thermodes. The positive correlation for the whole sample (r=0.78, n=40) becomes more distinct when the responses are grouped according to their Q_{10} (>2: r=0.96, n=17) (<2: r=0.94, n=13). For explanation see text (Eissel and Simon, 1980).

Hypothalamic interneurons with positive temperature coefficients were analyzed with the same technique in conscious ducks according to whether they were activated or inhibited during whole body cooling by heat extraction with an intestinal thermode. Among the thermoreactive interneurons with a local $Q_{10} > 1$, the sample of neurons activated by extrahypothalamic cooling exhibited a significantly greater average local Q_{10} than the sample of neurons inhibited by extrahypothalamic cooling (Lin and Simon, 1982). This observation could not only be reconciled with the observed positive hypothalamic Q_{10} of signal-to-effector coupling in cold defence of birds but, moreover, explained the general decrease in threshold temperatures by hypothalamic cooling in birds as due to a greater suppression of cold than warm signal transmission.

Attempts to relate the differences in Q_{10} between warm and cold signal transmission in the hypothalamus of ducks to the involvement of different transmitters have not been successful at the neuronal level, as far as biogenic amines are concerned (Sato and Simon, 1988). This corresponds to the comparatively large degree of uniformity in the mostly hypothermic effects elicited by intrahypothalamic application of various biogenic amines in birds (Hissa, 1988).

Neurophysiological evidence and effects of hypothalamic temperature on signal-to-effector coupling in temperature regulation of birds suggest differences in local temperature dependence among hypothalamic neurons conducting extrahypothalamic cold and warm signals, with a Q_{10} greater for cold than for warm signal transmission. This is opposite to the proposal made for mammals of a local $Q_{10} > 1$ of warm signal transmission that is greater than the $Q_{10} > 1$ of cold signal transmission (Hammel, 1968). In mammalian thermoregulation this proposal has received the most widespread attention in modelling the apparent thermosensory function of the mammalian hypothalamus. Apart from explaining the effects of hypothalamic thermal stimulation on threshold temperatures for effector activation, this model would also account for the more recently provided evidence for a positive hypothalamic temperature coefficient of the gain of input-to-effector coupling in mammals (Inomoto et al., 1982).

There are no *a priori* arguments against the assumption of a local
thermoresponsiveness of warm signal transmission greater than that of
cold signal transmission in the mammalian hypothalamus, and of a local
thermoresponsiveness of cold signal transmission greater than that of
warm signal transmission in the avian hypothalamus. In either case
experimental evidence is, however, far from suggesting the *necessity* of
these assumptions, especially in mammals in which, contrary to birds,
independent neurophysiological evidence for the proposed Q_{10} distribution
does not exist and for which, moreover, other models of hypothalamic
thermoreception have been put forward (Bligh, 1972). Thus, according to
these considerations each model has to be considered as heuristic. In
addition, the question has to be raised whether the proposed differences
in Q_{10} distribution between cold and warm signal pathways would be
sufficient to account for the characteristics of temperature regulation
in birds on the one hand and mammals on the other.

THERMORECEPTION VS. SYNAPTIC TEMPERATURE DEPENDENCE

In *birds*, effects of hypothalamic thermal stimulation on
thermoregulatory effectors have been observed which, although restricted
to few effectors and species, would be compatible with the assumption of
a limited thermosensory function of hypothalamic structures. For the
range of hypothalamic temperatures lower than normal, recent studies on
the willow ptarmigan have demonstrated that this particular bird
appropriately activated metabolic cold defence in response to modest
hypothalamic cooling, while inappropriate inhibition occurred during more
pronounced cooling (Mercer and Simon, 1987). Fig. 6 demonstrates the
changes in extrahypothalamic core temperature of various birds, as
compared to the typical mammalian reaction, induced by selective
discplacements of hypothalamic temperature from its normal level. A rise
in core temperature indicates activation of cold defence and/or
inhibition of heat defence, while a fall corresponds to inhibition of
cold defence and/or facilitation of heat loss. With displacements of
hypothalamic temperature below normal, the elevation of core temperature
in mammals typically indicates activation of cold defence, whereas birds
generally react with an inappropriate drop in core temperature. With
displacements of hypothalamic temperature above normal, however, the
response of the majority of birds corresponds to that of mammals. The
diagram indicates several notable exceptions. In response to lowering
hypothalamic temperature, the above-mentioned ptarmigan and possibly also
the emu (Jessen et al., 1982) display a weak, appropriate response in a
limited range of hypothalamic temperature, the same applies to behavioral
cold defence in pigeons which, in this respect, contrasts with the
inappropriate effect of hypothalamic cooling on autonomic cold defence.
In response to elevation of hypothalamic temperature, pigeons were, so
far, the only avian species in which panting was found to be inappropri-
ately inhibited, but this paradoxical response was limited to the
autonomic effector for heat defence, whereas the same animal responded
appropriately to hypothalamic warming with an activation of behavioral
heat defence (Schmidt and Simon, 1982). The summarized results give the
impression that, depending on the species, various degrees of transitory
temperature-response relationships may exist for hypothalamic thermal
stimulation ranging from purely "appropriate" to purely "inappropriate".
In addition, if one accepts a thermosensory function in the pigeon's
hypothalamus, the reactions of this birds represent striking examples for
differential sensor-to-effector coupling, with the behavioral effector
being coupled and the autonomic effectors not coupled to the hypothalamic
thermosensory input.

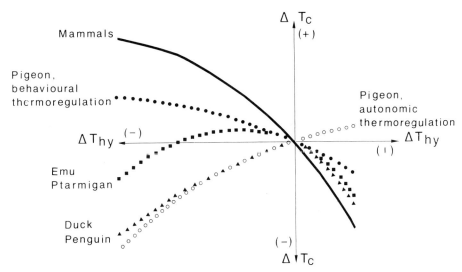

Fig. 6. Relationships between selective displacements of hypothalamic temperature from its normal value produced by means of chronically implanted thermodes and resulting deviations of core temperature from its normal value shown for a typical mammal and for various birds. Note that the influence of hypothalamic temperature changes in pigeons consists in "mammalian-like" changes in core temperature when thermoregulatory behavior is involved, but in "inappropriate" core temperature changes when autonomic thermoregulation is involved. Note also the partially "mammalian-like" response of core temperature to hypothalamic cooling in emu and ptarmigan.

Partially appropriate and partially inappropriate thermoregulatory responses to changes in hypothalamic temperature found in birds, however, cannot be explained alone by the Q_{10} distribution of intrahypothalamic cold and warm signal transmission suggested by the shifts in threshold temperatures and confirmed by neurophysiological analysis. Two explanations may be tentatively envisaged. First, there is a highly differentiated distribution of Q_{10} between the inhibitory and stimulatory hypothalamic outputs to the various effector systems of temperature regulation, partly of the type found in the duck, partly of the type proposed for mammals. This would correspondingly result in partly appropriate and partly inappropriate input-to-effector coupling. Second, and as a more likely explanation, it depends on the degree of involvement of specific, intrahypothalamic cold and warm receptive elements, whether the positive temperature dependence of synaptic signal transmission will dominate or not. In the avian hypothalamus, sensors appear to be generally of little importance and seem to be coupled differently to different effectors, depending on the effector and the species.

In combination with a $Q_{10} > 1$ of cold signal transmission that is greater than the $Q_{10} > 1$ of warm signal transmission, as it was demonstrated in birds, the idea of specific thermoreception existing in the hypothalamus is currently given preference as a general working hypothesis in assessing the thermal properties of central neurons in mammals and birds. However, specific thermosensory inputs originating from thermoreceptive elements within the central nervous system have long been a controversial issue, with the discussion reflecting a general problem faced by any concept which attributes enteroceptive functions to central nervous elements. In the case of specific hypothalamic osmoreception, for instance, osmosensitivity of the effector cell itself (Leng et al., 1982) has been considered as an alternative to the assumption of specific osmoreceptive elements, which have remained elusive, so far. With respect to the thermosensory function of the central nervous system, the discussion has remained centered on the hypothalamus of mammals. While the idea of an apparent thermosensory function due to a particular distribution of temperature dependence among cold and warm signal pathways has generally been given preference in models of mammalian thermoregulation, the analysis of intrinsic vs. synaptic temperature dependence at the neuronal level is still far from conclusive.

Following the discovery of thermosensory input generation along the entire spinal cord in both mammals and birds, the analysis of thermoreception within the central nervous system has not remained restricted to the hypothalamic region. Analysis of neuronal thermoresponsiveness at the spinal level has provided at least circumstantial evidence for the existence of central thermosensitive neurons subserving a specific thermoafferent function. Spinal anterolateral tract axons have been identified which transmit neuronal activation by either cooling or warming of the spinal cord itself. According to their location and their origin in spinal segments distal to the site of recording, only afferent functions could be attributed to these neurons (Simon, 1972). For mammals, Hensel (1973) has emphasized the close congruence of static responses in the temperature frequency relationships of cutaneous cold and warm receptors and, respectively, spinal cold and warm sensitive ascending neurons, of which a large fraction was, moreover, shown to share distinct dynamic response components with skin thermoreceptors (Simon, 1984). In birds spinal thermosensitivity was also shown to reside in the spinal cord itself (Necker and Rautenberg, 1975), and ascending spinal axons activated by either cooling or warming were demonstrated as well (Necker, 1975). Although their analysis has not been as detailed as in mammals, very similar static temperature response relationships were demonstrated. These similarities in the properties of spinal and cutaneous neuronal thermosensitivity in birds and mammals strongly suggest specific thermoreception in the spinal cord, if not by primary afferents, then by interneurons capable to produce, at specific sites of their somadendritic membrane, generator potentials as a function of their absolute temperature and with mechanisms analogous to those responsible for peripheral cold and warm reception.

By inference, the considerations applied to spinal cord thermosensitivity suggest the existence of specific thermoafferent elements in the brain stem as well, because there is no reason why they should be restricted to the extracephalic segments of the central nervous system. However, since the evaluation of this problem for the hypothalamus has proceeded from the idea of temperature dependent synaptic transmission (Nakayama et al., 1963), the lines of reasoning have been very different. Due to the complete lack of anatomical clues for the function of hypothalamic thermoresponsive neurons, discrimination of potentially receptive units from interneurons has remained more or less hypothetical.

Analysis of single unit thermoresponsiveness in the hypothalamus of birds has contributed negatively to the problem, if viewed from the concepts developed for mammals. Despite the lack of relevant hypothalamic thermosensory functions in the hypothalamus of ducks, thermoresponsive neurons were found to occur as frequently and with the same characteristics as in the hypothalamus of mammals (Simon et al., 1977). A recent, comparative *in vitro* approach carried out in hypothalamic slices from rats and ducks has, moreover, questioned neuropharmacological criteria which were thought to be discriminative with regard to intrinsic vs. synaptic thermoresponsiveness of hypothalamic neurons (Nakashima et al., 1987). Media low in Ca^{++} (and high in Mg^{++}) are used in many studies to suppress synaptic signal transmission. However, it was found that lowering Ca^{++} could change the temperature characteristics of central neurons, even from warm- to cold- responsive, as illustrated by the example of Fig. 7 obtained from a duck, but confirmed also for the rat. This negative result is especially important in that it questions the view derived for mammals from such studies that no intrinsic cold sensitivity exists at the hypothalamic level. In fact, most recent information obtained by intracellular recording *in vitro* on rat hypothalamic slices (Curras et al., 1988) may be interpreted as affirmative with regard to the existence of primary warm as well as cold responsiveness in the hypothalamus of mammals. This inference would be in accordance with the general extrapolation drawn from specific spinal to specific brain stem thermoreception.

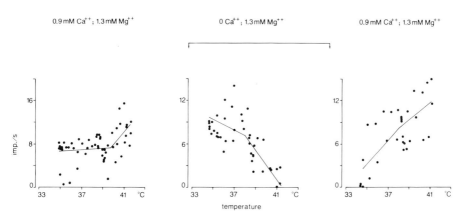

Fig. 7. Relationship between tissue temperature and discharge rate of a neuron recorded *in vitro* in a tissue slice taken from the preoptic and anterior hypothalamic region of a duck and superfused with oxygenated Krebs-Ringer solution. At normal Ca^{++} concentration in the medium the neuron exhibited warm responsiveness before (left diagram) and after (right diagram) an experimental period (middle diagram) in which the slice was superfused with Ca^{++}-free medium. In the Ca^{++}-free medium the neuron became reversibly cold responsive (Nakashima et al., 1987).

Although elucidation of the mode by which thermosensory inputs are generated in the central nervous system will remain decisive for the development of concepts of temperature regulation, the contribution of sensors located outside of the central nervous system seems to be quantitatively more important. In birds, the contribution of skin temperature to the control of autonomic and behavioral cold defence seems to exhibit considerable topical variability of input-to-effector coupling (Schmidt, 1982), but the neurophysiology of warm and cold reception in birds seems to be basically identical with skin thermoreception in mammals (Necker, 1981). The existence of specific thermoreception outside of the central nervous system in the body core, i.e., by "deep" peripheral thermoreceptors, must be inferred in birds, as in mammals, from the large discrepancy between the quantitative contributions of central nervous and cutaneous thermal stimulation to metabolic cold defence in comparison to its response to changes in temperature of the entire body core. The nervous elements responsible for deep body thermoreception outside of the central nervous system have, however, not been clearly identified and characterized.

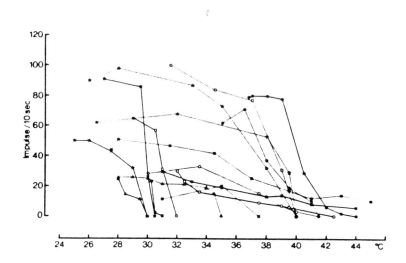

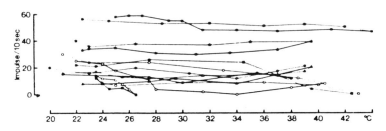

Fig. 8. Discharge rates of spinal motoneurons as a function of local spinal cord temperature in a bird (pigeon, upper diagram) and a lizard (iguana, lower diagram) (Görke, 1976).

The synopsis of the modes of thermoresponsiveness elucidated, so far, in birds as well as in mammals leads to the conclusion that temperature effects on nervous elements providing specific, afferent thermosensory input, as well as on synaptic transmission, both contribute to the relationships found between various profiles of body temperatures and states of activity of thermoregulatory effectors. Specific information from the skin is provided by a dual system of cold and warm receptors. By analogy, a similar dual system, which remains to be further elucidated, should account for thermoreception in deep body tissues outside of the central nervous system. For the central nervous system, the working hypothesis which proposes specific thermoreception, involving neurons with positive as well as negative temperature coefficients, has continued to develop. In addition, thermal modulation of synaptic transmission has to be taken into consideration at all levels of the central nervous system, acting upon afferent thermosensory transmission as well as on efferent transmission of the outputs controlling thermoregulatory effectors. At the hypothalamic level of both mammals and birds, the observed local temperature effects on the gain of input-to-effector coupling suggest that the net temperature coefficient of central nervous integration at the hypothalamic level is positive. Obviously, in the mammalian as well as the avian hypothalamus the influence of the more frequent warm responsiveness of central neurons dominates over neuronal cold responsiveness which is observed only in a minority of neurons. At the spinal level, however, the overall temperature coefficient might well be partially negative, e.g., in thermal control of shivering, due to the "cold responsiveness" of motoneurons in both mammals (Pierau et al., 1976) and birds (Görke et al., 1979). As illustrated by Fig. 8, the negative temperature coefficient of excitability found for spinal motoneurons of birds constitutes a qualitative difference to cold-blooded sauropsids as exemplified by studies on a lizard. In this particular property birds rather resemble mammals in which the temperature dependence of the excitability of spinal motoneurons exhibits bell-shaped characteristics with maxima in the hypothermic range

CONCLUSIONS

Exemplary differences in neuronal control of cold defence between mammals and birds have directed experimental work towards establishing criteria for discrimination between specific and non-specific thermoresponsiveness of central nervous neurons. Evidence derived from analysis of input-output relationships of the intact organism, as well as from analysis of neuronal activity presents circumstantial evidence for two different types of thermoresponsiveness.
1. Synaptic transmission may be temperature dependent, due to temperature-related changes in postsynaptic excitability and/or in presynaptic transmitter release. This property appears to be common to mammals and birds, with mostly positive, but in certain neuronal systems also negative temperature coefficients.
2. Central neurons may be intrinsically thermoresponsive, with positive as well as negative temperature coefficients, i.e., self-excitatory at rates which are a function of absolute temperature, with or without additional sensitivity to the rate of temperature change. Irrespective of whether such neurons are interneurons or primary afferents, they could provide for specific central cold and warm reception if they were appropriately connected to the input side of the central thermointegrative system.

Although birds seem to differ fundamentally in thermal hypothalamic control of cold defence, they seem to have in common with mammals the general temperature dependence of central signal transmission and the existence of specific thermoreceptors at multiple sites of the skin, the deep body tissues and the central nervous system. They seem to differ only quantitatively in the distribution of thermoreceptive elements along the central nervous axis. At the hypothalamic level, receptor density appears to be high and coupling of their signals to efferent control of effectors generally tight in mammals. Experimental evidence in birds suggests a low density of hypothalamic thermoreceptive elements, with less tight and variable coupling to thermoregulatory effectors, with the result of inappropriate effects of especially hypothalamic cooling on autonomic thermoregulatory effectors.

REFERENCES

Barre, H., Geloen, G., Chatonnet, J., Dittmar, A., and Rouanet, J.-L., 1985, Potential muscular thermogenesis in cold-acclimated muscovy duckling. Am. J. Physiol. 249, R533-R538.

Bligh, J., 1972, Neuronal models of mammalian temperature regulation, in: Essays on Temperature Regulation, J. Bligh and R. Moore, eds., North Holland, Amsterdam: p. 105-120.

Curras, M.C., Kelso, S.R., Boulant, J.A., 1988, Intracellular recordings of preoptic temperature sensitive and insensitive neurons. FASEB Journal, Vol. 2: A746.

Eissel, K., and Simon, E., 1980, How are neuronal thermosensitivity and lack of thermoreception related in the duck's hypothalamus? A tentative answer. J. Therm. Biol., 5: 219-223.

Görke, K., 1976, Der Einfluss von Temperatur-Änderungen im Rückenmark auf das Reflexverhalten und die Spontantätigkeit Spinaler Motoneurone bei Sauropsiden. Dissertation der Abteilung für Biologie an der Ruhr-Universität, Bochum, FRG.

Görke, K., Necker, R., and Rautenberg, W., 1979, Neurophysiological investigation of spinal reflexes at different temperatures of the spinal cord in birds and reptiles. Pflügers Arch., 359: 269-271.

Hammel, H.T., 1968, Regulation of internal body temperature. Ann. Rev. Physiol., 30: 641-710.

Hammel, H.T., Maggert, J.E., Simon, E., Crawshaw, L., and Kaul, R., 1977 Thermo- and osmo-regulatory responses induced by heating and cooling the rostral brain stem of the Adelie penguin, in: Adaptations Within Antarctic Ecosystems, G.A. Llano, ed., Smithsonian Institution, Washington D.C.: 89-500.

Hensel, H., 1973, Neural processes in thermoregulation. Physiol. Rev., 53: 948-1017.

Hissa, R., 1988, Controlling mechanisms in avian temperature regulation: a review. Acta Physiol. Scand., Vol. 132, Suppl. 567: 1-148.

Hori, T., Simon-Oppermann, Ch., Gray, D.A., and Simon, E., 1986, Thermally induced changes in neural and hormal control of osmoregulation in a bird with salt glands (Anas platyrhynchos). Pflügers Arch., 407: 414-420.

Inomoto, T., Mercer, J.B., and Simon, E., 1982, Opposing effects of hypothalamic cooling on threshold and sensitivity of metabolic response to body cooling in rabbits. J. Physiol. (Lond.), 322: 139-150.

Inomoto, T., and Simon, E., 1981, Extracerebral deep-body cold sensitivity in the Pekin duck. Am. J. Physiol., 241: R136-R145.

Jessen, C., Hales, J.R.S., and Molyneux, G.S., 1982, Hypothalamic thermosensitivity in an emu (Dromiceius novae-hollandiae). Pflügers Arch. 393: 278-280.

Leng, G., Mason, W.T., and Dyer, R.R., 1982, The supraoptic nucleus as an osmoreceptor. Neuroendocrinology, 34: 75-82.

Lin, M.T., and Simon, E., 1982, Properties of high Q_{10} units in the conscious duck's hypothalamus responsive to changes of core temperature. J. Physiol. (Lond.), 322: 127-137.

Mercer, J.B., and Simon, E., 1987, Appropriate and inappropriate hypothalamic cold thermosensitivity in Willow Ptarmigan. Acta Physiol. Scand., 131: 73-80.

Nakayama, T., Hammel, H.T., Hardy, J.D., and Eisenman, J.S., 1963, Thermal stimulation of electrical activity of single units of the preoptic region. Am. J. Physiol., 204: 1122-1126.

Nakashima, T., Pierau, Fr.-K., Simon, E., and Hori, T., 1987, Comparison between hypothalamic thermoresponsive neurons from duck and rat slices. Pflügers Arch., 409: 236-243.

Necker, J., 1975, Temperature-sensitive ascending neurons in the spinal cord of pigeons. Pflügers Arch. 353: 275-286.

Necker, R., 1981, Thermoreception and temperature regulation in homeothermic vertebrates, in: Progress in Sensory Physiology 2, Autrum, H., Ottoson, D., Perl, E., and R.F. Schmidt, eds., Springer-Verlag, Berlin: p. 1-47.

Necker, R., and Rautenberg, W., 1975, Effect of spinal deafferentation on temperature regulation and spinal thermosensitivity in pigeons. Pflügers Arch., 360: 287-299.

Nolte, P., 1988, Hypothalamische Kontrolle osmoregulatorischer Hormone, Untersuchungen mit Hypothalamuskühlung an der wachen Peking-Ente, M.D. Thesis, University of Giessen, FRG.

Pierau, Fr.-K., Klee, M.R., and Klussmann, F.W., 1976, Effect of temperature on postsynaptic potentials of cat spinal motoneurones. Brain Res., 114: 21-34.

Rautenberg, W., Necker, R., and May, B., 1972, Thermoregulatory responses of the pigeon to changes of the brain and the spinal temperatures. Pflügers Arch., 338: 31-42.

Saarela, S., and Vakkuri, O., 1982, Photoperiod-induced changes in temperature metabolism curve, shivering threshold and body temperature in the pigeon. Experientia, 38: 373-374.

Satinoff, E., 1978, Neural organization and evolution of thermal regulation in mammals. Science, 201: 16-22.

Sato, H., and Simon, E., 1988, Thermal characterization and transmitter analysis of single units in the preoptic and anterior hypothalamus of conscious ducks. Pflügers Arch., 411: 34-41.

Schmidt, I., 1982, Thermalstimulation of exposed skin area influences behavioral thermoregulation in pigeons. J. Comp. Physiol., 146: 201-206.

Schmidt, I., and Simon, E., 1979, Interaction of behavioral and autonomic thermoregulation in cold exposed pigeons. J. Comp. Physiol., 133: 151-157.

Schmidt, I., and Simon, E., 1982, Negative and positive feedback of central nervous system temperature in thermoregulation of pigeons. Am. J. Physiol., 243: R363-R372.

Scholander, P.F., Hock, R., Walters, V., Johnson, F., and Irving, L., 1950, Heat regulation in some arctic and tropical mammals and birds. Biol. Bull., 99: 237-258.

Simon, E., 1972, Temperature signals from skin and spinal cord converging on spinothalamic neurons. Pflügers Arch., 337: 323-332.

Simon, E., 1974, Temperature regulation: the spinal cord as a site of extrahypothalamic thermoregulatory functions. Rev. Physiol. Biochem. Pharmacol., 71: 1-76.

Simon, E., 1981, Effects of CNS temperature on generation and transmission of temperature signals in homeotherms: a common concept for mammalian and avian thermoregulation. Pflügers Arch., 392: 79-88.

Simon, E., 1984, Extrahypothalamic thermal inputs to the hypothalamic thermoregulatory network. J. Therm. Biol., 9: 15-20.

Simon, E., Hammel, H.T., and A. Oksche, 1977, Thermosensitivity of single units in the hypothalamus of the conscious Pekin duck. J. Neurobiol., 8: 523-535.

Simon, E., Pierau, Fr.-K., and Taylor, D.C.M., 1986, Central and peripheral thermal control of effectors in homeothermic temperature regulation. Physiol. Rev., Vol. 66: 235-300.

Simon-Oppermann, C., Simon, E., Jessen, C., and Hammel, H.T., 1978, Hypothalamic thermosensitivity in conscious Pekin ducks. Am. J. Physiol., 235, R130-R140.

THE SHIVERING RESPONSE IN COMMON EIDER DUCKS[*]

James B. Mercer

Department of Arctic Biology and Institute of Medical
Biology, University of Tromsø, P.O. Box 635
N-9001 Tromsø, Norway

INTRODUCTION

In homeotherms total body thermosensitivity (TBTS) can be determined at any given ambient temperature (T_a) by relating an induced fall in body core temperature (T_c) to the resulting increase in metabolic heat production (M). Experimental determinations of TBTS usually involve using heat exchangers (water perfused thermodes) to lower body core temperature (e.g. intravascular heat exchangers; Jessen et al., 1977, intestinal thermodes; Ionomoto & Simon, 1981). Some of the inherent problems related to the use of heat exchangers have been previously discussed (see Fig.2, Mercer & Simon, 1984). For example, in order to determine a correct TBTS it is important that temperature sensors are not directly influenced by the cooling thermode. i.e. there should be no hysteresis between the curves describing the relationship between T_c and M during body core cooling and in the recovery period following the end of body core cooling. In the avian class of homeotherms T_c can also be lowered by making use of the paradoxical effect resulting from local cooling (clamping) of the brain stem (Simon et al., 1986). In this case TBTS can be calculated by relating T_c to M during the recovery period following release of the thermal clamp.

In mammals TBTS has been determined in a large variety of species. On average M has been found to increase by about 4 to 6 W.kg^{-1} per degree fall in T_c (see Simon et al., 1986). By comparison only two studies designed specifically to determine TBTS have been carried out in birds. In these experiments TBTS was found to be 12 W.kg^{-1}.C$^{\circ -1}$ with the heat exchanger placed into the central body veins (domestic goose - Helfmann et al. 1981) and 6W.kg^{-1}.C$^{\circ -1}$ with the heat exchanger placed into the lower intestinal tract (Pekin duck - Ionomoto & Simon, 1981). In both avian species a lack of hysteresis indicated an accurate determination of TBTS. Estimations from previously published experiments in the Adelie penguin and in the pigeon resulted in TBTS values of -5.5W.kg^{-1} .C$^{\circ -1}$ for both species (for refs see Simon et al., 1986)

---

[*] Part of the results have been previously published in abstract form (Mercer, 1985).

The experiments reported in this study have been made in order to provide more information regarding TBTS values in a wild bird, the common eider duck (Somateria Mollissima).

METHODS

Animals

The experiments were carried out in 11 juvenile eider ducks (Somateria mollissima) whose body weights ranged between 1.4 and 2.0 kg. The birds were hatched from machine-incubated eggs and raised at the Department of Arctic Biology, University of Tromsø (69°40'N). The eggs were collected locally from the nests of wild incubating birds. Between experiments the animals were housed outdoors in an open pen and subjected to natural light and weather conditions.

Oesophageal thermode

In order to lower body temperature a perfusion thermode consisting of 3 U-shaped polyethylene tubes (pp 90) was introduced into the oesophagus via the oral cavity. The tip of the thermode lay approximately 2-3 cm cranial to the muscular part of the stomach. By perfusing the thermode with water of 20°C at a flow rate of about 60-70 ml.min^{-1}, heat could be extracted at rates of 3-4 times resting metabolic heat production.

Hypothalamic thermode

In two animals a hypothalamic perfusion thermode was chronically implanted. The thermode and the implantation procedure were similar to that previously described by Simon-Oppermann et al., (1978). The thermode consisted of four blind-ending stainless steel tubes (1.0 mm OD & 0.8 mm ID) located 4 mm apart from each other in a quadrangular arrangement, with a fifth centrally located thermocouple reenterant tube. The thermode was inserted under x-ray control during general anaesthesia (Equisitin) and was fixed to the skull with stainless steel wire and dental acrylic cement. Correct placement was assumed as the animals reacted to brain heating and cooling in a manner both qualitatively and quantitatively similar to that found in the pekin duck. i.e. the animals responded appropriately to hypothalamic heating and showed a clear paradoxical response to hypothalamic cooling (Simon-Oppermann et al., 1978).

Measurements

All experiments were carried out in a climatic chamber set to an ambient temperature (T_a) of 10°C. Metabolic heat production (M) was measured by indirect calorimetry using an open circuit system. A ventilated hood was placed over the animal's head and air was continuously withdrawn through a gas meter. Oxygen consumption and M in W.kg^{-1} were calculated from the airflow rate and oxygen extraction measured by means of an oxygen analyzer (Applied Electrochemistry Type S-3A, USA), assuming a caloric equivalent of 20.35 kJ/l O_2 (STPD). All temperatures were continuously recorded using copper-constantan thermocouples. Colonic temperature (T_{col}), carotid arterial wall temperature (T_{cw}), and right atrial temperature (T_{ra}) were measured to represent core temperature (T_c). T_{col} was measured at a depth of approximately 15 cm into the colon. In the series of experiments involving hypothalamic cooling oesophageal temperature (T_{oes}) was measured at a depth of about 20 cm deep in the oesophagus. T_{cw} was measured at the tip of a blind ending polyethylene tube (pp 90) surgically implanted with the end of the tube lying parallel to and, on the dorsal side of, the paired carotid arteries. T_{ra} was measured at the tip of a blind ending polyethylene tube (pp 90) with the tip in the right atrium. The tube was introduced through a small incision in the right jugular vein and the position determined by X-ray fluoroscopy. The surgical procedures were carried out under general anaesthesia (Equisitin).

Experimental Protocol

Oesophageal cooling. Following instrumentation the animals were allowed an equilibration period of about 1 hour in order to reach stable steady state, as judged by constancy of T_c and M. The animals were then subjected to a 20 min period of oesophageal cooling followed by a 30 min recovery period. When the animal had once again come into steady state conditions the next perfusion period was started. No more than 5 perfusion periods were performed in any one experiment.

In one set of experiments the effect of oesophageal cooling was examined in animals having their oesophagus anaesthetized. Prior to insertion of the thermode, Xylocain (2%) gel (ASTRA) was injected into the oesophagus via a piece of silastic tubing attached to a syringe. Following application the oesophagus was externally massaged. Prior to insertion the thermode was also well coated with the anaesthetic gel.

Combined oesophageal & hypothalamic cooling. In the two animals provided with hypothalamic perfusion thermodes a series of experiments were conducted in which use was made of the paradoxical effect of brain cooling in order to lower brain temperature (Ionomoto & Simon, 1981). In these experiments body core temperature was first lowered with combined oesophageal and hypothalamic cooling (a hypothalamic tissue clamp temperature of 26.0°C was employed). This technique usually resulted in a fall in T_c of about 3°C. The oesophageal cooling was then stopped and a period of 10 minutes elapsed before the hypothalamic clamp was released. During the period prior to release of the hypothalamic clamp M was usually very close to resting levels. The release of the hypothalamic clamp resulted in a rapid and large increase in M, which took about 15 to 20 minutes to return to the resting level as body core temperature increased. During this recovery period TBTS was calculated from the relationship between M and T_{col} or T_{oes}.

The relationship between T_c and M was evaluated by linear regression analysis with the least squares method. The slopes were taken as indices for total body thermosensitivity. Data was excluded from those periods during an experiment when the animals were not quiet.

RESULTS

The time course of a typical experiment showing the effects of a 20 minute period of oesophageal cooling is shown in Figure 1. There was an immediate fall in both T_{col} and T_{cw} followed, by a rapid increase in M to more than twice the control value. Within 2 to 3 minutes the increase in M prevented any further fall in T_{col} whereas, T_{cw} continued to fall until about the 8th minute of perfusion. (In no experiment was the induced fall in T_{cw} during the cooling period more than 0.5°C while the induced fall in T_{col} was always somewhat less). During the cooling period the animal achieved a new steady condition and a simple two point determination of TBTS could be made by relating the change in T_c to the change in M between the start and end of the 20 minute perfusion period. For T_{col} &

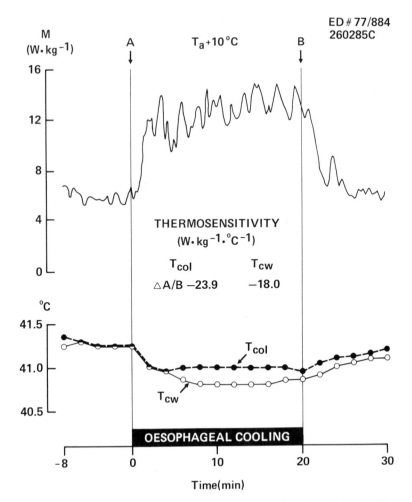

Fig.1 Time course of a single experiment showing the relationship between metabolic heat production (M) and body core temperature (colonic, T_{col}; carotid arterial wall, T_{cw}) before, during and after a 20 min. period of oesophageal cooling.

T_{cw} the resultant TBTS values were -23.9 & -18.0 W.kg^{-1}.°C^{-1} respectively. These values are approximately 4 to 5 times greater than those previously reported for the Pekin duck (Ionomoto & Simon, 1981). Following the end of the perfusion period M rapidly (within about 5 minutes) returned to pre-stimulus values. However, even after 10 minutes T_{col} & T_{cw} were still below the pre-stimulus values.

In Fig.2 the relationship between changes in M and T_{col} & T_{cw} resulting from 18 oesophageal cooling periods in 5 animals are shown. (Some of the mean values have fewer than 18 points due to either thermocouple breakage or due to periods when the animals were restless). In this figure (as also in figures 3 and 4) each of the data points shows the relationship between body core temperature and M as calculated from five different time

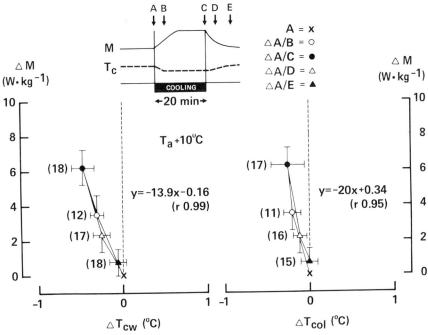

Fig. 2. The top part of the figure shows a schematic diagram of the time course for metabolic heat production (M) and body core temperature (T_C) during a single oesophageal cooling experiment. The points ABCDE represent different time points during the experiment. The mean results in the lower part of the figure shows the relationship between M and carotid arterial wall temperature (T_{cw}, left side of fig.) & M and colonic temperature (T_{col}, right side of fig.). Note that apart from the point marked with the symbol X (time point A), all other symbols represent differences between two time points. The error bars are standard deviations. Number of measurements in parenthesis.

periods before, during and after the perfusion period. These time periods are indicated in the schematic time course diagrams shown on the top of each figure. Thus, in these figures the symbol "X" represents the relationship between M & T_C at the start of the perfusion period and is designated the value zero (the mean absolute values ± SD for M, T_{col} & T_{cw} at the start of the perfusion period were 6.16 ± 0.61 W·kg^{-1}, 41.42 ± 0.41°C and, 41.21 ± 0.23°C, respectively, and at the end of the 20 min. perfusion period the mean absolute values ± SD were 12.64 ± 1.07 W·kg^{-1}, 41.11 ± 0.31 °C and, 40.73 ± 0.22 °C respectively); the open & closed circles represent the differences in M & T_C between the start and the 4th. minute of the perfusion period and the start and the end of the perfusion period, respectively. The open and closed triangles represent the differences in M and T_C between the end of the perfusion period and the 4th. minute of the recovery period and, between the end of the perfusion period and the 10th. minute of the recovery period, respectively. The curves were constructed by joining the mean data points in chronological order.

TBTS was calculated for T_{cw} & T_{col} using the five mean data points

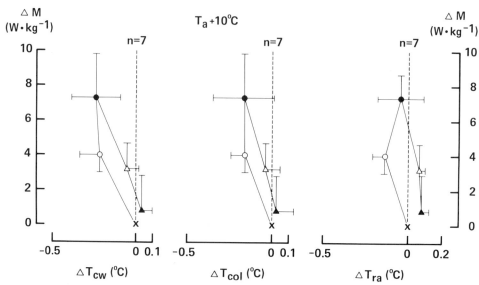

Fig. 3. The top part of the figure shows a schematic diagram of the time course for metabolic heat production (M) and body core temperature (T_c) during a single oesophageal cooling experiment. The points ABCDE represent different time points during the experiment. The mean results in the lower part of the figure shows the relationship between M and carotid arterial wall temperature (T_{cw}, left side of fig.), M and colonic temperature (T_{col}, centre of fig.) and M and right atrial temperature (T_{ra}, right side of fig.). Note that apart from the point marked with the symbol X (time point A), all other symbols represent differences between two time points. Error bars are standard deviations. Number of measurements in parenthesis.

and was -13.9W.kg^{-1}.°C^{-1} and -20.0W.kg^{-1}.°C^{-1}, respectively. The high correlation coefficients for the two linear regression analyses used to calculate TBTS (0.99 & 0.95 respectively) indicate a lack of hysteresis in the relationship between M and T_c.

In fig.3 the relationship between M and T_c, when measured simultaneously at three different sites (T_{cw}, T_{col} & T_{ra}), is shown. In these experiments, which are the results of 7 perfusion periods in two animals, TBTS was -10.33, -13.95 and, -7.97 W.kg^{-1}.°C^{-1}, respectively. The respective correlation coefficients for the linear regression calculations used to determine TBTS were 0.9, 0.88 and, 0.5 (the TBTS values using T_{cw} & T_{col} were slightly lower than those shown in Fig.2.). In the curves shown in Fig.3 a distinct hysteresis can be seen, particularly for the relationship between M and T_{ra}. In some of the experiments T_c had a value at the end of the cooling period which was the same as, or in some cases even greater than, that found prior to the cooling period. This was particularly evident

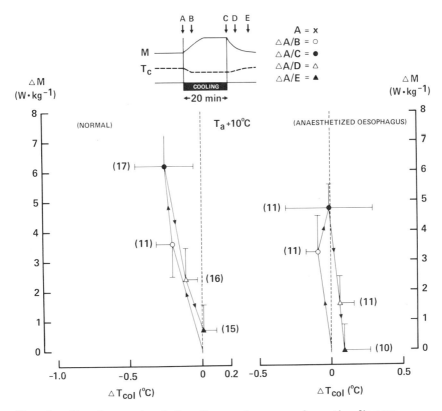

Fig. 4. The top part of the figure shows a schematic diagram of the time course for metabolic heat production (M) and body core temperature (T_c) during a single oesophageal cooling experiment. The points ABCDE represent different time points during the experiment. The mean results in the lower part of the figure shows the relationship between M and colonic temperature (T_{col}) in control animals (left side of Fig.) and in animals in which the oesophagus had been locally anaesthetized (right side of fig.). The control data is the same as that shown in the right hand side of Fig.2. Note that apart from the point marked with the symbol X (time point A), all other symbols represent differences between two time points. Error bars are standard deviations. Number of measurements in parenthesis.

when T_{ra} was used to represent T_c. A simple 2 point determination for TBTS (i.e. comparing the steady state values between the start and end of the 20 min. cooling period) resulted in TBTS values of -15.15, -22.03 and,-80.78 $W \cdot kg^{-1} \cdot °C^{-1}$ for T_{cw}, T_{col} and, T_{ra}, respectively.

In order to determine whether the high TBTS values found in these animals were due to direct stimulation of thermosensitive structures within the oesophagus itself, a series of 11 perfusion periods in 4 animals were made in animals in which the oesophagus was anaesthetized with 2% Xylocain. The results of these experiments are presented in the right hand side of Fig.4. For comparison the results for the same relationship in unanaesthetized (control group) animals (taken from Fig.2) are shown in the sensitivity was found, the curve describing the response having a slight hysteresis. A simple two point determination of TBTS in this group (comparing the start and end of the cooling period) resulted in a TBTS value of $-475 \text{W.kg}^{-1}.°\text{C}^{-1}$.

In the two animals provided with hypothalamic perfusion thermodes 9 experiments were made to determine TBTS from the recovery period following lowering T_c by combined hypothalamic and oesophageal cooling. Following release of the hypothalamic clamp the mean peak increase in M $\pm$ SD was $23.75 \pm 4.15 \text{W.kg}^{-1}$, the highest value obtained being 28.0W.kg^{-1}. The mean TBTS values calculated from the recovery period were $-8.99 \pm 2.41 \text{W.kg}^{-1}.°\text{C}^{-1}$ (n=9) and $-9.44 \pm 0.94 \text{W.kg}^{-1}.°\text{C}^{-1}$ (n=6) when T_{col} & T_{oes} were taken to represent T_c, respectively.

DISCUSSION

The results of this study demonstrate some of the problems that can arise when attempting to determine TBTS by the thermode technique. In this study TBTS varied according to factors such as the method used (i.e recovery versus oesophageal cooling) as well as on the body core site taken to represent the regulated variable. In addition, if a hysteresis occured between the cooling and recovery curves then the calculated thermo-sensitivities varied according to which data points were included in the calculations. The question naturally arises as to which of the determined thermosensitivities is the most valid.

Based on the arguments of Mercer & Simon (1984) the lack of a clear hysteresis in the results shown in Fig 2 would indicate that the two values presented in this figure are both valid. Of the two different TBTS values presented, that using T_{cw} to represent T_c was slightly lower than that using T_{col} and was probably related to the fact that the thermocouple used to measure T_{cw} was being influenced by the proximity of the oesophageal thermode. Such a thermal disturbance will results in an underestimation of TBTS (Mercer & Simon, 1984). Evidence that this thermocouple was being directly influenced by the cooling thermode was obtained from an observation in a separate study designed to test the effect of local tracheal cooling. In these experiments 3 small perfusion polyethylene thermodes (pp 60) were surgically attached to the outside wall of the trachea over a length of about 6cm. Cooling these thermodes resulted in a 50% increase in M. The increase in M was not associated with any detectable fall in T_{col} whereas T_{cw} decreased by an amount similar to that found using the more effective oesophageal thermode. For T_{cw} the mean calculated TBTS (two point determination) from three experiments was $- 4.76 \text{W.kg}^{-1}.\text{C}°^{-1}$. This sensitivity is half the assumed valid TBTS value for this species (see below), which indicates that during perfusion the tracheal thermodes were directly influencing the thermocouple measuring T_{cw}. As the oesophagus and trachea are in close association with each other it is assumed that during perfusion the oesophageal thermode was also having a direct influence on this thermocouple. Thus, with reference to T_{col} in Fig.2 a TBTS value of about $-20.0 \text{ W.kg}^{-1}.°\text{C}^{-1}$ would appear to be a valid value. However, in a second series of oesophageal cooling experiments in which T_{col} was also used to represent T_c (Fig.3) a slightly lower TBTS value (ca. $14 \text{W.kg}^{-1}.\text{C}°^{-1}$) was found. In these latter experiments a hysteresis was evident. The reason for the slightly different responses between these two groups is not known.

It is generally accepted that one of most reliable methods for determining TBTS is from the "recovery" method following lowering of body core temperature (Ionomoto & Simon, 1981). By taking advantage of the paradoxical affect of brain cooling in birds, this procedure can more easily be carried out in this class of homeotherms than in mammals. The technique has the advantage of eliminating thermal gradients within the body core with the consequence that temperature sensors in the body are not directly affected by the temperature of an implanted thermode. Based on this argument a TBTS value of about $9.0 W.kg^{-1}.°C^{-1}$, as determined using the recovery method following combined oesophageal and hypothalamic cooling, represents the most valid TBTS value for the eider duck. This value is slightly higher than that normally found in mammals ($4 - 6W.kg^{-1}.°C^{-1}$) and higher than that reported for the pekin duck ($6W.kg^{-1}.°C^{-1}$; Ionomoto & Simon, 1981). However, while it is slightly lower than that reported for the goose ($12W.kg^{-1}.°C^{-1}$; Helfmann et al., 1981), it is about half the sensitivity as calculated from the opesophageal cooling experiments. If one assumes that the TBTS value determined using the recovery method is a correct value then the values reported for the opesophageal cooling experiments are overestimations.

Apart from the high thermosensitivity resulting from the oesophageal cooling experiments there was some variation depending on whether T_{cw}, T_{col} or, T_{ra} were used to represent T_C (Fig.3). As explained above, the slightly lower sensitivity found when T_{cw} was used to represent T_C was probably related to the fact that the oesophageal thermode was directly influencing the thermocouple at this site. The exaggerated response found when T_{ra} was used to represent T_C may be due to the fact that this site represents a collecting point for "warm" venous return to the heart. Presumably most of the heat produced from the shivering response is derived from the pectoral muscles and that the higher temperature of the blood in the right atrium may simply reflects the effect of the increased heat content of the venous return from the shivering muscles.

It could be argued that the apparent overestimation of TBTS as determined from the oesophageal cooling experiments may reflect the fact that temperature sensors are being directly influenced by the thermode (see Fig.2; Mercer & Simon, 1984). Certainly, it seems unlikely that there are thermosensitive structures located within the walls of the oesophagus itself. The cooling response of the animals whose oesophagus had been locally anaesthetized would seem to negate this idea. In fact, the increased thermosensitivity following local application of the anaesthetic would rather indicate that the opposite effect had occurred. i.e. the oesophagus had become more thermosensitive. Since the local effect that this substance may have in birds is unknown, it is difficult to speculate as to the cause of the increased thermosensitivity.

If temperature sensors were being directly stimulated by the oesophageal thermode one would expect to find a hysteresis in the curve relating T_C to M. While there is no clear hysteresis in the curves presented in Fig.2, those shown in Fig.3 are in a reverse direction to that expected if temperature sensors were located close to the thermode (see Fig.2, Mercer & Simon, 1984). The hysteresis in the curves shown in Fig.3 may be the result of the fact that the animals cannot increase or decrease metabolic heat production rapidly enough to match the relatively rapid changes in heat extraction which occur at the onset and end of oesophageal cooling. Thus, during the initial phase of oesophageal cooling there may be an imbalance between the amount of heat being extracted by the thermode and the amount being produced metabolically. Consequently, assuming that dynamic effects can be excluded, the initial fall in T_C is probably greater than would normally occur if the heat extraction had occurred more gradually. Likewise, there may also be an imbalance between the amount of heat extracted and produced at the end of the perfusion period. In this case the rate of decrease in metabolic heat production is slower than the relatively rapid termination of heat extraction via the thermode, resulting in a slight increase in T_C. Under these circumstances a simple two point determination of TBTS (comparing the changes in M and T_C between the changes in M and T_c between the start and the end of the cooling period) should give the most reliable result. However, two point determination of TBTS using the data presented in Fig.3 resulted in rather high values e.g. 22 $W.kg^{-1}.C^{\circ-1}$ when T_{col} was taken to represent T_b. This value is about 2.5 times greater than that found using the "recovery" method.

The "reverse" hysteresis response resulting from the oesophageal cooling experiments, in which T_c at the end of the perfusion period approaches or was even greater than its value at the start of the perfusion period gives the impression of a positive feed-back response. Since the overestimations of TBTS resulting from these experiments cannot be readily explained on the basis of thermal inputs, it is possibile that non-thermal factors are involved. However, the present data is insufficient to confirm this idea.

REFERENCES

Helfmann, W., Jannes, P., and Jessen, C., 1981, Total body thermosensitivity and its spinal and supraspinal fractions in the conscious goose, Pflugers Arch., 391:60.

Ionomoto, T., and Simon, E., 1981, Extracerebral deep-body cold sensitivity in the Pekin duck, Am. J. Physiol., 241:R136

Jessen, C., Mercer, J.B., and Puschmann, S., 1977, Intravascular heat exchanger for conscious goats, Pflugers Arch.,368:263.

Mercer, J.B., 1985, Use of an oesophageal thermode for the determination of total body thermosensitivity in conscious eider ducks (Somateria mollissima), Acta Physiol Scand., 124:S542.

Mercer, J.B., and Simon, E., 1984, A comparison between total body thermosensitivity and local thermosensitivity in mammals and birds, Pflugers Arch., 400:228.

Simon, E., Pierau, Fr-k., and Taylor, D.C.M., 1986, Central and peripheral thermal control of effectors in homeothermic temperature regulation, Physiol. Rev., 66:235.

Simon-Oppermann, C., Simon, E., Jessen, C., and Hammel, H.T., 1978, Hypothalamic thermosensitivity in conscious pekin ducks, Am. J. Physiol. 235:R130.

ON THE THERMOSENSITIVITY OF THE SPINAL CORD IN PIGEONS

Randi Eidsmo Reinertsen and Claus Bech

Department of Zoology
University of Trondheim
N-7055 Dragvoll, Norway

INTRODUCTION

Since spinal thermosensitivity seems to be much more important than that of the hypothalamus in evoking appropriate thermoregulatory reflexes in birds, this has formed the subject of a number of studies. Appropriate thermoresponses during spinal cooling have been demonstrated for several species, e.g. the pigeon (Rautenberg et al., 1972), the Adelie penguin (Hammel et al., 1976), the goose (Helfmann et al., 1981) and the Pekin duck (Inomoto and Simon, 1981). The local thermosensitivity of the thermosensory areas that have been identified has been determined by studying the maximum response of a thermoregulatory effector to an experimentally induced temperature change. The spinal cord of the pigeon, in particular, is highly sensitive with changes in heat production being linearly related to changes in the spinal cord stimulus. Values of between -0.9 and -2.05 W kg^{-1} °C^{-1} for the spinal sensitivity have been recorded for this species (Graf, 1980; Rautenberg, 1969). This vide range of values may be partly due to the use of different experimental methods. A general problem in assessing effector sensitivity to thermal stimulation of the spinal cord using the thermode technique, has been the difficulty in defining the stimulation temperature. The exact location of the temperature-transducing elements within and/or close to the spinal cord is not yet known wherefore their actual temperatures cannot be determined precisely.

Studies on spinal cord sensitivity provide information that is then utilized in building up analytical and descriptive models of the neuronal organization of thermoregulation. It is therefore important that the sensitivity value be quantitatively correct. Local or general changes in body temperature that act as load errors in homeothermic thermoregulation correspond to local or general changes in tissue heat content. It thus follows, as Simon et al. (1986) have pointed out, that effector responses that are dimensionally identical with the rate of change in heat content (W or W kg^{-1}) are the most suitable for any direct comparison of stimulus intensity with thermoregulatory effector activity.

Spinal cord thermosensitivity, however, has not generally been estimated in the above manner, but rather by relating the maximum activation of metabolic heat production to the experimentally induced decreases in local temperature. We have attempted to estimate the local thermosensitivity by relating the response in terms of metabolic heat production, to the heat subtracted by cold stimulation of the spinal cord. This study was made to gain some basic data for the thermosensitivity of the spinal cord using an experimental procedure that enabled a quantitative estimate to be made irrespective of the actual spinal cord temperature.

METHODS

Animals

The experiments were made on two adult pigeons (Columba livia) whose body weights varied between 400 g and 420 g during the experimental period. Between each experiment the animals were housed in individual cages and exposed to a diurnal rhythm of 8 hours light and 16 hours darkness, i.e. approximately the conditions in winter. The room temperature varied between 15°C and 20°C. They were given a special pigeon maintenance seed mixture, both this and water being available ad libitum.

Spinal cord thermode and implantation procedure

The thermode and implantation procedure was similar to that previously described by Rautenberg (1969). Under a general anaesthesia an aperture was made in the third thoracic vertebra from the dorsal side. The thermode, a 15 cm long, hairpin-shaped, piece of polyethylene tubing was inserted into the peridural space dorsal to the spinal cord, and thrust cranially approximately 8.0 cm. A second polyethylene tube was then inserted along the thermode tube for a distance of about 7.0 cm (fig. 1).

Before the start of each experiment, a thermocouple was also inserted into the colon, to measure the colonic temperature. Two thermocouples were inserted into the re-entrance tube to measure the temperature of the spinal cord (T_{sc}) at two different spots (fig. 2). Spinal cord temperature was measured at 1 cm, 2.5 cm, and 5.5 cm from the water inlet point of the thermode. The thermodes were perfused with water from a thermostatically-controlled water-bath which enabled us to keep the vertebral temperature at a constant, selected level. The tube connections to the water-bath were equipped with Y-tubes. Through these tubes we inserted thermocouples that measured water inlet temperature and water outlet temperature from the thermode. The flow rate was determined by collecting the water that was passed through the thermode during each stimulation period (fig. 1).

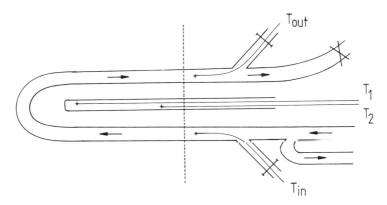

Fig. 1. Position of the spinal cord thermode and reentrant tube used for measuring spinal cord thermode temperature at two different positions (T_1 and T_2). Water inlet temperature (T_{in}) and water outlet temperature (T_{out}) from the thermode were measured at the point of entering and leaving the spinal cord area, respectively (dotted line).

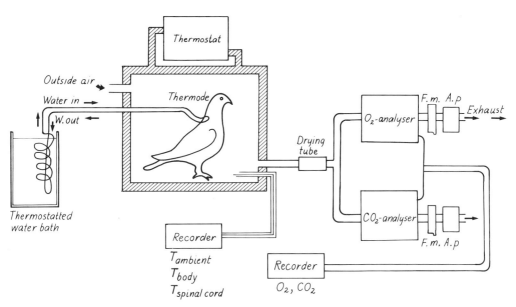

Fig. 2. Apparatus for measuring temperature sensitivity of the spinal cord and thermoregulatory responses to changes in spinal cord temperature. Spinal cord temperature is altered by means of a water-perfused thermode connected to a thermostatted water bath. Metabolic rate are measured by analyzing the effluent air for O_2 and CO_2 content. Ambient temperature, body temperature and spinal cord temperature are measured together with water inlet and outlet temperatures. The metabolic chamber is kept at a constant temperature. (F.m.=flowmeter; A.p.=air pump).

O_2-consumption measurements

During each of the above experiments the pigeon was kept inside a 7-litre brass chamber (fig. 2), surrounded by circulating water that kept the ambient temperature of the chamber (T_a) constant at 25°C. Oxygen consumption (VO_2) was measured by the open circuit method. Air was sucked through the chamber at a rate of approximately 1500 ml per minute. The oxygen and carbon dioxide concentrations of the effluent air were measured and recorded. The oxygen consumption values were corrected to STPD conditions and converted to heat production, using 1 ml O_2 (STPD) $g^{-1} h^{-1}$ equal to 5.582 W kg^{-1}, which is assuming a respiratory exchange ratio of 0.8.

Experimental procedures

After a control period, the spinal cord thermode of the pigeon being studied was perfused, for periods of 10 minutes duration, with water at a constant (thermostatically controlled) inlet value that varied between 35 and 15°C. The various perfusion periods were separated by at least 20 minute intervals during what no perfusion occured. Six different water inlet temperature values were used during each experimental series. Spinal cord temperature was directly measured each time at two different sites; either at 1 cm and 2.5 cm, 1 cm and 5.5 cm, or 2.5 cm and 5.5 cm. Spinal cord thermosensitivity was then calculated in terms of the measured increase in metabolic heat production (in Watts per kilo body weight) in relation to the induced loss of heat (also in Watts per kilo body weight) during each cold stimulation of the vertebral canal.

RESULTS

Figure 3 shows the effect of the experimental cooling of the spinal cord on oxygen consumption and body temperature during two successive stimulation periods in one of the experimental series. The ambient temperature was 25°C. During the first stimulation period with a water inlet temperature of 25°C, the spinal cord temperature decreased to 34.5°C at a distance of 2.5 cm from the inlet point, but decreased to only 38°C at 5.5 cm from the inlet point. An inlet temperature of 16°C resulted in decreases in the respective spinal cord temperatures to 29°C and 37°C. Thus we see that the position of the thermocouple has a great influence on the temperature fall induced by a particular stimulation temperature value. The spinal cooling experiments also induced corresponding increases in oxygen consumption, and in the colonic body temperature, which rose during the final stages of each stimulation period, due to the excess heat production.

The values obtained for the two pigeons are shown in figs. 4-6, the changes in metabolic heat production being plotted against the measured temperature of the spinal cord at 5.5 cm from the water inlet point (fig. 4), at 2.5 cm (fig. 5) and at 1.0 cm (fig. 6). Fig. 4 shows that there was a close linear relationship between the stimulaton intensity of the spinal cord and the metabolic response. A linear regression analysis, based on the 36 average ΔM values obtained for the two pigeons for the various perfusion temperatures, resulted in the line: $\Delta M = 75.31 - 1.85 \times T_{sc}$ with $r^2 = 0.77$. The slope of the regression

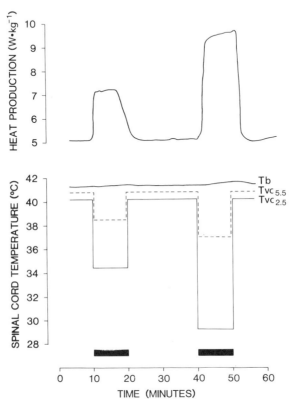

Fig. 3. Courses of body temperature measured in the colon (T_b) and vertebral canal temperature measured at 5.5 and 2.5 cm from the water inlet point ($Tvc_{5.5}$, $Tvc_{2.5}$) during cooling of the vertebral canal in a consious pigeon. The effect on heat production is shown in the upper part of the figure. Ambient temperature was 12°C; bars mark the thermal stimuli.

line indicates a local thermosensitivity of -1.85 W kg^{-1} °C^{-1} when related to the temperature of the spinal cord. When a regression analysis was made to find the relationship between $\triangle M$ and the measured spinal cord temperature at a point 2.5 cm from the water inlet point (fig. 5), the $\triangle M$ values was = $22.59 - 0.59 \times T_{sc}$, with $r^2 = 0.77$. The slope of the regression line in this case indicates a local thermosensitivity of -0.59 W kg^{-1} °C^{-1}. When the final regression analysis was made, for the relationship between $\triangle M$ and the spinal cord temperature at a point 1.0 cm from the water inlet point (fig. 6), the result was $\triangle M = 13.54 - 0.38 \times T_{sc}$ with $r^2 = 0.65$. The slope of the regression line in this case indicates a local thermosensitivity of -0.38 W kg^{-1} °C^{-1}.

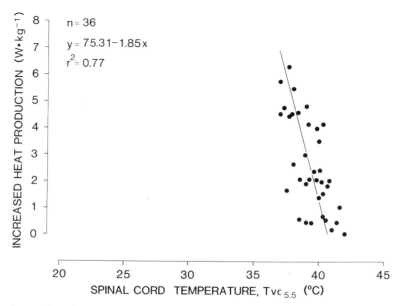

Fig. 4. Relation of metabolic responses to temperature of vertebral canal measured at a point 5.5 cm from the water inlet point (Tvc$_{5.5}$). Values are averages of stimulation periods of at least 10 minutes.

The above results have revealed the existence of an appreciable thermal gradient within the spinal cord during cold stimulation. The cold sensitivity of the spinal cord, when estimated in terms of the increase in metabolic heat production in relation to the induced decrease in spinal cord temperature, is seen to be strongly dependent on the actual spinal cord temperature, i.e. on the point at which the measurement is made. Our results indicate spinal sensitivities between -0.38 and -1.85 W kg^{-1} °C^{-1}. It is interesting that these values are in the same range as those previously published, viz. -0.9 to

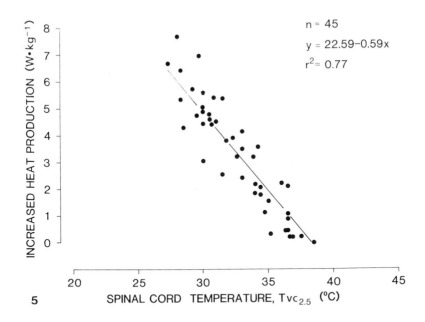

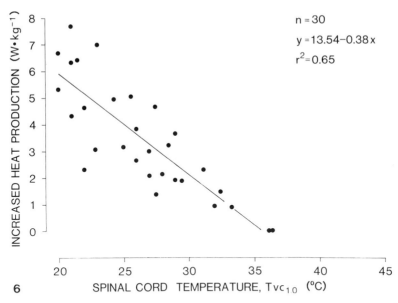

Fig. 5 and 6. Relation of metabolic responses to temperature of vertebral canal measured at a point 2.5 cm from the water inlet point ($Tvc_{2.5}$) (fig. 5, upper) and at a point 1.0 cm from the water inlet point ($Tvc_{1.0}$) (fig. 6, lower). Values are averages of stimulation periods of at least 10 minutes.

-2.05 W kg^{-1} °C^{-1}. Thus, the variation in cold sensitivity of the spinal cord reported in previous studies, may well be due to the variation in the sites at which the spinal cord temperature was measured. This makes the use of such values less suitable in accessing the contribution of the individual thermosensitive areas to the proportional control of thermoregulation.

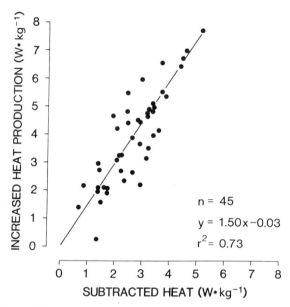

Fig. 7. Relationship between heat producton and heat subtracted by cooling of the vertebral canal.

Figure 7 shows the changes in metabolic heat production (ΔM) in relation to the measured amount of heat (ΔH) removed from the vertebral canal during the same experimental series shown in figs 4-6. A linear regression analysis, based on all the 45 average ΔM values obtained for the two pigeons for the various experimental perfusion temperatures, resulted in the relationship: $\Delta M = -1.50 \times 0.03 \Delta H$, with $r^2 = 0.73$. The slope of the regression line indicates a thermosensitivity for the particular stimulated part of the spinal cord of 1.50 W kg^{-1}/W kg^{-1}. In regard to the cold sensitivity of the spinal cord, Mercer and Simon (1984) have pointed out that the uncertainty that exists about the actual temperature at the unknown sites at which the spinal thermosensors were located, will result in a greater degree of uncertainty about the "true" sensitivity of this thermosensitive area as a whole. The way in which we estimated spinal sensitivity does not yield information about the position of the thermode relative to the location of the thermosensors. However, since the above calculation does not involve spinal cord temperature, the sensitivity value, expressed in terms of the increase in metabolic heat production in relation to the induced heat loss in the vertebral canal, is independent of the actual measurements of spinal cord temperature. In addition, with such a method it is possible to avoid situation AIII which is suggested in Fig. 8. In that situation the thermode would have been correctly placed, but the thermometer was situated away from the stimulated area; if anything, an inappropriate thermosensitivity would be disclosed, because the body temperature outside the cooled area will change in the opposite direction to the stimulation.

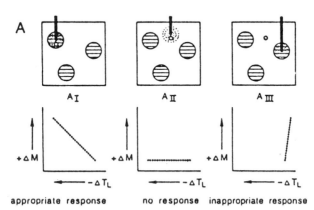

Fig. 8. Theoretical effects of local cooling in the body attempting to evaluate the sensitivity of a circumscribed area with a high density of thermosensory structures by determining relationship between the measured change in local temperature with the metabolic response. The black rod is the cooling thermode, the hatched circles are sites of high thermosensory density. In AIII the thermode is correctly placed, but the thermometer is placed outside of the cooled area. After Mercer and Simon (1984), reproduced with permission of the authors.

CONCLUSION

We think that this method does seem to provide a reliable tool for estimating spinal cord thermosensitivity. By allowing controlled variation in both the flow rate and the temperature of the water perfusing the thermode, this method also makes separate studies of the static and the dynamic thermoregulatory responses to cold stimulation of the spinal cord possible. Such studies may, in their turn, yield valuable information about the neuronal organization of thermoregulation.

ACKNOWLEDGEMENT

This study was supported by The Norwegian Research Council for Science and the Humanities (D.65.46.124).

REFERENCES

Graf, R., 1980, Diurnal changes of thermoregulatory functions in pigeons. II Spinal thermosensitivity. Pflügers Arch., 386:1.

Hammel, H.T., Maggert, J., Kaul, R., Simon, E. and Simon-Oppermann, C., 1976, Effects of altering spinal cord temperature on temperature regulation in the adelie penguin, Pygoscelis adeliae, Pflügers Arch. 362:1.

Helfmann, W., Jannes, P., and Jessen, C., 1981, Total body thermosensitivity and its spinal and supraspinal fractions in the conscious goose, Pflügers Arch., 391:60.

Inomoto, T., and Simon, E., 1981, Extracerebral deep-body cold sensitivity in the Pekin duck, Am. J. Physiol., 241:R136.

Jessen, C., Helfmann, W., and Jannes, P. 1980, Effects of cooling preferantially the head of the conscious goose. In: "Contribution to thermal physiology", Z. Szelényi and M. Székely eds., Adv. Physiol. Sci. Vol. 32, Pergamon Press.

Mercer, J.B., and Simon, E., 1984, A comparison between total body thermosensitivity and local thermosensitivity in mammals and birds, Pflügers Arch. 400:228.

Rautenberg, W., 1969, Die Bedeutung der zentralnervösen Thermosensitivit die Temperaturregulation der Taube. Z. vergl. Physiol., 62:235.

Rautenberg, W., Necker, R., and May, B., 1972, Thermoregulatory responses of the pigeon to changes of the brain and the spinal cord temperatures, Pflügers Arch. 338:31.

Simon, E., Pierau, F.-K., and Taylor, D.C.M., 1986, Central and peripheral thermal control of effectors in homeothermic temperature regulation, Physiol. Rev. 66:235.

SHIVERING AND NONSHIVERING THERMOGENESIS IN BIRDS:

A MAMMALIAN VIEW

Eamonn Connolly, Jan Nedergaard and Barbara Cannon

Department of Metabolic Research
The Wenner-Gren Institute, Biologihus F3
The University of Stockholm, S-106 91 Stockholm

INTRODUCTION

Homeothermia, the ability to maintain a stable body temperature which is higher than that of the environment, is only found in two classes of animals, mammals and birds. Both of these groups of animals must have well-developed and accurately controlled thermoregulatory mechanisms allowing them to survive not only in a wide variety of habitats, but also in regions where diurnal or seasonal temperature fluctuations may be large. Whereas regulatory thermogenesis in mammals has been the object of intense investigation, the relative importance of shivering and nonshivering thermogenesis in birds has been much less studied. This review will attempt to draw attention to several basic principles learned from the mammalian studies which may be applied in order to obtain a further understanding of thermoregulation in birds. Several comprehensive reviews have appeared dealing with other aspects of thermoregulation in birds (Dawson et al., 1983; Hissa, 1988).

NONSHIVERING THERMOGENESIS AND ACCLIMATION TO COLD

Nonshivering thermogenesis consists of both an obligatory component (i.e. the heat production associated with general metabolism) and a facultative component which is a specific increase in heat production in response to a change in environmental temperature. In the following discussion we will confine ourselves to nonshivering thermogenesis which is facultative, cold-induced and, of course, not associated with muscle contraction.

Some features of cold acclimation in the mammal are illustrated in Fig. 1. A mammal that has been living at thermoneutral temperature has a metabolic rate which is the minimal or basal metabolic rate (BMR). This is the heat production involved in maintenance metabolism and this rate is generally unchanged by adaptation to different environments. If the warmacclimated animal is exposed to a markedly cold environment ($5^{o}C$), there is an immediate rise in the metabolic rate which is accompanied by a vigorous activity in the muscles of the animal (Fig. 1). Thus, in the first phase of cold exposure, shivering thermogenesis is essential for maintenance of body core temperature.

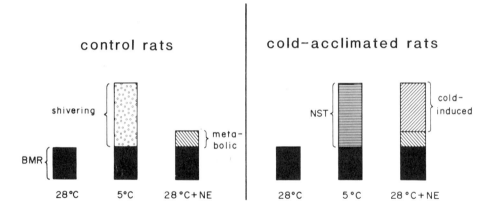

Fig. 1. Principles of mammalian cold acclimation. "NE" indicates norepinephrine administration. "NST" indicates nonshivering thermogenesis

If the animal is now exposed to the cold environment for a period of several weeks so that it becomes cold-acclimated, the basal metabolic rate (measured at thermoneutrality) remains unchanged. However, in the cold, the metabolic rate remains high even though the shivering activity of the muscles has fully disappeared (Fig. 1). The progress of cold acclimation can be followed by examining a gradual replacement of shivering thermogenesis by nonshivering thermogenesis (Hart et al., 1956; Fig. 2).

Thus, chronic cold exposure causes the animal to adapt to the new environment by recruiting an increased capacity for nonshivering thermogenesis. As this capacity increases, the need for the uncomfortable shivering thermogenesis is reduced and finally removed. The conclusion of such observations is, therefore, that nonshivering thermogenesis is the only source of extra heat in the cold-adapted rodent.

To fully understand the mechanisms of regulatory thermogenesis in birds, several important questions must be answered. Is there in birds a nonshivering thermogenesis which can be recruited by acclimation to cold? If this is so, what is the physiological mediator of the heat production? Which organ or tissue in the bird is responsible for the cold-induced nonshivering thermogenesis and what is the mechanism by which the heat is produced? From the mammalian point of view, do birds have brown fat or a brown fat-like tissue? We will now review the evidence presently available and thereby attempt to answer these questions.

IS NONSHIVERING THERMOGENESIS PRESENT IN BIRDS?

El Halawani et al. (1970) examined leghorn chickens which were acclimated to a warm (21°C) or cold (5°C) temperature for up to nine months. During the acclimation period the metabolic rates of the cold-acclimated group were higher than those of the warm-acclimated group when both were measured at 5°C, suggesting an increased capacity for

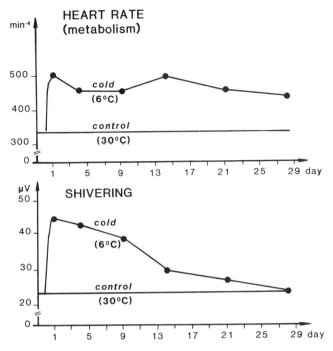

Fig. 2. Shivering and nonshivering thermogenesis during cold-acclimation in the mammal. Animals were exposed to the indicated temperatures on day 0. Adapted from the data of Hart et al. (1956).

thermogenesis in the cold-acclimated birds. The most striking effect described was the almost complete disappearance of muscular shivering activity after cold acclimation, despite the consistently raised metabolic rate (Fig. 3A). The warm-acclimated birds however, continued to show the shivering response when exposed to cold (Fig. 3A). Thus, the cold-acclimated birds must have recruited a capacity for nonshivering thermogenesis as a response to the cold stress. One feature of this study was the very long acclimation times (5 months) necessary to achieve the disappearance of shivering, which can be compared with only 3 weeks in the rat (Fig. 2).

In muscovy ducklings that had been warm- ($25^{o}C$) or cold- ($4^{o}C$) acclimated for 5 weeks, Barré et al. (1986a) were able to show that there was a large increase in the metabolic rate of the cold-acclimated birds with decreasing temperature which could not be related to shivering activity (Fig. 3B). No such cooling-induced nonshivering thermogenesis could be observed in the warm-acclimated ducklings.

Other than these two examples, there are no studies where the parameters necessary to show nonshivering thermogenesis in birds have been fully examined, i.e. both shivering and metabolic rate (total thermogenesis) in both warm- and cold-acclimated birds.

Based on the evidence available, however, nonshivering thermogenesis is probably present in birds, but such work should be extended and confirmed in more rigorous examinations of other bird species.

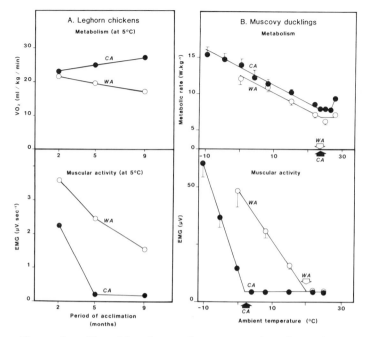

Fig. 3. Nonshivering thermogenesis in birds. Adapted from the data of (A) El-Halawani et al. (1970) and (B) Barré et al. (1986a). "WA" and "CA" indicate warm- and cold-acclimated.

THE MEDIATOR OF NONSHIVERING THERMOGENESIS

In the mammal, the involvement of the sympathetic nervous system and adrenergic stimulation in nonshivering thermogenesis is undisputed. When a mammal which has been acclimated to thermoneutrality is injected at thermoneutrality with norepinephrine, a small but significant rise in metabolism occurs. This response can be attributed to both a small increase in the activity of the minimal amount of brown adipose tissue that is present in the warm-acclimated animal and to "metabolic" effects of norepinephrine on other tissues in the body that are not directly involved in cold-induced thermogenesis (Fig. 1). In the cold-acclimated animal, the increase in metabolic rate in response to norepinephrine markedly increases (Fig. 1). This response to norepinephrine consists of the small response seen in the warm animals combined with a large cold-induced capacity for nonshivering thermogenesis.

On this basis, one can define a simple test to demonstrate this increased nonshivering thermogenic capacity. Both warm-acclimated (preferably thermoneutral) and cold-acclimated animals are taken and their metabolic rates are measured. Then norepinephrine is injected and the increase in metabolic rate observed. If the cold-adapted animal has an increased capacity for nonshivering thermogenesis, then the response to the injection will be much greater than that of its warm-acclimated counterpart (Fig. 4). However, in order to do this simple test, it is necessary to know the physiological inducer of nonshivering thermogenesis in the animal under study.

It is thus of interest to examine the effects of catecholamines in birds. Indeed norepinephrine has been shown under certain circumstances to elevate metabolic rate in the pigeon (Hissa et al., 1975). This obser-

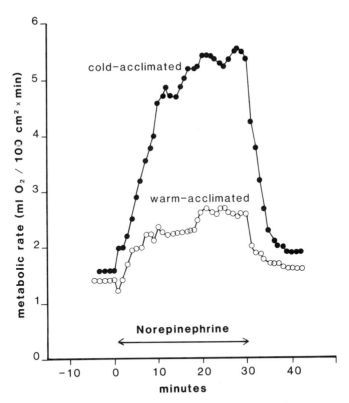

Fig. 4. Recruitment of brown adipose tissue-mediated nonshivering thermogenic capacity after cold acclimation of the rat. Data adapted from Himms-Hagen et al. (1975).

vation in itself, however, merely shows that norepinephrine can induce a metabolic effect (as in the warm-acclimated rat, Figs. 1 and 3). To imply the existence of cold-induced nonshivering thermogenesis, the observation must be made in both warm- and cold-acclimated birds at the same temperature (preferably thermoneutral), with the cold birds showing an elevated response.

The peptide hormone glucagon has also been implicated as a mediator of heat production in birds. King penguin chicks respond to a glucagon infusion by increasing their metabolic rate (Barré and Rouanet, 1986). Furthermore, when the birds were placed in the cold, their metabolic rate increased as shivering started and the subsequent administration of glucagon induced a smaller increase in oxygen consumption and transiently eliminated shivering (Barré and Rouanet, 1986). Whereas this might seem to suggest that a glucagon-mediated response was responsible for the elevation of metabolic rate in the cold, a similar result would also be obtained if glucagon could elevate non-regulatory thermogenic or metabolic processes, which by producing heat allowed the animal to stop shivering for a short time. Again, without a demonstration of an increased response to the hormone in cold-acclimated birds, the data are not conclusive. Unfortunately, such studies have not been performed in birds and until they are, the very existence of nonshivering thermogenesis as well as its physiological mediator in birds will remain a matter of debate.

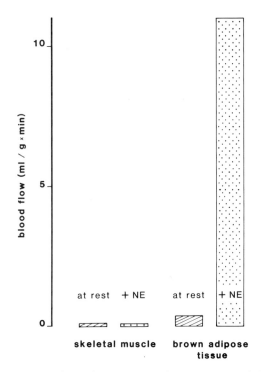

Fig. 5. Blood flow in cold-acclimated rat. Adapted from the data of Foster and Frydman (1978).

THE SITE OF NONSHIVERING THERMOGENESIS

Until the late 1970's it was by no means clear which tissue was responsible for cold-induced nonshivering thermogenesis in the mammal. However, Foster and Frydman (1978, 1979) investigated the problem by examining the changes in blood flow to different tissues of the cold-acclimated rat, compared to its warm-acclimated counterpart. The tissue responsible for producing the large amounts of heat necessary for keeping the body temperature stable must be receiving a copious blood supply, both to provide oxygen for respiration and to transport away the heat to the rest of the body. Using radiolabelled microspheres, these authors showed that norepinephrine could induce a large increase in the percentage of cardiac output that was directed to the brown adipose tissue and the blood flow through the tissue was elevated about 10-fold (Foster and Frydman, 1978, 1979; Fig. 5). No other tissue showed such an elevation in blood flow and, largely based on this work, it is now generally accepted that brown adipose tissue is solely responsible for norepinephrine-mediated (cold-induced) nonshivering thermogenesis in the small mammal.

Current research on bird thermogenesis in some ways resembles the situation in mammalian research just before the blood flow studies of Foster and Frydman in that there seems to be no direct evidence implicating a specific tissue in bird nonshivering thermogenesis. There are no blood flow studies performed on birds, although it would appear that experiments of this type could pinpoint any tissue involved.

The effector organ or tissue in the bird responsible for nonshivering thermogenesis could be (1) skeletal muscle, (2) brown adipose tissue or

(3) none of these. We will present and discuss the data implicating these alternatives below.

Skeletal muscle

In the cold-acclimated rat there are well-documented changes in mitochondria induced by cold acclimation (Himms-Hagen, 1975; Fig. 6A). Furthermore, deacclimation studies (Himms-Hagen, 1975) clearly showed that the disappearance of these adaptations in the muscles correlated fully with the falling metabolic rate of the animal. Thus, skeletal muscle was heavily implicated as the site of nonshivering thermogenesis in the cold-acclimated rat.

However, although there is an increased blood flow to brown adipose tissue in the cold-acclimated rat during cold stress, there is very little change in the blood flow to the muscles (Foster and Frydman, 1978; Fig. 5) and these adaptive changes in skeletal muscle are no longer thought to reflect nonshivering thermogenesis.

Changes in the muscles of birds have also been described and, taking the example of mitochondrial parameters, a complete parallel to the rat can be seen in bantam chicks that have been acclimated to cold for 2 weeks (Aulie and Grav, 1979; Fig. 6B). Other changes in muscle mitochondrial metabolism have been demonstrated (Barré et al., 1986b; Barré and Nedergaard, 1987). It is of course possible that the changes in muscle are a consequence of "training" induced by shivering and that in neither the rat nor the bird are they in themselves indices of nonshivering thermogenic capacity.

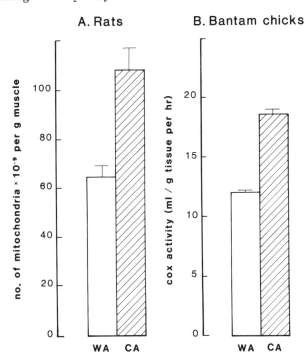

Fig. 6. Adaptive changes in muscle after cold acclimation. "WA", warm-acclimated, "CA", cold-acclimated. Adapted from (A) Himms-Hagen et al. (1975) and (B) Aulie and Grav (1979).

Brown adipose tissue

The importance of brown adipose tissue for nonshivering thermogenesis in the mammal has naturally led to the idea that this tissue may also be present in birds. One might suggest that the possession of brown adipose tissue may be a prerequisite for homeothermy. To date there are several articles which claim to have identified possible deposits of brown adipose tissue in birds (Lockenbill and Cohen, 1966; Oliphant, 1983; Barré et al., 1986a) but also one which claims the complete absence of the tissue (Johnston, 1971).

In order to identify brown adipose tissue (in the mammalian sense), three major criteria must be fulfilled which are known characteristics of the active tissue:
(1) the cells of the tissue must have multilocular fat deposits,
(2) the cells must have abundant mitochondria with densely packed cristae, and
(3) most importantly, the tissue must be shown to possess a hormonally-controlled mechanism which allows large amounts of heat production to occur.

We can now examine the evidence for brown adipose tissue in birds using these guidelines.

Multilocular adipocytes. It is certainly true that adipose tissue in the bird can contain multilocular cells which have an appearance similar to mammalian brown adipocytes (Luckenbill and Cohen, 1966; Oliphant, 1983; Barré et al., 1986a). However, multilocularity is in itself only an indication of high fat turnover and can therefore be observed even in white adipose tissue under certain conditions.

Mitochondria. Multilocular adipocytes containing reasonably densely packed mitochondria have been reported in two bird species, the chickadee and the grouse (Oliphant, 1983). The mitochondria shown in this study had distinct cristae and were reminiscent of those seen in active mammalian brown adipose tissue (see Suter, 1969; Thomson et al., 1969). Thus, on the basis of the first two criteria, some birds appear to possess brown adipose tissue. However, in the absence of a confirmation that this tissue is thermogenically active and contains a mechanism for thermogenesis, there is no reason to assume that the "brown adipose tissue" seen in the chickadee and the grouse is not merely metabolically active white fat.

Thermogenic mechanism. In the mammal, the mechanism of heat production in response to norepinephrine stimulation has been almost fully elucidated (for reviews see Nedergaard and Lindberg, 1982; Cannon and Nedergaard, 1985ab; Trayhurn and Nicholls, 1986). Thermogenesis occurs in response to adrenergic stimulation of the brown adipose tissue. Although the tissue possesses both $alpha_1$- and beta-adrenergic receptors, it is mainly the action of norepinephrine on the beta-adrenergic receptor that gives rise to thermogenesis. Occupancy of the beta-receptor leads to the activation of adenylate cyclase and the consequent elevation of cAMP levels within the cells. This leads, via protein kinase activity, to a stimulation of hormone-sensitive lipase and mobilization of stored triglyceride within the brown fat cell.

The free fatty acids released are not only substrates for mitochondrial respiration, but are also thought to act directly, or via fatty-acyl-CoA derivatives, on the inner mitochondrial membrane uncoupling protein thermogenin which is unique to brown fat mitochondria.

NORMAL MITOCHONDRIA

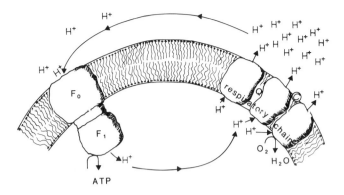

BROWN FAT MITOCHONDRIA

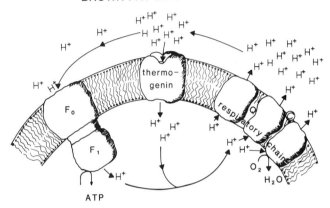

Fig. 7. The action of the uncoupling protein thermogenin in brown fat mitochondria.

In normal mitochondria, respiration leads to the build-up of a transmembranal proton gradient which can only be dissipated by re-admission of the protons into the mitochondrial matrix through the F_1-ATPase (Fig. 7). The movement of protons is thus <u>coupled</u> to the synthesis of ATP. In brown fat mitochondria, thermogenin provides an alternative pathway for protons to re-enter the mitochondria without passing through the F_1-ATPase (Fig. 7). Thus, when thermogenin (which functions as a "proton-equivalent" transporter) is opened by the presence of fatty acids or their acyl-CoA derivatives, respiration is partially <u>uncoupled</u> from ATP synthesis and the energy stored in the fatty acid substrate is dissipated as heat. This mechanism is unique to brown fat, making this tissue a specialized thermogenic organ.

Thermogenin also has the capacity to bind purine nucleotides and it is thought that ATP and ADP bind to the protein in-vivo in order to maintain coupled respiration in times of low thermogenic demand. After adrenergic stimulation, fatty acids become available which can overcome the inhibitory action of purine nucleotide binding to thermogenin, thereby inducing uncoupling.

In most physiological situations, the thermogenic activity of brown adipose tissue is directly proportional to the thermogenin content of the tissue. The thermogenin content of brown fat can be conveniently measured by examining the binding of guanosine-5´-diphosphate (GDP) to isolated mitochondria from the tissue. Thus, the amount of GDP binding to brown fat mitochondria (per mitochondrial protein) can be recruited to a higher level after chronic exposure of the animal to the cold, indicating an elevated nonshivering thermogenesis in the cold animal. Thermogenin can also be quantified immunologically.

It is important to point out that without this well-defined mechanism for heat production, the role of mammalian brown adipose tissue in nonshivering thermogenesis would be difficult to justify.

In birds, no such functional studies have been performed on the so-called "brown fat". If active brown fat is postulated in the birds, a demonstration of a mechanism for thermogenesis in the tissue is essential.

Other sites of nonshivering thermogenesis

It is of course possible that birds have some other tissue which is brown adipose tissue-like in being thermogenic. Since fatty acids seem to provide the thermogenic substrate in cold-exposed birds (see Dawson et al., 1983), it is an intriguing possibility that a mechanism whereby fatty acids themselves can induce uncoupling in tissues, as described by Andreyev et al. (1988), could be operative in birds. In this model, the free fatty acids are transported specifically on the adenine nucleotide translocase and are not suggested merely to permeate the lipid bilayer. Such a mechanism could in principle be occurring in the liver, muscles or any other tissue containing significant amounts of mitochondria.

CONCLUSIONS

To date there are very few studies that clearly demonstrate the presence of cold-induced, facultative nonshivering thermogenesis in birds. It is important to stress that such demonstrations must involve an enhanced response in an a bird that has been acclimated to a cold temperature, compared to its warm-acclimated counterpart. Based on the available data, however, we consider it likely that nonshivering thermogenesis exists in birds.

The neurohumoral mediator of nonshivering thermogenesis in birds has yet to be identified. The mere fact that a hormone can induce thermogenesis is, in itself, not sufficient; rather, the demonstration of a clear hormone response which can be specifically recruited when the bird is cold-acclimated is necessary.

The site of nonshivering thermogenesis in birds is still a matter of debate. The blood flow studies performed in cold-acclimated mammals were instrumental in proving the unique role of brown adipose tissue in nonshivering thermogenesis. Similar studies might also shed light on the role of the various tissues in bird nonshivering thermogenesis.

The occurrence of an adipose tissue bearing some of the characteristics of mammalian brown fat, has been occasionally reported in birds. However, in the absence of detailed biochemical data to confirm that this tissue possesses a thermogenic mechanism which is functionally recruited in the cold-acclimated animal, the role of such adipose deposits in nonshivering thermogenesis will remain mere speculation.

REFERENCES

Andreyev, A. Y., Bondareva, T. O., Dedukhova, V. I., Mokhova, E. N., Skulachev, V. P. and Volkov, N. I., 1988, Carboxyatractylate inhibits the uncoupling effect of free fatty acids, FEBS Lett., 226:265.

Aulie, A. and Grav, H. J., 1979, Effect of cold acclimation on the oxidative capacity of skeletal muscles and liver in young bantam chicks, Comp. Biochem. Physiol., 62A:335

Barré, H., Cohen-Adad, F., Duchamp, C. and Rouanet, J. L., 1986a, Multilocular adipocytes from muscovy ducklings differentiated in response to cold acclimation, J. Physiol., 375:27.

Barré, H., Nedergaard, J. and Cannon, B., 1986b, Increased respiration in skeletal muscle mitochondria from cold-acclimated ducklings: uncoupling effects of free fatty acids, Comp. Biochem. Physiol. 85B:343

Barré, H. and Nedergaard, J., 1987, Cold-induced changes in Ca^{2+} transport in duckling skeletal muscle mitochondria, Am. J. Physiol., 252:R1046.

Barré, H. and Rouanet, J. L., 1986, Calorigenic effect of glucagon and catecholamines in king penguin chicks, Am. J. Physiol., 244:R758.

Cannon, B. and Nedergaard, J., 1985a, The biochemistry of an inefficient tissue: brown adipose tissue, Essays Biochem., 20:110.

Cannon, B. and Nedergaard, J., 1985b, Brown adipose tissue. The molecular mechanisms controlling activity and thermogenesis, in: "New perspectives in adipose tissue," A. Cryer and R. Van, eds., Butterworth, London.

Dawson, W. R., Marsh, R. L. and Yacoe, M. E., 1983, Metabolic adjustments of small passerine birds for migration and cold, Am. J. Physiol., 245:R755.

El Halawani, M. E., Wilson, W. O. and Burger, R. E., 1970, Cold-acclimation and the role of catecholamines in body temperature regulation in male leghorns, Poultry Sci., 49:621.

Foster, D. O. and Frydman, M. L., 1978, Nonshivering thermogenesis in the rat. II. Measurements of blood flow with microspheres point to brown adipose tissue as the dominant site of the calorigenesis induced by noradrenaline, Can. J. Physiol. Pharmacol., 56:110.

Foster, D. O. and Frydman, M. L., 1979, Tissue distribution of cold-induced thermogenesis in concious warm- or cold-acclimated rats reevaluated form changes in tissue blood flow: The dominant role of brown adipose tissue in the replacement of shivering by nonshivering thermogenesis, Can. J. Physiol. Pharmacol., 57:257.

Hart, J. S., Heroux, O. and Depocas, F., 1956, Cold acclimation and the electromyogram of unanesthetized rats, J. Appl. Physiol., 9:404.

Himms-Hagen, J., Behrens, W., Muirhead, M. and Hbous, A., 1975, Adaptive changes in the calorigenic effect of catecholamines: role of changes in the adenyl cyclase system and of changes in the mitochondria, Mol. Cell. Biochem., 6:15.

Hissa, R., 1988, Controlling mechanisms in avian temperature regulation: a review, Acta Physiol. Scand., 132 suppl. 567:1.

Hissa, R., Saarela, S. and Pyörnilä, A., 1975, Thermoregulatory effects of peripheral injections of monoamines on the pigeon, Comp. Biochem. Physiol., 51C:235.

Johnston, D. W., 1971, The absence of brown adipose tissue in birds, Comp. Biochem. Physiol., 40A:1107.

Luckenbill, L. M. and Cohen, A. S., 1966, The association of lipid droplets with cytoplasmic filaments in avian subsynovial adipose cells, J. Cell. Biol., 31:159.

Nedergaard, J. and Lindberg, O., 1982, The brown fat cell, Int. Rev. Cytol., 74:187.

Oliphant, L. W., 1983, First observations of brown fat in birds, Condor, 85:350.

Suter, E., 1969, The fine structure of brown adipose tissue. I Cold-induced changes in the rat, J. Ultrastruct. Res., 26:216.

Thomson, J. F., Habeck, D. A., Nance, S. L. and Beetham, K. L., 1969, Ultrastructural and biochemical changes in brown fat in cold-exposed rats, J. Cell. Biol., 41:312.

Trayhurn, P. and Nicholls, D. G. eds., 1986, "Brown adipose tissue", Edward Arnold Ltd., London.

MUSCULAR NONSHIVERING THERMOGENESIS IN COLD-ACCLIMATED DUCKLINGS

Hervé Barré, Claude Duchamp, Jean-Louis Rouanet,
André Dittmar and Georges Delhomme

Lab. Thermorégulation et Energétique de l'exercice C.N.R.S.
8, avenue Rockefeller 69373 Lyon cedex 08, France

INTRODUCTION

In small terrestrial mammals, the existence of nonshivering thermogenesis (NST) is fully demonstrated in new-borns, cold-acclimated and hibernating species. The activation of the sympathetic catecholaminergic system by cold or by diet is largely documented. Norepinephrine is considered as the mediator of NST and brown adipose tissue as the major site of this NST.

In adult birds, since the publication of the works of Steen and Enger (1957), Hart (1962) and West (1965), it was believed that shivering thermogenesis constitutes the only thermogenic defence mechanism against cold. Likewise more recent studies on winter-acclimation in adult pigeons by Saarela and Vakkuri (1982) and in juvenile black grouses by Rintamäki et al. (1983), have considered shivering thermogenesis to be the only mechanism of heat production in these birds when ambient temperature decreases below thermoneutrality.

The first clear demonstrations of the existence of a nonshivering thermogenesis in birds were made by El Halawani (1970) in 5- and 9-month-old cold-acclimated chickens and more recently by Barré et al. (1985 and 1986a) in cold-acclimated muscovy ducklings reared at 4°C Ta from the age of 1 wk by comparison with controls reared at 25°C Ta.

DEMONSTRATION OF A NST IN COLD-ACCLIMATED DUCKLINGS

Resistance to cold (Fig. 1). Acclimation to cold in ducklings is demonstrated by their better resistance to cold. Despite their smaller size and unfavorable surface-to-volume ratio, the 4 wk-old cold-acclimated (CA) ducklings can better maintain their body temperature on prolonged exposure to - 15°C than the controls of the same age (TNa) or of the same body mass (TNm).

Direct demonstration of NST in 4 wk-old CA ducklings. Nonshivering thermogenesis was assessed by the simultaneous measurement of metabolic rate and integrated electromyographic activity (EMG) at constant Ta ranging from - 10 to + 28°C (Fig. 2). Metabolic rate (MR) was measured

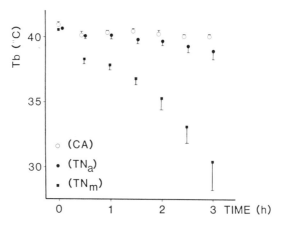

Fig. 1 Change in cloacal temperature (Tb) during 180 min. exposure to cold (Ta - 15°C). (See text for legend).

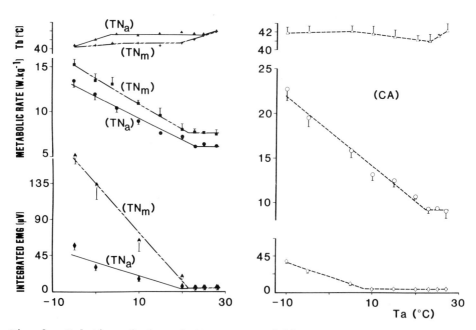

Fig. 2 Relation of cloacal temperature (Tb), metabolic rate and integrated muscle electrical activity (EMG) to ambient temperature (Ta) in 4 wk-old cold-acclimated ducklings (CA) and controls of same age (TNa) or body mass (TNm). n = 6 per Ta in each group.

by indirect calorimetry with an open-circuit system. Shivering was measured as integrated EMG activity of the gastrocnemius leg muscle, after systematic comparison showing parallel electrical activity in pectoral and gastrocnemius muscles. In cold acclimated ducklings, the shivering threshold temperature (STT) was 14°C lower than the lower critical temperature (LCT), whereas in controls, these thresholds were not significantly different. An interval of Ta observed between the STT and the LCT demonstrates per se a NST in a cold-acclimated young bird, a fact hitherto generally rejected. The capacity for NST in the CA ducklings estimated by MR in the cold at 8°C, just at the shivering threshold was 60 % above resting MR.

A potentiated shivering thermogenesis (PST). In 4 wk-old CA ducklings, the NST was ascribed to skeletal muscle itself on the grounds of the increased thermogenic efficiency of muscular activity. The responses of CA and control ducklings in the cold zone are compared in the figure 3, where both parameters (metabolic rate and integrated EMG activity) appear to be linked whatever the acclimation temperature. CA ducklings and controls exhibit a linear correlation of shivering with MR, but the slopes of the regression lines are significantly different between CA and control of the same age (TNa), and between CA and control of the same body mass (TNm), but not between both controls. Since the regression line for CA ducklings has not shifted in parallel, but is steeper than those of TN ducklings, the extra-heat production in CA ducklings is not merely additional to the thermogenesis level of controls, but increases with increasing shivering. In other words, a regulatory thermogenesis of muscular origin may be considered in each group of ducklings, but the thermogenic efficiency of this muscular activity as judged by integrated EMG appears to be increased by cold acclimation. In 4 wk-old CA ducklings, the NST is in fact a potentiated shivering thermogenesis (PST).

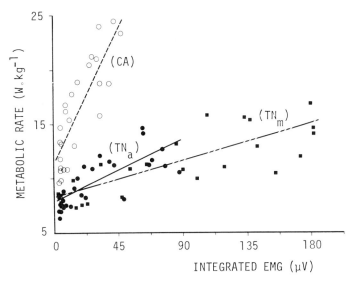

Fig. 3 Relation of metabolic rate to integrated EMG activity in 4 wk-old cold-acclimated (CA) and control ducklings of same age (TNa) or body mass (TNm).

A true NST. In a later stage of cold acclimation, a true NST develops in response to cold. In 6 wk-old CA ducklings (Fig. 4) instead of 4 wk-old CA ducklings, the range of cold ambient temperature where MR increases without any shivering activity is wider (21°C between LCT and shivering threshold versus only 14°C at the age of 4 wks) and the maximal amplitude of the NST is larger (74 % above resting MR versus only 60 % in 4 wk-old CA ducklings).

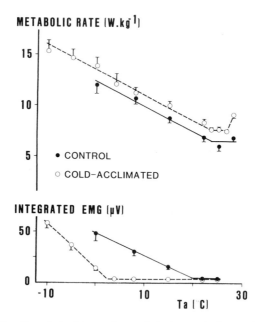

Fig. 4 . Relation of metabolic rate and integrated muscle electrical activity (EMG) to ambient temperature (Ta) in 6 wk-old cold-acclimated ducklings and in controls. n = 6 per Ta in each group.

GLUCAGON, A MEDIATOR OF NST IN BIRDS

The role of the glucagon in the control of heat production, particularly as a possible mediator of NST in birds, was first suggested by Freeman (1970) on the grounds of the potent glycogenolytic and lipolytic action of this hormone, both in vivo and in vitro. An increase in plasma glucagon level was found in CA ducklings by Barré et al. (1986b). In addition, a considerable increase in oxygen consumption was reported after glucagon injection in Japanese quails by Krimphove and Opitz (1975) and in winter-acclimatized king penguin chicks by Barré and Rouanet (1983), in birds exposed to thermoneutral ambient temperature. If glucagon is a potential mediator of NST in birds, chronic treatment with glucagon should mimic the state of cold adaptation, as does treatment with norepinephrine in rats. We have tried to induce artificially cold acclimation by repeated daily injections of glucagon to ducklings chronically glucagon-treated (GT) and maintained at thermal neutrality for 6 wks (Barré et al. 1987a).

NST induced by a glucagon-test injection in GT ducklings. In the thermoneutral zone at 25°C Ta, after the glucagon test injection in GT ducklings, there was a sudden increase in MR. Its maximum value (98 % above the saline control value) was reached 40 min after this injection. A plateau was maintained for the next 40 min, and then MR decreased progressively. In control ducklings, the glucagon-induced change in MR followed the same pattern as in GT ducklings, but the maximum value was only 29 % above the saline control value. In GT ducklings, an hyperthermia developed after glucagon injection, sometime up to death despite an active polypnea. The glucagon-induced increase in Tb reached 2.4°C in GT ducklings.

In the cold at 4°C Ta, no significant change in MR was observed in GT ducklings after the glucagon test injection, although within 1 min after the injection, shivering activity was completely suppressed. In control ducklings, inhibition of shivering activity followed the same pattern as in GT ducklings, but was concomitant with a 25 % decrease in MR and a 2.5°C decrease in Tb after the glucagon test injection. Under these conditions, in GT ducklings, glucagon was responsible for a true NST that may be estimated to be 55 % above resting MR. Since shivering is suppressed at 4°C Ta, this powerful calorigenic effect of glucagon was observed despite the absence of shivering at thermoneutrality and even despite the suppression of shivering in the cold. One is justified in considering it is a NST comparable with that described in CA small mammals receiving norepinephrine. Because a similar effect has been observed in juvenile penguins, glucagon can be considered as a mediator of NST in young birds during the period of growth.

THE PINK ADIPOSE TISSUE : A SPECIFIC DIFFERENTIATED ADIPOSE TISSUE IN COLD-ACCLIMATED DUCKLINGS

Since BAT is lacking in birds, the effector tissue responsible for NST must be different from that of mammals. Our results on 6 wk-old CA ducklings (Barré et al. 1986a) revealed the differentiation of a pink adipose tissue with multilocular adipocytes in response to cold acclimation, contrary to control ducklings with unilocular adipocytes. In CA ducklings, the vessel lumen of the pink adipose tissue with the presence of microvesicles for the transport of lipids through the endothelium, suggests the existence of an active lipolysis in this tissue and underlines the difference with the white adipose tissue (WAT) of controls. The significance of the profound structural changes observed in the pink adipose tissue of CA ducklings appears to be irrelevant to respiration itself : indeed the oxidative capacity judged from cytochrome oxydase activity of the pink adipose tissue, although much higher than that of WAT, remained very low as compared with the oxidative capacity measured in other tissues such as muscles or even liver from CA ducklings and such as BAT from CA rats (Barré et al. 1987b). Consequently, the important NST exhibited by 6 wk-old CA ducklings can be hardly attributed to the pink adipose tissue. But the liver and skeletal muscles would be some possible sites of such a thermogenesis.

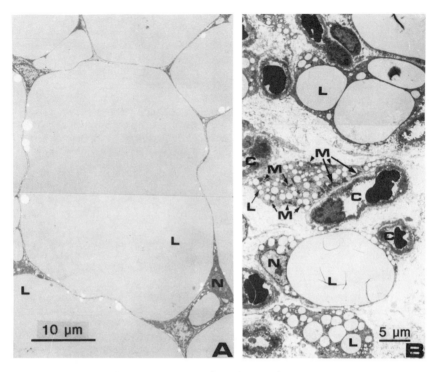

Fig. 5 Electron micrographs of A) white adipose tissue from control duckling (typical unilocular white adipocyte), B) pink adipose tissue from 6 wk-old cold-acclimated duckling (multilocular adipocytes with numerous lipid droplets (L), capillaries (C), mitochondria (M) and relatively large nuclei (N).

EVIDENCE FOR THE PRESENCE OF NST FROM MUSCULAR ORIGIN IN CA DUCKLINGS

Temperature change in an organ do not give by itself suitable information about thermogenesis in this organ since temperature also depends on blood flow that is responsible for heat transfer by convection. Moreover, if a comparison is made between organs, it is necessary to consider heat transfer by conduction taking the environment of each organ into account. But oxygen consumption of an organ can be deduced from simultaneous measurement of blood flow and local oxygen tension if arterial oxygen tension remained constant. Strong evidence for the presence of NST in skeletal muscle of muscovy ducklings was based upon the simultaneous tissular blood flow (TBF), tissular oxygen tension (PO2) and tissue temperature measurements in order to estimate continuously and in vivo the gastrocnemius muscle metabolism of CA and control ducklings. NST was induced by a glucagon test injection at 25°C Ta in the thermoneutral zone. Test injections were made via a polyethylene catheter, intraperitoneally implanted before the beginning of the experiment and fixed with adhesive tape. Metabolic rate of the duckling was measured by indirect calorimetry and cloacal temperature (Tb) using copper-constantan thermocouples.

Tissular blood flow and local temperature measurements. Tissular blood flow measurement was made in an awake bird by a thermal clearance technique using a single probe including a combined heating/sensing thermistor designed and developed in our laboratory. This measurement is done with one thermistor from two points : - a first point without heating is used for the measurement of the tissue temperature, the value of this temperature is stored in an analog memory and used as reference temperature, - a second point with a heating device is used for the measurement of heat loss by conduction and convection, this measurement being the effective thermal conductivity of the tissue. In order to determine the changes in thermal conductivity of the tissue resulting from heat loss by conduction only, the zero blood flow was determined at the end of the experiment immediately after the sacrifice of the bird. The thermal conductivity increment above the zero level corresponding to heat loss by convection has been shown to be linearly related to tissular blood flow in biological tissues. The transient impulsional heating protocol (6 measurements per minute) permitted simultaneous measurement of local temperature in the tissue and tissular blood flow. The accuracy of the analog function circuits determines the measurement limits of the system : presently temperature may be measured to 0.03°C, power to 0.1 mW and thermal conductivity to 1.5 %.

Oxygen tension measurements. Tissular PO2 was measured by a potentiostatic three-electrode system embedded in a single percutaneous sensor (700 µm in diameter). The arterial blood gases were also measured. Blood samples were collected through a polyethylene catheter inserted in the right carotid artery. Blood PO2 was determined by microelectrodes thermostated at 40°C.

NST from muscular origin. Measured on the whole organism a considerable glucagon-mediated calorigenic effect (+ 45 % and + 31 % above the saline control value in CA and TN ducklings respectively) was induced in the ducklings exposed at thermoneutrality. After the glucagon injection, Tb and muscle temperature increased in both control and CA ducklings, these increases being more important in CA ducklings than in controls. The increases in Tbs were respectively 1.15°C in CA and 0.68°C in control ducklings and the increases in muscle temperatures were respectively 0.70°C in CA and 0.40°C in control ducklings. These results do not give alone any information about the amount of heat produced by the gastrocnemius muscle under these conditions, because they do not take blood flow and oxygen consumption into account.

Thermal conductivity id es tissular bloow flow to the gastrocnemius muscle increased by 160 % after the intraperitoneal glucagon injection in CA ducklings, but did not change significantly in TN ducklings. Muscular PO2 decreased by 34 % in CA ducklings, but did not change significantly in TN ducklings after the glucagon test injection. The fall in local oxygen tension after glucagon injection despite the increase of blood flow un-equivocally points an increase in oxygen consumption of the gastrocnemius muscle.

The decrease in muscular oxygen tension was not due to a decrease in arterial blood oxygen tension, since instead PaO2 increased in relation with hyperventilation. Thus these results indicating a larger thermogenic effect due to glucagon in gastrocnemius muscle of CA ducklings than in controls point out to a significant role of the skeletal muscle in NST.

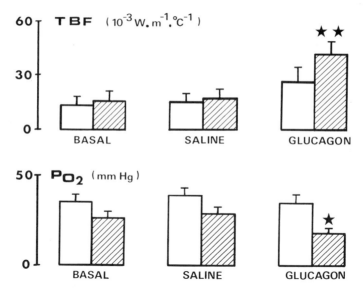

Fig. 6 Tissular blood flow (expressed in thermal conductivity units) and tissular oxygen tension from gastrocnemius muscle of 6 wk-old control (open bars) and cold-acclimated (shaded bars) ducklings (basal values, after saline injection and 60 min. after a glucagon test injection). *, ** : significantly different from basal values (P< 0.05, P < 0.01, respectively).

CONCLUSION

Our results show the existence of a nonshivering thermogenesis in a young bird acclimated to cold : the muscovy duckling. Glucagon is a mediator of this NST and because its calorigenic effect was shown in the gastrocnemius muscle, the skeletal muscle itself is a possible site of this NST.

ACKNOWLEDGMENTS

This work was supported by grants from Département de Biologie Humaine, Université Claude Bernard, Lyon I and Centre National de la Recherche Scientifique (Département des Sciences de la Vie).

REFERENCES

Barré, H. and Rouanet, J.L., 1983. Calorigenic effect of glucagon and catecholamines in king penguin chicks. Am. J. Physiol., 244: R758-R763.

Barré, H., Géloen, A., Chatonnet, J., Dittmar, A. and Rouanet, J.L., 1985. Potentiated muscular thermogenesis in cold-acclimated muscovy duckling. Am. J. Physiol., 249: R533-R538.

Barré, H., Cohen-Adad, F., Duchamp, C. and Rouanet, J.L., 1986a. Multilocular adipocytes from muscovy ducklings differentiated in response to cold aclcimation. J. Physiol. (London), 375: 27-38.

Barré, H., Géloen, A., Mialhe, P. and Rouanet, J.L., 1986b. Effects of glucagon on bird thermogenesis. In "Endocrine regulations as adaptive mechanisms to the environment", I. Assenmacher and J. Boissin Ed., Editions du CNRS, Paris, pp. 395-401.

Barré, H., Cohon Adad, F. and Rouanet, J.L., 1987a. Two daily glucagon injections induce nonshivering thermogenesis in muscovy ducklings. Am. J. Physiol., 252: E616-E620.

Barré, H., Bailly, L. and Rouanet, J.L., 1987b. Increased oxidative capacity in skeletal muscles from cold-acclimated ducklings : a comparison with rats. Comp. Biochem. Physiol., 88B: 519-522.

El Halawani, M.E., Wilson, W.O. and Burger, R.E., 1970. Cold acclimation and the role of catecholamines in body temperature regulation in male leghorns. Poultry Sci., 49: 621-632.

Freeman, B.M., 1970. Thermoregulatory mechanisms of the neonate fowl. Comp. Biochem. Physiol., 33: 219-230.

Hart, J.S., 1962. Seasonal acclimatization in four species of small wild birds. Physiol. Zool., 35: 224-236.

Krimphove, M. and Opitz, K., 1975. Untersuchungen der calorigenen Wirkung von Glucagon. Arch. Intern. Pharmacodyn., 216: 328-350.

Rintamäki, H., Saarela, S., Marjakangas, A. and Hissa, R., 1983. Summer and winter temperature regulation in the black grouse Lyrurus tetrix. Physiol. Zool., 56(2) : 152-159.

Saarela, S. and Vakkuri, O., 1982. Photoperiod-induced changes in temperature metabolism curve, shivering threshold and body temperature in the pigeon. Experientia, 38: 373-374.

Steen, J. and Enger, P.S., 1957. Muscular heat production in pigeons during exposure to cold. Am. J. Physiol., 191: 157-158.

West, G.C., 1965. Shivering and heat production in wild birds. Physiol. Zool., 38: 111-120.

NONSHIVERING THERMOGENESIS IN WINTER-ACCLIMATIZED KING PENGUIN CHICKS

Claude Duchamp, Hervé Barré, Didier Delage, Gilles Berne,
Pierre Brebion and Jean-Louis Rouanet

Lab. Thermorégulation et Energétique de l'exercice C.N.R.S.
8, avenue Rockefeller 69373 LYON cedex 08 (FRANCE)

Nonshivering thermogenesis (NST) has long been denied in birds as refered in the works of Steen & Enger (1957), Hart (1962), West (1965), and more recently Saarela & Vakkuri (1982), Rintamaki et al. (1983). NST was regarded as a feature of mammals because of the lack of the thermogenic brown adipose tissue, the most exclusive site of this heat production. A multilocular pink adipose tissue differentiated in response to cold acclimation in young ducklings was found in our laboratory, but its biochemical properties led it rather to a role in substrate supply (Barré et al. 1986a).

If such a thermogenesis exists in birds, it is most likely to be found in the time of life when the bird is most exposed to the cold and/or in cold-climate species. Thus in the 4 week-old muscovy ducklings acclimated to 4°C from the age of 1 week, a potentiated shivering thermogenesis was demonstrated by Barré et al. (1985). In a later stage of cold-acclimation the 6-wk-old ducklings developed a true NST (Barré et al. 1986a).

The king penguin chick (<u>Aptenodytes patagonicus</u>) appeared to be an interesting model for the study of such a thermogenesis developed by thermal acclimatization in a wild bird. Indeed, this animal species normally lives in the subantarctic zone, i.e. in a cold climate with ambient temperatures ranging from - 3 to + 15°C. During their breeding the king penguin chicks fed irregularly by their parents, stand alone in the colony. The chicks have to huddle together in order to reduce the stress of the cold during the subantarctic winter. Then after molting, during the passage from shore to marine life in the cold seawater, the juvenile chicks have to face an intensive and prolonged energetic demand to maintain homeothermy. Thus, in attempt to demonstrate metabolic acclimatization and nonshivering thermogenesis in king penguin chicks we studied different stages during their breeding period.

EXISTENCE OF METABOLIC ACCLIMATIZATION

In juvenile chicks never immersed (NI), evidence for progressive thermal acclimatization to marine life was sought by measurement of metabolic rate (MR) and body (Tb) temperature in water. Metabolic rate was measured by indirect calorimetry using an open circuit as described by Depocas and Hart (1957) (fig. 1).

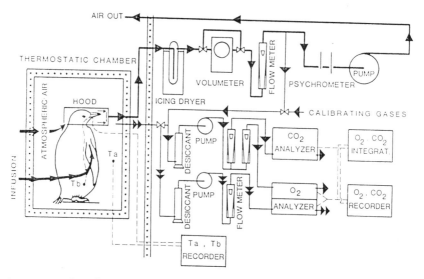

Fig. 1   Open-circuit system used for O2 consumption and CO2 production measurement.

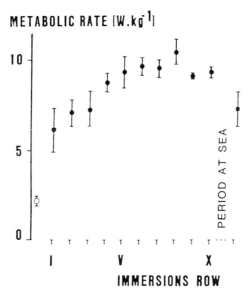

Fig. 2 . Evolution of resting metabolic rate in never-immersed (NI) king penguin chicks. (Open circle), in air at thermoneutrality (filled circles), in water (Tw = 7°C) during 10 successive immersions ; (closed squares), in water (Tw = 7°C) in acclimatized chicks after a long period at sea.
(means ± SE, n = 4 in each group)

A comparison was made between NI chicks and water acclimatized by a long period (2-3 months) of marine life. In never immersed chicks a simulation of seawater adaptation was performed by 10 successive immersions (fig. 2). As the tests proceeded, MR at 7°C water temperature (Tw) progressively rose from 6.0 to 9.4 W.kg-1 becoming

3 to 5 times higher than in air and then remained at this high level up to the 10th immersion experiment. At the same time Tb increased from 37.6 to 38.4°C from the first to the 4th immersion and up to the 10th experiment. In comparison MR during immersion in water acclimatized chicks after a long period at sea (2-3 months) was 23 % lower than the maximum measured in NI chicks between the 5th and 10th immersion, whereas Tb remained at a high level (Tb = 38.9°C).

The progressive increase in thermogenic capacity in cold water illustrated here, allows the king penguin chicks to maintain their body temperature efficiently during long exposures in cold water. In chicks acclimatized to marine life (A), the same exposure in water (7°C Tw) requires less regulatory thermogenesis, perhaps due to improved insulation. Progressive increase in thermogenic capacities and reinforcement of internal insulation demonstrated that the passage from shore to marine life consisted of a true cold acclimatization.

EXISTENCE OF NONSHIVERING THERMOGENESIS

The enhancement of metabolic capacities in response to a stressfull cold environment may suggest the development of calorigenic response without shivering for regulatory thermogenesis. In attempt to determine whether nonshivering thermogenesis was involved in metabolic adaptation of king penguin chicks to cold environment, NST was sought in these birds. Barré suggested in 1984, that winter-acclimatized chicks developed a NST, but in this study, sensing for shivering by touch alone could not detect electrical activity associated with a tonus increase and other experiments with electromyographic activity (EMG) recordings were required to confirm these results.

The direct demonstration of the existence of NST was performed using the classical method based on the comparison of the ambient temperature eliciting the first increase in metabolic rate in the cold (lower critical temperature : LCT) with the threshold of ambient temperature for shivering (STT) (Jansky 1973 ; Barré et al. 1985, 1986a). A significant gap between LCT and STT should demonstrate per se the existence of NST. Simultaneous measurements of resting metabolic rate (RMR) and integrated EMG were performed in winter-acclimatized (WA) chicks and in controls of the same age, deacclimated chicks reared at 25°C Ta for at least 3 weeks.

WA chicks exhibited a higher RMR than controls (2.4 vs 2.1 W.kg-1, P < 0.01) (Table 1). This increase of the obligatory thermogenesis shifted downward their LCT (- 9.6 vs + 4.4°C, P < 0.001).

Table 1 Characteristic parameters of energy metabolism, insulation, and critical temperatures of 14 winter-acclimatized (WA) and 8 control king penguin chicks.

	WA	CONTROLS	
Body mass, kg	7.91 ± 0.12	7.64 ± 0.16	
RMR, W.kg-1	2.43 ± 0.06	2.16 ± 0.06	**
Conductance, W.m-2.°C-1	1.06 ± 0.04	1.57 ± 0.11	***
LCT, °C	- 9.6 ± 2.5	4.4 ± 1.0	***
STT, °C	- 18.5 ± 2.3	4.6 ± 2.51	***
Δ(LCT - STT), °C	8.9	- 0.2	

Values are mean ± SE. **, *** comparisons between WA and CONTROLS (P < 0.01 and P < 0.001 respectively) (See text for other legends).

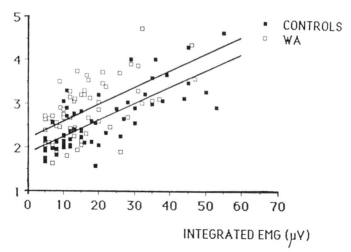

Fig. 3 Relation of metabolic rate to integrated EMG activity in winter-acclimated (WA) and control king penguin chicks. Respective equations of regression lines were in WA y = 0.039 x + 2.194 (r = 0.60) and in controls y = 0.038 x + 1.860 (r = 0.76).

Winter-acclimatization resulted in an improvement of thermal insulation. Thermal conductance was only 1.06 vs 1.57 $W.m^{-2}.°C^{-1}$ in WA and control chicks respectively.
There was no significant increase in EMG activity between 10 and 20°C in CA chicks and between + 25 and + 5°C in controls.
The statistically determined STT (Barré et al, 1985) of WA chicks was 9°C lower than LCT (- 18.5 vs - 9.6°C, $P < 0.05$) whereas STT and LCT of controls were similar (4.6 vs 4.4°C, NS). Consequently within this range of 9°C Ta in WA chicks, regulatory thermogenesis was independent of any shivering. The capacity for NST in these chicks estimated by MR in the cold at - 20°C Ta was 0.7 $W.kg^{-1}$ i.e. 28 % above RMR.
The relationship of shivering and thermogenesis in the cold in WA and control chicks showed (fig. 3) that both parameters appear to be linked despite the acclimation temperature. Since the regression lines of shivering with MR for WA and control chicks shifted in parallel, the extra heat production by shivering is simply additional, it cannot be considered as a potentiated shivering thermogenesis but rather as a true nonshivering thermogenesis as observed by El Halawani et al. (1970) in long-term cold adapted domestic fowls and by Barré et al. (1986a) in 6 wk-old cold-acclimated ducklings.
The extremely low LCT in WA chicks due to an enhanced obligatory thermogenesis, the increased thermal insulation, the development of NST and the reversibility of the phenomenon, demonstrate that breeding in the subantarctic climate consists of a true cold acclimatization in king penguin chicks.

GLUCAGON AS A MEDIATOR OF THIS NST

The calorigenic action of perfused hormonal agents have been evaluated in king penguin chicks by the changes in metabolic rate using indirect calorimetry as described above. The role of the sympathetic catecholaminergic nervous system in the control of heat production in mammals, is largely documented as regards NST (Himms-Hagen, 1976).

It is also well known that catecholamines and glucagon share many pharmacological properties generally mediated by cAMP. In spite of the similarities the calorigenic action of glucagon in mammals and even more so in birds, remains a matter of controversy (Alexander, 1979). Nevertheless, the role of glucagon as a possible mediator of NST in birds was first suggested by Freeman (1970) and a considerable (70 %) increase in metabolic rate was later reported in 3-mo-old japanese quails exposed to thermoneutral ambient temperature (Krimphove & Opitz, 1975). Recently, Barré et al. (1987) have shown that a chronic glucagon treatment in 6-wk-old ducklings reproduced the artificial cold-acclimation of rats chronically treated by norepinephrine and in these birds, a glucagon injection induced a true NST. Thus glucagon was then proposed as a potential mediator of NST in birds.

In king penguin chicks, the calorigenic action of glucagon and catecholamine infusion was evaluated both at thermoneutrality, and in the cold (fig.4).

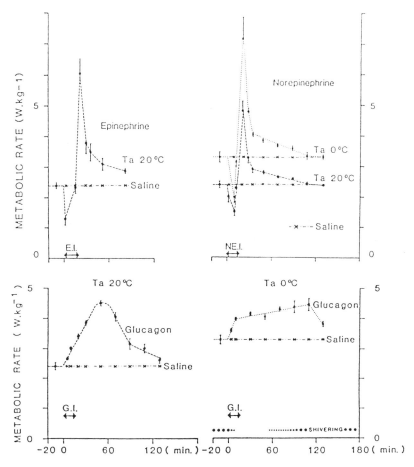

Fig. 4 . Comparison of effects of glucagon infusion (G.I.), epinephrine infusion (E.I.) and norepinephrine infusion (NE.I.) on metabolic rate in 7 king penguin chicks at neutral ambient temperature (●, Ta 20°C) or in cold (■, Ta 0°C). Control penguin chicks receive saline solution infusion. Shivering activity in cold is represented just below the graph (after Barré & Rouanet, 1983).

At Ta = 20°C, the mean increase in metabolic rate was
0.73 W.kg-1 for epinephrine infusion (4 µg.kg-1.min-1 iv), 0.42 W.kg-1
for norepinephrine infusion (10 µg.kg-1.min-1 iv) and 1.16 W.kg-1 for
glucagon infusion (0.05 µg.kg-1.min-1 iv) i.e. 30, 17 and 47 % of the
control value, respectively. Thus the calorigenic effect of glucagon is
much more potent than that of catecholamines and is able to take part
in the regulatory thermogenesis in the cold. Thus at 0°C Ta for the
same glucagon infusion, the mean increase in specific MR was
0.84 W.kg-1, 27 % above control rate. This less apparent effect in the
cold could be explained by the substitution of the glucagon calorigenic
action for part of the regulatory thermogenesis, inhibition of
shivering thermogenesis and perhaps an endogenous hormonal reaction. In
contrast the light metabolic response to intravenous infusion of
catecholamines resulted first from reflex apnea and polypnea, following
the hypertension effect of these hormones at neutral ambient
temperature. These results agree with the slight calorigenic effect
(12 %) reported in adult wild pigeons by Hissa et al. (1975) or no
effect as seen by Chaffee & Roberts (1971).

A lipolytic effect. Blood samples collected among the glucagon
infusion were analyzed to measure plasma free fatty acids concentration
(fig. 5). Glucagon infusion induced significant higher level of plasma
FFA (220 and 150 % above control value at 20 and 0°C Ta respectively)
whereas norepinephrine infusion induced no significant variations of
plasma FFA. The potent lipolytic action of glucagon already reported by
Freeman (1970) and its powerful calorigenic action in comparison with
catecholamines would strongly suggest its role as a potent mediator of
NST in birds.

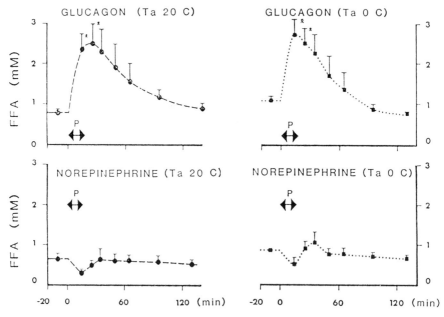

Fig. 5   Effects of glucagon (0.05 µg.kg-1.min-1 iv) and norepinephrine
(10 µg.kg-1.min-1 iv) perfusion (P) on plasma free fatty acid in 4 king
penguin chicks at neutral ambient temperature (●, Ta 20°C) and in cold
(■, Ta 0°C). * P < 0.05. (mean ± SE).

SKELETAL MUSCLE AS THE POSSIBLE SITE OF THIS NST

Like other birds king penguin chicks lack the thermogenic brown adipose tissue that is the main support of NST in neonate and cold-acclimated mammals. This absence lead us to attribute the origin of the observed regulatory thermogenesis to other organs, among them skeletal muscles or the liver, as possible sites of NST in birds. To evaluate the changes in the capacities of different organs in energy expenditure, we measured the cytochrome oxidase activity, generally used as an index of the aerobic oxidative capacity of organs (Jansky, 1973). Thus, an increased oxidative capacity of skeletal muscles has been shown in intermittently cold-exposed bantam chicks in relation to shivering thermogenesis (Aulie & Grav, 1979) and in relation to the development of NST in cold-adapted ducklings (Barré et al., 1987) with the additional involvement of the liver. It was of interest to see if the development of NST in winter-acclimatized king penguin chicks led also to an increase of this oxidative capacity in various tissues, especially in skeletal muscle (Table 2).

The cytochrome oxidase activity per unit wt respectively was 27.5 and 50.3 % higher in both gastrocnemius and pectoral muscles from WA chicks than from controls, whereas no significant changes occured in that of the liver, although the organ masses were not different from one group to the other. On the basis of these results it is apparent that skeletal muscles (gastrocnemius and pectoral) may contribute to the increased capacity for thermogenesis developed in WA chicks as in ducklings, but in penguin chicks the liver does not appear to be directly involved.

The question remains to find out whether this capacity is developed for shivering or for NST. But, since a NST is observed in these birds it has to be sought in muscles rather than in the liver.

<u>In mammals</u> cold acclimation induces proliferation and hypertrophy of BAT, accompanied by increases in total protein and amount of mitochondria (Desautels & Himms-Hagen, 1979).

<u>In cold-acclimated ducklings</u>, mitochondria of skeletal muscle have an altered morphology suggesting a differentiated mitochondrial function (Rouanet, 1983).

<u>In WA king penguin chicks</u> mitochondrial protein content (MPC) per unit wt (Table 3) was not significantly increased in the liver as compared with controls. In pectoral and gastrocnemius muscles, intermyofibrillar MPC were not changed by winter-acclimatization, but in subsarcolemmal mitochondria it was increased from 37 and 65 % above control values in pectoral and gastrocnemius respectively.

As intermyofibrillar mitochondria are directly involved in the process of contraction, the increase of protein content of subsarcolemmal mitochondria alone would implicate the mitochondria in muscle NST.

Table 2   Cytochrome oxidase activity (nAt O2.min-1.mg tissue-1) in some possible thermogenic tissues of 6 winter-acclimatized (WA) penguins chicks and 6 controls.

	CONTROLS	WA	
Liver	53.0 ± 2.8	52.6 ± 2.5	
Gastrocnemius muscle	23.8 ± 0.9	30.3 ± 1.6 **	+ 27.5 %
Pectoral muscle	20.5 ± 1.1	30.8 ± 2.0 **	+ 50.3 %

Values are mean ± SE. ** Comparison between WA and control chicks, $P < 0.01$.

Table 3. Yield of mitochondrial protein per g net weight (mg.g tissue-1) of some possible thermogenic tissues of 6 winter-acclimatized (WA) king penguin chicks and 6 controls

	CONTROLS	WA		
LIVER	21.0 ± 1.7	24.7 ± 1.3		
GASTROCNEMIUS MUSCLE				
Subsarcolemmal mitochondria	0.8 ± 0.1	1.2 ± 0.05	***	+ 65 %
Intermyofibrillar mitochondria	2.7 ± 0.2	2.4 ± 0.2		
PECTORAL MUSCLE				
Subsarcolemmal mitochondria	1.2 ± 0.02	1.7 ± 0.04	*	+ 37 %
Intermyofibrillar mitochondria	2.9 ± 0.1	3.0 ± 0.2		

Values are mean ± SE. *, *** Comparisons between WA and TN chicks, $P < 0.05$ and $P < 0.001$ respectively.

CONCLUSION

We have demonstrated the existence of metabolic and insulative adaptations in king penguin chicks at different stages of their breeding period, the passage from shore to marine life and during the subantarctic winter. In the winter-acclimatized chicks we have demonstrated the existence of a true nonshivering thermogenesis by simultaneous recording of metabolic rate and electromyographic activity at different temperatures. The developed NST in penguin chicks appears to be mediated by glucagon via an increase of the rate of lipolysis in adipose tissue, providing the cells with more available fuel. It must be pointed up that the sensitivity to the uncoupling effect of FFA on muscle mitochondria is increased in cold-acclimated ducklings (Barré et al. 1986b). The results in the king penguin chicks support the idea that the origin of the observed regulatory nonshivering thermogenesis lies in skeletal muscle and in particular in the subsarcolemmal mitochondria.

ACKNOWLEDGMENTS

This work was supported by grants from Terres Australes et Antarctiques Françaises (Mission de Recherche), from Comité National Français des Recherches Antarctiques and from Centre National de la Recherche Scientifique (GR 001).

REFERENCES

Alexander, G., Cold thermogenesis. In : Environmental Physiology III, edited by D. Robertshaw. Baltimore, MD : University Park, 1979, vol. 20, p. 43-155 (Int. Rev. Physiol. Ser.).

Aulie, A. and Grav H.J., 1979, Effect of cold acclimation on the oxidative capacity of skeletal muscles and liver in young bantam chicks, Comp. Biochem. Physiol., 62A: 335-338.

Barré, H., 1984, Metabolic and insulative changes in winter- and summer-acclimatized chicks, J. Comp. Physiol., B154: 317-324.

Barré, H., Cohen-Adad, F., Duchamp, C. and Rouanet, J.L., 1986a, Multilocular adipocytes from muscovy ducklings differentiated in response to cold acclimation, J. Physiol. (London), 375: 27-38.

Barré, H., Cohen-Adad, F., and Rouanet, J.L., 1987, Two daily glucagon injections induce nonshivering thermogenesis in muscovy ducklings, Amer. J. Physiol., 252: E616-E620.

Barré, H., Géloën, A., Chatonnet, J., Dittmar, A. and Rouanet, J.L., 1985, Potentiated muscular thermogenesis in cold-acclimated muscovy duckling, Amer. J. Physiol., 249: R533-538.

Barré, H., Nedergaard, J. and Cannon, B., 1986b, Increased respiration in skeletal muscles mitochondria from cold-acclimated ducklings: uncoupling effect of free fatty acids, Comp. Biochem. Physiol., 244 : 343-348.

Barré, H. and Rouanet, J.L., 1983, Calorigenic effect of glucagon and catecholamines in king penguin chicks. Amer. J. Physiol.,244 : R758-R763.

Chaffoo, R.R.J. and Roberts, J.G., 1971, Temperature acclimation in birds and mammals, Ann. Rev. Physiol., 33: 155-202.

Depocas, F. and Hart, J.S., 1957, Use of the Pauling oxygen analyzer for measurement of oxygen consumption of animals in open-circuit systems and in a short lag, closed-circuit apparatus, J. Appl. Physiol., 10: 388-392.

Desautels, M. and Himms-Hagen, J., 1979, Roles of noradrenaline and protein synthesis in the cold-induced increase in purine nucleotide binding by brown adipose tissue mitochondria, Can. J. Biochem., 57: 968-976.

El Halawani, M.E., Wilson, W.O. and Burger, R.E., 1970, Cold-acclimation and the role of catecholamines in body temperature regulation in male leghorns, Poultry Sci. 49: 621-632.

Freeman, B.M., 1970, Thermoregulatory mechanisms of the neonate fowl. Comp. Biochem. Physiol., 33: 219-230.

Hart, J.S., 1962, Seasonal acclimatization in four species of small wild birds, Physiol. Zool., 35: 224-236.

Himms-Hagen, J., 1976, Cellular thermogenesis. Ann. Rev. Physiol. 38: 315-351.

Hissa, R., Pyörnila, A. and S. Saarela, 1975, Effect of peripheral noradrenaline on the thermoregulation in temperature acclimated pigeon, Comp. Biochem. Physiol., 51C: 243-247.

Jansky, L., 1973, Nonshivering thermogenesis and its thermoregulatory significance. Biol. Rev., 48: 85-132.

Krimphove, M., and Opitz, K., 1975. Untersuchungen der calorigenen Wirkung von Glucagon, Arch. Int. Pharmacodyn. Ther. 216: 328-350.

Rintamaki, H., Saarela, S., Marjakangas, A. and Hissa, R., 1983, Summer and winter temperature regulation in the black grouse Lyrurus tetrix, Physiol. Zool., 56(2), 152-159.

Rouanet, J.L., 1983, Arguments morphologiques et fonctionnels en faveur d'une participation du muscle squelettique à la thermogenèse sans frisson chez le caneton acclimaté au froid, Thèse 3ème Cycle, Université Cl. Bernard, Lyon.

Saarela, S., and Vakkuri, O., 1982, Photoperiod-induced changes in temperature metabolism curve, shivering threshold and body temperature in the pigeon, Experientia, 38: 373-374.

Steen, J., and Enger, P.S., 1957, Muscular heat production in pigeons during exposure to cold, Amer. J. Physiol., 191: 157-158.

West, G.C., 1965, Shivering and heat production in cold birds, Physiol. Zool., 38: 111-120.

APPARENT NON-MUSCULAR THERMOGENESIS

IN COLD-EXPOSED PHASIANID BIRDS

        Esa Hohtola, Ritva Imppola and Raimo Hissa

        University of Oulu
        Departmnet of Zoology
        SF-90570 Oulu, Finland

INTRODUCTION

    Birds maintain homeothermy during cold exposure by two mechanisms: heat loss is curtailed by behavioral/insulative adjustments and internal heat production is elevated through an increase in metabolic rate. According to the well-accepted contention, the increase in heat production in birds is accomplished by muscular shivering. This contention is based on the following facts: 1) The cold-induced increase of metabolic rate is accompanied by an increase in the electrical activity (EMG) of skeletal muscles (Steen and Enger, 1957; West, 1965), 2) A simultaneous increase in muscle-body temperature gradient occurs (Steen and Enger, 1957; Hohtola, 1982), 3) Birds lack both the anatomical substrate and the physiological mechanisms for brown-adipose-tissue-type thermogenesis (see Hissa, 1988).

    A survey of literature reveals, however, that this view is based on observations made in a small number of bird species and taxa. Although electromyographic measurements on shivering have been performed in a variety of birds (see Hohtola and Stevens, 1986) using a detailed quantitative approach in some cases (West et al., 1968; Hohtola, 1982), actual muscle-body temperature gradients have been measured only in the pigeon (Steen and Enger, 1957; Hohtola, 1982). Furthermore, there is now evidence of a thermogenic mechanism in the skeletal muscle of young birds that is independent of muscle electrical activity (Barré et al., 1985). It is thus clear that studies encompassing a wider variety of species and careful recordings of muscle electrical activity and temperature gradients between tissues are needed to further elucidate the role of shivering and other possible modes of thermogenesis in birds.

    In this work, we studied the thermoregulatory capabilities in two winter-acclimatized phasianid birds, the pheasant and the partridge. Routine measurements of heat production, body temperatures and EMG in cold-exposed birds revealed an aberration in the expected coupling between metabolic rate and muscle electrical activity and prompted a more detailed analysis of muscle-body temperature gradients.

MATERIALS AND METHODS

Animals

Pheasants (Phasianus colchicus) and partridges (Perdix perdix) of both sexes were obtained from breeders in Central Finland. All experiments were performed on winter-acclimatized birds maintained in large outdoor aviaries. The mean mass of the birds were: 975 g (female pheasants), 1245 g (male pheasants), 371 g (partridges). Commercial poultry food and snow as a water source were available ad libitum. Each bird was fasted for 14 hours before measurements.

Measurements

Measurements were performed in metabolic chambers that allowed variation of ambient temperature (T_a) between -30°C and +30°C. Metabolic rate was determined by measuring oxygen consumption in an open-flow system. Dry, CO_2-free air was metered through the chamber at a steady rate between 2.0 and 4.6 l/min depending of the size of the bird. Dried, CO_2-free outlet air was passed through an oxygen analyzer (Beckman E2 or Applied Electrochemistry S-3A) for measuring the oxygen consumption (VO_2).

All temperatures were measured with copper-constantan thermocouples. Skin temperature (T_s) was measured by taping a thin thermocouple on the featherless area of the leg. Colonic temperature (T_{cl}) was measured by introducing a lubricated thermocouple 7 cm beyond the cloacal opening and by fixing it to the retrices with adhesive tape. Muscle temperatures (T_m) were measured using 0.20 mm thermocouple leads introduced into a depth of 10-15 mm using a hypodermic needle, which was removed after insertion. The thermocouple was bended and fixed to muscle surface with adhesive tape. All temperatures were recorded on a Leeds-Northrup multipoint recorder.

Muscle electrical activity was routinely recorded with uninsulated pin electrodes which were inserted 5 to 8 mm deep in the muscle (see Hohtola, 1982). Other electrode types were also used occasionally to vary the EMG pickup area. The signal from the electrodes was amplified, filtered (50-1000 Hz) and fed through an integrator to obtain a quantitative index of shivering, while the raw EMG was monitored on an oscilloscope (see Hohtola, 1982).

RESULTS

The thermoregulatory responses of the birds to changes in ambient temperature can be seen from Figures 1 and 2. Oxygen consumption showed a well-defined increase with decreasing T_a in all experiments. The maximum VO_2 at -30°C ranged from two times BMR in male pheasants to 3 times BMR in partridges. No clear-cut lower critical temperature could be defined for either species. The drop in leg skin temperature showed a normal peripheral vasoconstriction during a decrease in T_a. Surprisingly, colonic temperature often increased when T_a was lowered. This phenomenon was especially pronounced in female pheasants (see Fig. 1).

Contrary to our experience from cold-exposed birds and other reports from similar experiments, we were not able to detect electromyographic signs of shivering in breast or leg muscles of

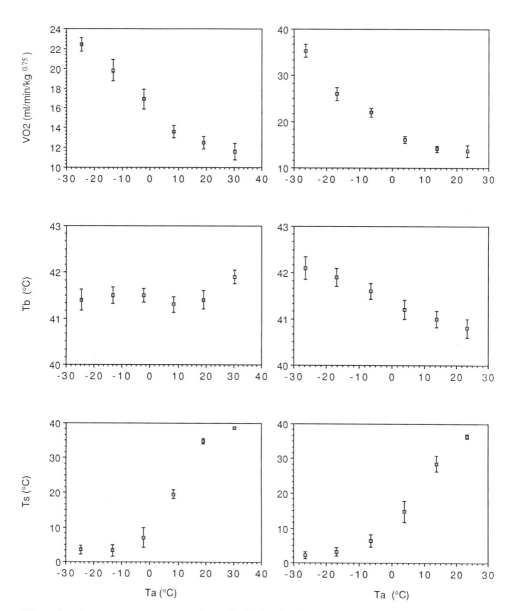

Fig. 1. Oxygen consumption (VO_2), body temperature (T_b) and leg skin temperature (T_s) in male (left panel, N=5) and female (right panel, N=6) as a function of ambient temperature (T_a). Symbols depict means ± S.E.M.

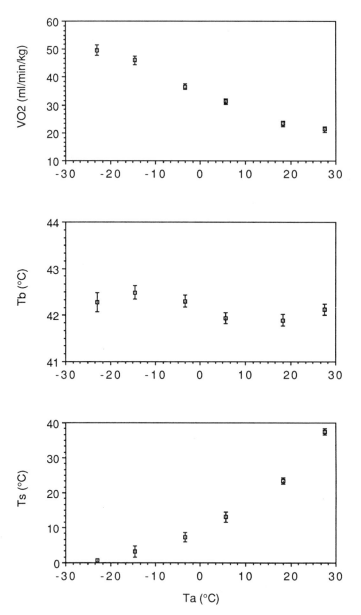

Fig. 2. Oxygen consumption (VO$_2$), body temperature (T$_b$) and leg skin temperature (T$_s$) in partridges as a function of ambient temperature (T$_a$). Symbols depict means ± S.E.M. N=10.

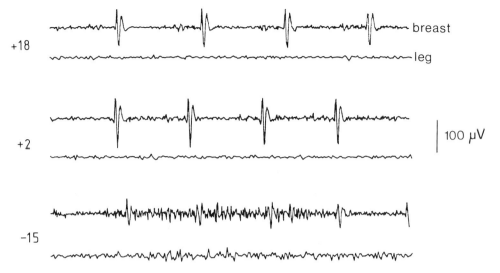

Fig. 3. A representative tracing of electrical activity in the pectoral and leg muscles of a pheasant at three ambient temperatures. The ECG is visible in the breast muscle recording because widely spaced safety pin electrodes were used.

either species. An example, obtained using safety pin electrodes, is shown in Fig. 3. It is apparent that the increase in metabolic rate seen in Figures 1 and 2 occurs well before any signs of EMG in either muscle group. As an electrical reference, an increase in the heart rate can be detected from the ECG signal, which is readily picked up by this electrode type. The maximal response seen at the lowest T_a in Fig. 3 consisted of occasional EMG bouts having an amplitude that was an order of magnitude lower that what is known from the pigeon, for example (Hohtola 1982). In fact, the EMG-activity was so low that no deflection from the baseline could be seen in the integrated EMG. As a control experiment, we measured shivering from cold-exposed pigeons using exactly the same equipment and electrodes and were able to record the normal EMG response.

The lack of the normal electromyographic shivering response prompted us to measure breast muscle and colonic temperatures in a separate series of experiments. It appeared that in both species, muscle temperature was always lower than colonic temperature. Furthermore, this temperature difference increased linearly with ambient cooling (Figures 4 and 5). In male pheasants, we measured also leg muscle temperature, which showed the same response, but was even lower than breast muscle temperature (Fig. 5). Similarly, the mean difference between colonic and breast muscle temperature at -30°C was approximately 3°C in partridges (data not shown).

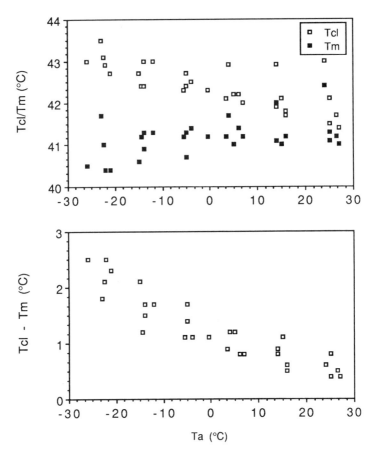

Fig. 4. Cloacal (T_{cl}) and breast muscle (T_m) temperatures in female pheasants as a function of ambient temperature (T_a). The upper panel shows the actual temperatures, while the lower panel shows the temperature difference. N=6.

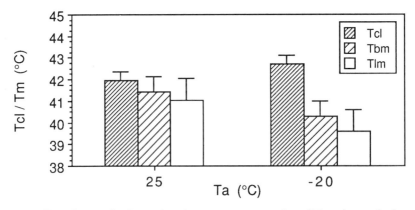

Fig. 5. Cloacal (T_{cl}), breast muscle (T_{bm}) and leg muscle (T_{lm}) temperatures in male pheasants at two ambient temperatures (T_a). The bars show means and S.E.M. N=4.

DISCUSSION

The present results differ from previous studies on cold-exposed birds in two respects: 1) No electromyographic signs of true shivering could be detected and 2) The temperature gradient between muscle and body was reversed. However, as inappropriate methods for detecting and quantifying shivering have been used in many studies (see Hohtola, 1982), it is difficult to say whether these species actually represent an exception among the birds studied so far. An even more striking observation was the reversed temperature gradient between skeletal muscle and abdomen. In this case, generalisations are also questionable due to the small number of species studied in this respect. Similar experiments have been performed only in the pigeon before this (Steen and Enger, 1957; Hohtola 1982). In one previous study, however, leg muscle temperatures were found to be lower than body temperature in cold-exposed pheasants (Ederstrom and Brumleve, 1964).

Before analyzing other explanations for the present findings we make the following conclusions: 1) The increase in oxygen consumption shows that thermoregulatory heat production occurs in cold-exposed pheasants and partridges. 2) The lack of electrical activity in the two major muscle groups suggests that shivering thermogenesis is not the source of the extra heat production. 3) The "reversed" temperature gradient between body and muscle shows that other mechanisms of muscular heat production can also be excluded as a source of heat. 4) It must therefore be concluded that a non-muscular, regulatory source of heat exists in cold-exposed, winter-acclimatized pheasants and partridges.

An alternative explanation for the lack of EMG in these muscles of the cold-exposed birds could be a high degree of localization in the shivering response in species showing a clear distinction between aerobic and anaerobic muscles. Shivering in birds is usually a generalized response in all skeletal muscles (Hohtola, 1982; Hohtola and Stevens, 1986), but such a localization has been observed recently in the Bantam hen (Aulie and Tøien, this volume), where the iliotibularis and gastrocnemius muscles in the leg shiver during moderate cold exposure and shivering becomes more generalized only when the cold exposure is severe. Phasianid birds are usually poor flyers and good runners and obviously possess both pure aerobic and and anaerobic muscles. However, it is not probable that a muscle outside the major muscle groups studied in the present work could account for the heat production since the lowest ambient temperatures tested (-30°C) should have evoked shivering in all muscle groups.

The temperature gradient between muscle and colon poses an even greater problem. There is no evidence of any visceral thermogenic tissues that could account for the high colonic temperatures in cold. Gallinaceous birds do have a well-developed gut with extremely long caeca in some cases, both of which undergo seasonal changes in length (Moss, 1972; Pulliainen and Tunkkari, 1983). Among these birds, however, the phasianid species have the shortest caeca and guts (Pulliainen, 1984). Existing estimates (Thompson and Boag, 1975) of caecal heat production in birds indicate that it is not a major component of total energy expenditure. Furthermore, it is difficult to envision a mechanism for regulatory heat production in these tissues. The exceptionally high colonic temperatures (e.g. female pheasants) may also be restricted to the inside of the gut wall due to local microbial activity and possible intestinal vasoconstriction in cold. Temperature measurements at other abdominal sites are therefore necessary to clarify this phenomenon.

At the present, the results of the this study seem to differ so much from previous work done in cold-exposed birds, that no apparent explanations can be obtained from the literature. Therefore, as a tentative conclusion, it is suggested that a non-muscular, possibly visceral, thermogenic mechanism exists in winter-acclimatized phasianid birds.

REFERENCES

Barré, H., Geloen, A., Chatonnet, J., Dittmar, A. and Rouanet, J-L., 1985, Potentiated muscelar thermogenesis in cold-acclimated Muscovy duckling, Am. J. Physiol., 249:R533.

Ederstrom, H.E. and Brumleve, S.J., 1964, Temperature gradients in the legs of cold-acclimatized pheasants, Am. J. Physiol., 207:457.

Hissa, R., 1988, Controlling mechanism in avian temperature regulation: a review, Acta Physiol. Scand., 132(Suppl. 567):1-148.

Hohtola, E., 1982, Thermal and elctromyographic correlates of shivering thermogenesis in the pigeon, Comp. Biochem. Physiol., 73A:159.

Hohtola, E. and Stevens, E.D., 1986, The relationship of muscle electrical activity, tremor and heat production to shivering thermogenesis in Japanese quail, J. Exp. Biol., 125:119.

Moss, R., 1983, Gut size, body weight, and digestion of winter foods by grouse and ptarmigan, Condor, 85:185.

Pulliainen, E., 1984, On the gut size and chemical composition of the food of the partridge (Perdix perdix) in Finland, Suomen Riista, 31:13.

Pulliainen, E., and Tunkkari, P., 1983, Seasonal changes in the gut length of the willow grouse (Lagopus lagopus) in Finnish Lapland, Ann. Zool. Fenn., 20:53.

Steen, J. and Enger, P.S., 1957, Muscular heat production in pigeons during exposure to cold, Am. J. Physiol., 191:157.

Thompson, D.C. and Boag, D.A., 1975, Role of the caeca in Japanese quail energetics, Can. J. Zool., 53:166.

West, G.C., 1965, Shivering and heat production in wild birds, Physiol. Zool., 38:111.

West, G.C., Funke, E.R. and Hart, J.S., 1968, Power spectral density and probability analysis of electromyograms in shivering birds, Can. J. Physiol. Pharmacol., 46:703.

SHIVERING IN AEROBIC AND ANAEROBIC MUSCLES

IN BANTAMS (GALLUS DOMESTICUS)

Arnfinn Aulie and Øivind Tøien*

Veterinary College of Norway
Department of Physiology
Oslo, Norway

INTRODUCTION

Since shivering seems to be the main thermoregulatory heat producing mechanism in resting birds (Palogangas and Hissa, 1971; Calder and King, 1974; Hohtola, 1982), shivering should be present at ambient temperatures (Ta) below the lower critical temperature (Tc). In willow ptarmigan, the pectoral muscle shivered at Tc (West, 1972; Aulie, 1976). In the bantam hen, however, Ta had to be decreased to 20°C, or 4° C below Tc, before any EMG-activity could be detected in the pectoral muscle (Aulie, 1976; Tøien et al., 1986). This difference could be due to the fact that the ptarmigan pectoral muscle contains only aerobic fibers (Type IIA) (Grammeltvedt, 1978) while in the bantams it contains anaerobic fibers (Type IIB) (McNaughtan, 1974).

In the present study it was tested if a muscle containing mainly aerobic fibers was activated at higher Ta than an anaerobic muscle during cold stress. EMG-activity has been recorded from m. iliotibialis (aerobic) and m. pectoralis (anaerobic) in bantam cocks at different Ta. In order to study the shivering pattern during a sudden heat loss, EMG was recorded from the same muscles in bantam hens incubating thermoregulated eggs.

MATERIALS AND METHODS

Two bantam cocks (Gallus domesticus) and three incubating bantam hens were used. For experiments, the bird was placed in a ventilated Plexiglass chamber situated in a

---

* Present adress: Division of General Physiology, Department of Biology, University of Oslo, Oslo, Norway.

temperature controlled chamber (ACS, model TY110). For the brooding experiments, the Plexiglass chamber was equipped with a plastic bowl containing 4 eggs circulated with thermoregulated water (Tøien et al. 1986). Temperatures were measured with thermocouples connected to a multipoint potentiometer. The body temperature (Tb) was measured with a thermocouple inserted 4 cm into the cloaca. Ta was measured close to the air outlet in the Plexiglass chamber and the egg temperature (Te) was measured in the center of each egg. As EMG-electrode was used 2 insulated stainless steel wires, 0.2 mm diameter, inserted into a 15 mm long piece of teflone tube. The tube was equipped with two 1 mm diameter holes 4 mm apart. The wires were placed in such a way that a 1mm uninsulated section of each wire was situated underneath one of the two holes. One electrode was inserted under the skin on the surface of the cranial part of the left m.iliotibialis (approximately 2 cm anterior to the head of femur). The other electrode was placed on the left m.pectoralis. The bird was grounded through a safety-pin placed on the back. The electrodes were connected to two Beckman 9852A EMG-averaging couplers in a Dynograph recorder (R511A). The rectified EMG was recorded in the couplers average mode with a time constant of 0.1 s. After the bird had been equipped with thermocouple and EMG-electrodes, it was allowed to calm down for at least half an hour at Ta of 30-32°C. At the beginning of the incubation experiments, Te was kept at 40°C. Results are given as mean plus minus standard deviation.

RESULTS

At Ta of 25°C there was sporadic EMG-activity in the iliotibiales muscle but not in the pectoral muscle of the cocks (Fig. 1). When Ta was increased, the EMG-activity decreased and vanished between 30.5 and 34°C. When Ta was lowered from 34°C, the EMG-activity reappeared at approximately the same Ta as it had disappeared. This threshold Ta for shivering is defined as Tsh. The mean Tsh was 32.0+1.1°C (N=12), while the mean Tb was 41.3+0.4°C (N=10) at this ambient temperature. While the EMG-activity of the iliotibialis muscle increased at Ta below Tsh, the pectoral muscle was normally inactive until Ta was lowered to about 20°C. Below this temperature the EMG-activity increased in both muscles (Fig. 1).

Hens incubating 4 eggs at Te=40°C had a Tsh between 30.5 and 31.5°C and a Tb of 41.2+0.2°C (N=27). When Te was changed to 25°C, Tb decreased to 40.6+0.2°C (N=14) in 14.1+3.6 min. Tb increased when Te was returned to 40°C and normal Tb (41.2°C) was reached after 16.5+8.0 min (N=10) (Fig. 2). When Te was changed from 40 to 25°C, EMG-activity was detected in the iliotibialis muscle after 28+17 sec (N=11). In the pectoral muscle the EMG-activity appeared after 193+90 sec (N=10). The muscle activity increased until stable Tb was reached, but the pectoral muscle never reached the same activity level as iliotibialis. When Te was returned to 40°C, the EMG-activity

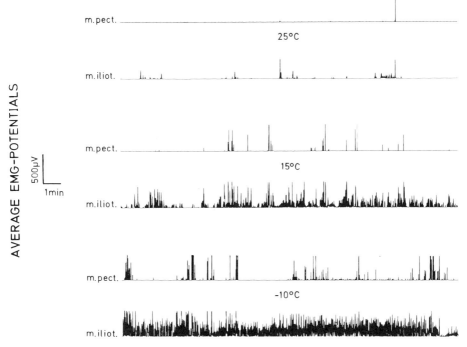

Fig. 1. EMG-activity recorded simultaneously from m.pectoralis and m.iliotibialis in a resting bantam cock exposed to different ambient temperatures.

in the pectoral muscle normally disappeared within a few seconds, while the activity in the iliotibialis muscle decreased more slowly until normal Tb was reached (Fig. 2).

DISCUSSION

The ambient temperature had to be increased to above 32° C (Tsh) before the EMG-activity of the iliotibialis muscle vanished. Between Tc and Tsh the iliotibialis muscle showed burst like EMG-activity which is typical for thermally induced tremor (Hohtala and Stevens, 1986). The pectoral muscles become active several degrees below Tc. Since Ta seldom is higher than 32°C, the thigh muscles in these birds seems to be active most of the time. The iliotibialis muscle is well suited for such a continuous activity since it is a typical aerobic muscle (Bass et al., 1970). Although the non-postural pectoral muscle has been regarded as the most important site for thermogenesis in cold (Steen and Enger, 1957; Hart, 1962), it does not participate in the thermoregulation at room temperatures in the bantam. The reason may be that

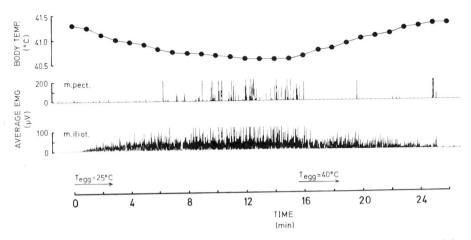

Fig. 2. Body temperature and EMG-activity from m.pectoralis and m.iliotibialis in a bantam hen incubating water circulated eggs. Ta=31°C.

in bantams these muscles mainly depend on anaerobic metabolism (Bass et al., 1970) and therefore are not suited for continuous activity. The pectoral muscle of the Japanese quail (another Galliformes) is a typical aerobic muscle (Kaiser and George, 1973). This muscle shivers at 26°C and exhibits an EMG-pattern similar to that recorded from the iliotibialis muscle in the bantam (Hohtala and Stevens, 1986).

The iliotibialis muscle in the bantam shivered at ambient temperatures above the lower critical temperature estimated from oxygen consumption measurements (Tøien et al., 1986). This indicates that the heat production does not reach its minimum at Tc but at Tsh. According to the "Newtonian Model" for thermoregulation, the heat production of birds is lowest within the thermoneutral zone and increase linearly at ambient temperatures below Tc (Scholander et al., 1950). Nichelmann (1983), however, suggests that the relationship between Ta and heat production best can be presented by a parabola. The heat production should thereby reach a minimum at a particular Ta, termed the thermoneutral temperature (Tn). In turkey Tn is about 30°C. When Ta was raised from 14 to 30° C, both evaporative and passive heat loss increased while the heat production decreased. Thus, the bird is both heated and cooled simultaneously in order to maintain a constant Tb within this temperature range (Nichelmann, 1983). In the bantam Tn is probably 32°C or the same as Tsh.

When a resting bantam is exposed to a Ta just below 20° C, shivering appears in the pectoral muscle. This is surprising since the heat producing capacity of the aerobic muscles hardly is utilized at this Ta. The metabolism is less than 10% above resting level (Tøien et al., 1986). The delayed activation of the pectoral muscle was more evident during the incubation experiments. When the egg temperature suddenly was changed from 40 to 25°C, the EMG-activity in the iliotibialis muscle appeared within 1/2 min, which is about the time it takes to cool the egg from 40 to 25°C. The pectoral muscle activity, however, appeared about 3 min thereafter. It seems as if the heat loss from the bird has to reach a certain level before the pectoral muscles get involved in the thermoregulation. The shivering starts in the aerobic muscles and only when the heat loss gets severe does the anaerobic muscles join in the shivering thermogenesis. The same recruitment pattern has been found during voluntary contractions of heterogeneous mammalian muscles. The motor units are recruited in the following sequence: Slow twitch (type I fibers)- fast twitch fatigue resistant (type IIA fibers) and fast twitch fatiguable (type IIB fibers) (Burke, 1981). The iliotibialis muscle of bantam contain 7 % type I fibers and 93 % type IIA fibers (McNaughtan, 1974). If the recruitment sequence is the same during shivering as during voluntary contractions, one would expect only the type I fibers to be active at Ta close to Tsh. From our present data it is not possible to decide at which Ta the type IIA fibers are activated. However, the fact that the EMG activity in the iliotibialis muscle increased considerably as Ta approached 20°C, indicates that type IIA fibers are activated at a higher ambient temperature than the type IIB fibers in the pectoral muscle.

When the egg temperature was changed from 25 to 40°C, the EMG-activity in the pectoral muscle disappeared within seconds, while the iliotibialis muscle shivered until Tb reached 41.2°C, or normal body temperature. Thus, although Tb was as low as 40.6°C after the cold egg exposure the pectoral muscle shivering ceased as soon as the broodpatch was exposed to 40°C. This indicates that in order to maintain shivering in the pectoral muscle, the total thermal input has to be high. In this case the input from the peripheral cold receptors in the broodpatch seems to be vital. The iliotibialis muscle, on the other hand, shivers as long as the bird is hypothermic, which means that stimulation of central receptors may play a more important role.

REFERENCES

Aulie A., 1976, The shivering pattern in an artic (willow ptarmigan) and a tropical bird (bantam hen). Comp. Biochem. Physiol., 53A:347-350.
Bass A., Lusch G., and Pette D., 1970, Postnatal differentiation of the enzyme activity pattern of energy-supplying metabolism in slow (red) and fast (white) muscles of the chicken. Europ. J. Biochem., 13:289-292.

Burke R.E., 1981, Motor units: anatomy, physiology, and functional organization, in: "Handbook of Physiology. The Nervous System II, part 1.", V. B. Brooks, ed., Am. Physiol. Soc., Bethesda.

Calder W. A., and King J. R., 1974, Thermal and caloric relations of birds, in: "Avian biology IV", D. S. Farner and J. R. King, eds., Academic Press, New York.

Grammeltvedt R., 1978, Atrophy of breast muscle with a single fiber type (m.pectoralis) in fasting willow grouse, Lagopus lagopus (L). J. Exp. Zool., 205:195-204.

Hart J. S., 1962, Seasonal acclimation in four species of small wild birds. Physiol. Zool., 35:224-236.

Hohtola E., 1982, Shivering thermogenesis in birds. Acta Univ. Oul. A, 139, Biol. 17.

Hohtola E., and Stevens E. D., 1986, The relationship of muscle electrical activity, tremor and heat production to shivering thermogenesis Japanese quail. J. Exp. Biol., 125:119-135.

Kaiser C. E., and George J. C., 1973, Interrelationship amongst the avian orders Galliformes, Columbiformes, and Anseriformes as cnvinced by the fiber types in the pectoral muscle. Can. J. Zool., 51:887-892.

Mc Naughtan A. F., 1974, An ultrastructure and histochemical study of fiber types in the pectoralis thoracica and iliotibialis muscle of the fowl (Gallus domesticus). J. Anat., 118:171-186.

Nichelmann M., 1983, Some characteristics of the biological optimum temperature. J. Therm. Biol., 8:69-71.

Palokangas R., and Hissa R., 1971, Thermoregulation in young black-headed gull (Larus ridibundus L). Comp. Biochem. Physiol., 38A:743-750.

Scholander P. F., Hock R., Walters V., Johnson F., and Irvin L., 1950, Heat regulation in some artic and tropical mammals and birds. Biol. Bull., 99:237-258.

Steen J. B., and Enger P. S., 1957, Muscular heat production in pigeons during exposure to cold. Am. J. Physiol., 191:157-158.

Tøien Ø., Aulie A., and Steen J. B., 1986, Thermoregulatory responses to egg cooling in incubating bantam hens. J. Comp. Physiol., B156:303-307.

West G.C., 1972, Seasonal differences in resting metabolic rate of Alaska ptarmigan. Comp. Biochem. Physiol., 42A:867-876.

METABOLIC ACCLIMATIZATION TO COLD AND SEASON IN BIRDS[*]

William R. Dawson and Richard L. Marsh

Museum of Zoology and Department of Biology
The University of Michigan
Ann Arbor, MI 48109, U.S.A.

Department of Biology
Northeastern University
Boston, MA 02115, U.S.A.

INTRODUCTION

With the prominent role augmented rates of thermogenesis play in the regulation of body temperature by most birds in winter cold, it is of interest to examine the extent to which this process is affected by metabolic acclimatization. Such adjustment affects not only the cold resistance of these animals, but also their energy requirements in a season when food supplies are declining and the time to locate them minimal. We shall emphasize naturally occurring forms of metabolic acclimatization in wild birds, but some consideration also will be given to that associated with exposure to cold in the laboratory. Smaller birds are of primary concern because of their limited capacities for insulative acclimatization, though indications of metabolic acclimatization in larger forms also will be considered. Particular attention will be devoted to indications of acclimatization involving metabolic level, thermogenic capacity, endurance in the cold, extent of energy reserves, and the biochemical correlates of cold resistance. Our goal is to characterize metabolic acclimatization by birds to cold and season to an extent commensurate with current knowledge of this form of compensation.

ACCLIMATIZATION IN GENERAL METABOLIC LEVEL

Early work on metabolic acclimatization in birds emphasized changes in basal metabolic level occurring after exposure of the animals to new experimental temperatures. Attention has also been given to natural seasonal changes in general metabolic level for these animals. Interpretation of some of the earlier results reported is hindered by limited sample sizes and lack of rigorous statistical analysis.

Typically, a rise in metabolic level, defined by basal metabolic rate (BMR) or resting metabolic rate in or near the zone of thermal neutrality

---
[*]Research of W. R. Dawson and associates supported by grants from the National Science Foundation, currently BSR 84-07952.

(RMR), occurs within one to four weeks after transfer of birds from moderate (ca. 25°-30° C) to intermediate (ca. 12°-22° C) or cold (ca. -14° to +10° C) ambient temperatures. The opposite effect occurs in a similar time frame with transfer in the reverse direction (Gelineo, 1964). In the domestic fowl (Gallus gallus), domestic pigeon (Columba livia), collared turtle dove (Streptopelia decaocto), and several passerines, the level of BMR is generally 10-55% greater following maintenance at cool or cold temperatures than at warm ones (Dontcheff and Kayser, 1934; Gelineo, 1934, 1955; Arieli et al., 1979). The largest percentage change was noted in the siskin (Carduelis spinus) whose BMR was 85% greater following maintenance at Ta of -2° to +8° C than at 29°-32° C (Gelineo, 1955). Not unexpectedly, the shifts in level of BMR are accompanied by changes in the positioning of the zone of thermal neutrality, the boundaries of this zone being one to several degrees lower in birds maintained at cool or cold ambient temperatures than in those at warm ones (Gelineo, 1934, 1955, 1964).

Whether the fluctuations in metabolic level observed following experimental manipulation of thermal conditions have an important counterpart in nature appears controversial. In addition to questions concerning adequacy of sampling and statistical analysis, comparisons are complicated by various circumstances. For example, seasonal fluctuations in body mass (see below) produce uncertainties as to the proper basis (mass-specific, total, or some other) on which to compare metabolic values obtained at different times of the year. Dontcheff and Kayser (1934) described an annual rhythm in the BMR of the pigeon. Mean values for the three individuals studied were elevated by approximately 10% in winter, but they were also relatively high during the summer months, leading the authors to conclude that temperature was not the only stimulus affecting metabolic level. Miller (1939) found a much more pronounced annual rhythm in the daytime metabolic rate of house sparrows (Passer domesticus) in Iowa. Mass-specific RMR of sparrows studied at 28° C within 1-3 days of capture was approximately 50% higher in November through February than in April through June. On the other hand, Hart's (1962) comparison of representatives of the house sparrow in summer and winter reveals no difference in metabolic level at ambient temperatures in the vicinity of 30° C. However, in line with Miller's observation, RMR in the zone of thermal neutrality during winter exceed summer values in the serin, Serinus canarius, European goldfinch, Carduelis carduelis, and in the mute swan, Cygnus olor (Gelineo, 1969; Bech, 1980).

The absence of significant seasonal change in BMR or RMR in or just below the zone of thermal neutrality has been reported for several species in addition to the house sparrow (see above): yellow bunting, Emberiza citrinella (Wallgren, 1954); northwest crow, Corvus caurinus, and black brant, Branta nigricans (Irving et al., 1955); cardinal, Cardinalis cardinalis (Dawson, 1958); starling, Sturnus vulgaris, pigeon (cf. Dontcheff and Kayser, 1934), and evening grosbeak, Coccothraustes vespertinus (Hart, 1962); great tit, Parus major (Hissa and Palokangas, 1970); American goldfinch, Carduelis tristis (Dawson and Carey, 1976); house finch, Carpodacus mexicanus (Dawson, et al., 1985); greenfinch, Carduelis chloris, and siskin (Saarela, 1988). As noted above, many of these comparisons may be complicated by seasonal changes in body mass. Such changes commonly involve winter fattening (see King, 1972; Dawson et al., 1983b) and the addition of presumably inert adipose tissue could distort mass-specific values. For example, the American goldfinch is significantly heavier in winter than in spring and summer, primarily as a result of fat deposition (Dawson and Carey, 1976; Carey et al., 1978). However, BMR does not vary between winter and spring on either a total (per bird) or mass-specific (per g) basis (Dawson and Smith, 1986). Also, Hart's (1962) results for RMR of pigeons, starlings, evening grosbeaks, and house sparrows at 30° C do not vary seasonally on either basis.

Complicating matters further, the RMR of chicks of the king penguin (Aptenodytes patagonicus) is lower in winter than in summer Barré (1986). The functional significance of this difference is unclear, but it may reflect ontogenetic changes rather than metabolic acclimatization per se.

The adaptive value of shifts in general metabolic level described above is not immediately obvious. A rise in this level in association with exposure to cold would seem unnecessary at first glance, given the capacities of birds for regulatory thermogenesis. Perhaps it reflects an emergency response linked with protection of peripheral tissues from injury in captive birds abruptly transferred into a cold environment. Improved thermogenic capacity and cold resistance have also been found to accompany the rise in general metabolic level associated with maintenance of birds in the cold (Gelineo, 1955; Gelineo, 1964). Despite these considerations, only a small fraction of birds naturally acclimatized to winter shows rises in BMR or RMR at or near thermoneutrality. The difference has been laid to the fact that seasonal acclimatization is a highly integrated response that develops gradually in association with long term environmental trends in temperature and photoperiod and short term thermal oscillations, whereas the experimental form of cold acclimatization, as noted above, represents an emergency response to abrupt exposure to cold, such as that described for rodents by Hart (1964). Perhaps it also reflects the fact that a rise in general metabolic level could increase energy costs, which would be an important disadvantage for wild birds during winter, as discussed by King and Farner (1961). Where seasonal changes in general metabolic level do occur, it is not always clear that they reflect adjustments to cold. RMR of white-browed scrub wrens (Sericornis frontalis) resident in an arid part of Australia averaged 19% lower in summer than in winter and Ambrose and Bradshaw (1988) regard this as an adjustment serving to reduce energy and water loss and thermoregulatory problems in the hot, dry portion of the year. White-browed scrub wrens in more mesic regions lacked such a seasonal shift in metabolic level.

AVIAN THERMOGENIC CAPACITY AND THERMOGENIC ENDURANCE

Cold Induced Thermogenesis in Birds

The role, if any, of regulatory non-shivering thermogenesis in avian cold defense is currently unclear (Cannon and Nedergaard, 1988), and we shall not consider it here. Shivering appears to be the major way in which birds effect thermogenesis in the cold (see, for example, West, 1965; Hohtola, 1982; Hohtola and Stevens, 1986). In most adult birds, the flight muscles (pectoralis and supracoracoideus), which comprise 15-25% of body mass (Hartman, 1961), are prominently involved in this activity. The biochemical data presented here pertain to these muscles, particularly the pectoralis.

Initial understanding of the scope of thermogenic responses of homeotherms to cold was largely the product of J. Giaja's efforts. The European goldfinch and great grey shrike (Lanius excubitor) were among the first species he studied (Giaja, 1925). Subsequently, he reported values for summit metabolism (the maximal metabolic rate attained in response to cold) for several other birds (see, for example, Giaja, 1931). In these studies, scope of the cold-induced thermogenic response was assessed in terms of the metabolic quotient, i.e., the ratio of summit metabolism to BMR (Giaja, 1925). Comparisons indicated that maximal levels of cold-induced thermogenesis in birds range between ca. 3 and 5 times BMR (see Brody's, 1945:284, summary of Giaja's results; Scholander et al., 1950).

Giaja's studies can probably be criticized on technical grounds, e.g., the cold challenges for the birds measured were intensified by removing part of their plumage and, in some instances, by immersing them in cold water

Table 1. Cold-induced Metabolic Expansibility in Some Small Birds[a]

Species	Body Mass (g)	Metabolic Quotient	Reference
Starling	85	5.1	Hart (1962); BMR, Johnson and McTaggart Cowan (1975)
Dark-eyed Junco	21.9	6.3	W. R. Dawson, J. M. Olson, and B. K. Smith unpublished
House Finch			Dawson et al. (1983a)
(California)	21.6	5.8	
(Colorado)	21.7	6.4	
Common Redpoll	14	5.6	Rosenmann and Morrison (1974)
Pine Siskin	13.8	4.7	Dawson and Carey (1976)
American Goldfinch	14.4	5.6	Dawson and Smith (1986)
Evening Grosbeak	62	5.6	Hart (1962); BMR, Dawson and Tordoff (1959)
House Sparrow	25.6	5.2	Koteja (1986); BMR, Kendeigh (1944)

[a]Values presented are for winter acclimatized birds, except in cases of common redpolls and house sparrows. Scientific names of common redpoll and pine siskin are *Carduelis flammea* and *C. pinus*, respectively. Those of other species are given in text.

prior to testing, thus risking abnormally low peripheral temperatures in key thermogenic tissues such as the flight muscles. Nonetheless, Giaja's efforts provide a conceptual framework important for assessing the extent and role of metabolic acclimatization of birds to cold. Curiously, only a few recent studies (Table 1) have attempted to pursue further the matter of cold-induced metabolic scope in these animals. These indicate somewhat higher metabolic quotients than those generally obtained by Giaja (see Brody, 1945:284). Beyond the values reported in Table 1 (4.7-6.4), poorwills (*Phalaenoptilus nuttallii*) are notable in reaching metabolic rates in arousal from torpor that are 6.3-8.9 times BMR (Withers, 1977).

Given the importance of skeletal muscle in avian regulatory thermogenesis, it is noteworthy that the metabolic quotients for cold defense in birds (Table 1) tend to be lower than the factors relating BMR and metabolic rate during maximal activity. These latter can reach 14 (flight) and 12 (running), though they typically lie between 5 and 10 (Brackenbury, 1984). Some problems exist with making precise comparisons between maximum metabolic levels attained during cold defense and exercise (see Marsh and Dawson, 1988a). However, the discrepancy in metabolic factors implies that birds cannot use the full aerobic potential of their muscles in chemical thermoregulation. Nevertheless, the thermogenic capacities of winter-acclimatized individuals impart substantial levels of cold resistance (see, for example, Dawson and Carey, 1976; Saarela, 1988).

Evidence for acclimatization of thermogenic capacity in birds comes from studies dealing respectively with: 1) experimental temperature conditioning; 2) effects of cold water immersion on juvenal penguins; and 3) seasonal comparisons. As noted in the comments concerning acclimatization effects on BMR or RMR, the shifts in metabolic level associated with experimental exposure of birds to cold is accompanied by improved cold resistance.

For example, the summit metabolism of European goldfinches maintained at -3° to +10° C is at least 35% greater than that of those maintained at 18°-22° C (Gelineo, 1955).

Thermogenic capacity also is enhanced in juvenal penguins with repeated immersion in cold (7° C) water (Barré and Roussel, 1986). Juvenal king penguins upon their first immersion only increase metabolism to 3 times the rate observed when they are maintained in air at thermoneutrality. Subsequent immersions improve this performance and, by the fourth one, this factor has increased to ca. 4-5, staying at this high level, which is 72% greater than that observed initially, through the 10th immersion. Adjustment of macaroni penguins (Eudypta c. chrysolophus) to sea water also involves an increase in thermogenic capacity, amounting to at least 36%. The apparent role of cold exposure in these adjustments suggests that the transition from land to sea by juvenal penguins of these species actually represents a form of metabolic acclimatization (Barré and Roussel, 1986).

Seasonal metabolic acclimatization of thermogenic performance has been noted in several species of birds. In the domestic pigeon and house sparrow the level of summit metabolism appears to increase during winter (Hart, 1962). On the other hand, several small birds, unlike small mammals, show only minor (<10-15%) seasonal variation in this function (Hart, 1962; Dawson et al., 1983a; Dawson and Smith, 1986; Saarela, 1988). Enzymatic measures also indicate a seasonally stable cellular aerobic capacity (Carey et al., 1978; Marsh and Dawson, 1982; Yacoe and Dawson, 1983). However, despite relatively stable thermogenic capacities, winter representatives of species such as the American goldfinch (Table 2) can show dramatically increased endurance during severe cold challenges (Hart, 1962; Dawson and Carey, 1976; Dawson et al., 1983a). The development of seasonal changes in ability to sustain high rates of thermogenesis can differ geographically in widely distributed species, based on results of Dawson et al. (1983a) for the house finch. Representatives of this species from Riverside, California, and Boulder, Colorado, localities having mild and moderately severe winters, respectively, differ in the extent of seasonal variation in cold resistance. Such resistance is limited throughout the year in the California birds, whereas it is increased significantly during winter in the Colorado individuals (Table 2). California house finches can acclimatize to winter conditions, for transfer to Ann Arbor, Michigan, a location having cold winters, and maintenance there during fall and winter resulted in cold resistance matching the maximal level noted in Colorado house finches (Dawson et al., 1983a).

FUNCTIONAL CORRELATES OF METABOLIC ACCLIMATIZATION IN BIRDS

Introduction

The fact that changes in ability to sustain elevated rates of thermogenesis during cold challenges constitute the primary feature of seasonal metabolic acclimatization in most birds studied has prompted the suggestion that this form of compensation is associated with alterations in the storage, mobilization, and/or metabolism of energy substrates (Dawson et al., 1983b). Material balance studies of American goldfinches and house finches indicate that fatty acids comprise the main energy substrates in fasted birds during moderate and severe cold challenges (Carey et al., 1978; Dawson et al., 1983a). Some glycogen may also be catabolized, especially in severe cold, but its role in thermogenesis seems minor (Thomas and George, 1975; Carey et al., 1978; Marsh and Dawson, 1982). On the other hand, glycogen has been suggested as the primary energy substrate supporting shivering in pigeons (see George and John, 1986), but this only pertained in partially defeathered birds during exposure to cold. Under these conditions muscle

Table 2. Seasonal Variation in Thermogenic Endurance of Two Finches

Species (season)	Body Mass (g)	Period at <-60° C over which thermogenesis maintained (min)	Reference[a]
American Goldfinch			1
(winter)	14.5	>180	
(summer)	12.8	<60	
House Finch			2
Colorado (winter)	21.7	98	
Colorado (spring)	21.8	9	
California (winter)	21.6	16	
California (spring)	21.0	6	

[a]References: 1, Dawson and Carey (1976); 2, Dawson et al. (1983a)

fibers with very low aerobic capacity were recruited and hypothermia rapidly ensued. Overall respiratory exchange ratios for birds fasted in the cold approximate 0.70-0.75, further suggesting the importance of fat oxidation in thermogenesis (Dawson and Marsh, 1988a). Also, estimates of glucose turnover in birds engaging in metabolic cold defense show that this substrate accounts for only a small fraction of the energy expended (Riesenfeld et al., 1979; Marsh and Dawson, 1982; Marsh et al., 1984). However, the presence of ingested carbohydrate in replete birds might alter the primary reliance on fatty acids in thermogenesis, though data are lacking on this point.

The mechanisms responsible for added mobilization of fat during cold exposure of birds are not well known. Glucagon appears to be the major lipolytic hormone in these animals, whereas catecholamines are much less effective, unlike the situation in mammals (Pearce, 1977; McCumbee and Hazelwood, 1978; LeClerq, 1984). It is of interest that glucagon is increased in cold-exposed ducklings (unpublished data cited in Barré et al., 1987) and active adult mallards, Anas platyrhynchos (Harvey et al., 1985).

Winter Fattening in Wild Birds

Small birds that remain at middle or high latitudes during the colder months of the year often show winter fattening (see reviews by King, 1972; Blem, 1976; Dawson et al., 1983b). Fat reserves are increased during late fall and maintained at a high level through early spring in species such as the American goldfinch (Carey et al., 1978; Dawson and Marsh, 1986). These reserves play an important role both in sustaining the birds through the long cold nights of winter and also during periods of inclement weather occurring at this season. However, they do not provide a large safety margin and it has been estimated for the American goldfinch fasting in the cold that they would only support a metabolic rate twice BMR for a couple of days (Carey et al., 1978). Not all small birds show winter fattening (Blem and Pagels, 1984) and the response can vary intraspecifically in species that do (Dawson and Marsh, 1986). The variation superficially appears correlated with climate (Blem, 1973; Dawson and Marsh, 1986) and ambient temperature over the preceding 2-10 days seems to be the main proximate factor affecting fat levels in dark-eyed juncos (Junco hyemalis) wintering in the northcentral

United States (Rogers et al., 1988). Additionally, American goldfinches show winter fattening in southeastern Michigan, whereas those in southern California, where conditions are milder, do not. However, goldfinches wintering in eastern Texas, which had average winter temperatures similar to those at Riverside, California (the locality where the California birds were studied) showed a similar degree of winter fattening to that noted in the Michigan ones. In reporting these results, Dawson and Marsh (1986) suggested that this pattern reflects two circumstances: 1) a high degree of interchange occurs among populations of goldfinches wintering in different areas of the eastern half of the United States; and 2) regulatory mechanisms controlling winter fattening appear independent of the proximate effects of temperature in this species. The complexity of winter fattening and its control is further indicated by the fact that ecologically similar species wintering together may maintain quite different levels of fat (Steube and Ketterson, 1982; Dawson and Marsh, 1986). The selective forces and proximate limits that determine the degree of winter fattening are as yet poorly understood. Scaling effects on heat loss and metabolism probably should be included among the ultimate factors influencing these relationships (Calder, 1984), just as photoperiod should be regarded as a potentially major proximate factor (Evans, 1969; Dawson and Marsh, 1986). The correlation between winter fattening and increases in ability to sustain high rates of thermogenesis has led to the former's being seen as a component of metabolic acclimatization (Dawson and Carey, 1976; Carey et al., 1978; Dawson et al., 1983b). This correlation may involve multiple considerations. Added fat storage probably increases cold tolerance by providing a larger total energy reserve that can serve to extend endurance at a given rate of thermogenesis. However, winter fattening also may be associated with other metabolic changes effective in sustaining high metabolic rates. Findings for two cardueline finches illustrate such an association. Most American goldfinches tested in midwinter can tolerate severe cold challenges for substantial periods; but no individuals do so by late spring. Although this change in cold resistance is correlated with mean fat reserves (Dawson and Carey, 1976; Carey et al., 1978), the distributions of fat contents of goldfinches in these two seasons overlap (Dawson and Marsh, 1986). Similarly, house finches show seasonal cycles of cold resistance and fattening at Boulder, Colorado, but these are absent in birds of this species at Riverside, California, where winters are milder (Dawson et al., 1983a). However, the amount of fat in summer and winter birds is quite variable, just as in the American goldfinches. Additionally, the accumulation of superabundant fat reserves by captive house finches from California and Colorado did not significantly extend their cold resistance beyond that of free-living birds in the latter state (Dawson et al., 1983a). The data thus far establish a role for winter fattening in metabolic acclimatization, but one that appears merely a part of a suite of adjustments permitting sustained mobilization and oxidation of fatty acids.

Changes in Mobilization, Delivery, and Use of Energy Substrates as Components of Metabolic Acclimatization

Summer-acclimatized American goldfinches and house finches became hypothermic under severe cold challenges, before significant amounts of depot fat were used (Carey et al., 1978; Dawson et al., 1983a). Inadequate mobilization from adipose tissue or insufficient uptake and catabolism of plasma free fatty acids could be responsible. Plasma titers of free fatty acids did not differ significantly between summer and winter in cold stressed individuals (Marsh and Dawson, 1982; Marsh et al., 1984), suggesting that mobilization was not limiting, though such a conclusion must remain provisional because it is based on static measures of these titers. Liver glycogen reserves of goldfinches and house finches are small and do not provide sufficient mobilizable substrate to sustain high levels of thermogenesis (Carey et al., 1978; Marsh and Dawson, 1982; Marsh et al., 1984). Gluconeogenic capacity of the liver in these birds may be augmented during winter (Marsh and Dawson, 1982; Marsh et al., 1984), but the extent and role of this in cold resistance remain to be defined. Marsh and Dawson (1988a) suggest that metabolic endurance during cold stress is not limited by the mobilization of energy substrates from storage depots and their delivery to shivering muscles. However, this needs to be tested.

Fatty acids apparently provide most of the energy for cold-induced thermogenesis in birds and the ability to oxidize these compounds may involve a variety of adjustments in substrate metabolism. In this connection, Marsh and Dawson (1982) studied carbohydrate use in American goldfinches during cold stress in summer and winter. This was prompted by results for exercising mammals showing: a) correlation between fatigue and the depletion of glycogen in liver and muscles (see, for example, Karlsson et al., 1974) or hypoglycemia (Clark and Conlee, 1979; Callow et al. 1986); and b) the prolonging of endurance by experimental treatments that reduce carbohydrate depletion in these tissues (e.g., Rennie et al., 1976). As noted earlier in this discussion, thermogenic endurance shows pronounced seasonal variation in the American goldfinch (Table 2). Winter-acclimatized representatives of this species had less depletion of glycogen and lower turnover of plasma glucose in the cold than summer-acclimatized ones (Marsh and Dawson, 1982). Moreover, the former increased metabolic rate substantially without increasing glucose turnover or glycogen breakdown. The observations on glucose turnover contrast markedly with those on mammals (e.g., Depocas, 1962; Minaire, 1973) and the domestic fowl (Riesenfeld et al., 1979), in which turnover rates vary directly with metabolic rate in the cold. Of possibly greater immediate relevance to the matter of avian metabolic acclimatization, house finches, which show far less change in thermogenic endurance between winter and summer (Table 2), did not differ seasonally in either glucose turnover or glycogen depletion during cold exposure (Marsh et al., 1984).

Specific activity of some catabolic enzymes changes seasonally in certain finches. Despite a decrease between summer and winter in carbohydrate use by cold-challenged American goldfinches, their pectoralis muscles showed a modest (14-28%) increase in the activity of phosphofructokinase (PFK) during the latter season (Marsh and Dawson, 1982; Yacoe and Dawson, 1983). This was accompanied in these muscles by an increase of 50-100% in the specific activity of β-hydroxyacyl-CoA dehydrogenase (HOAD), an indicator of capacity for β-oxidation of fatty acids (Marsh and Dawson, 1982; Yacoe and Dawson, 1983). Pectoralis HOAD activity also was significantly higher in winter than in spring in house finches in Colorado (Carey et al., 1988). Despite the changes in HOAD in goldfinches, the capacities for oxidation of fatty acids by crude mitochondrial preparations isolated from the muscles of winter- and summer-acclimatized individuals appeared similar (Yacoe and Dawson, 1983). However, these preparations showed rather low rates of oxygen consumption in comparison with values estimated from Dawson and Carey's (1976) in vivo metabolic measurements.

Understanding of the regulation of glycolysis and β-oxidation is crucial for interpreting the information summarized above. Marsh and Dawson (1982) hypothesized that the increase (as indicated by levels of HOAD activity) in β-oxidative capacity in the pectoralis muscles of winter-acclimatized American goldfinches could result in inhibition of glycolysis, despite the rise in glycolytic capacity (as indicated by measurements of PFK activity). This suggestion is consistent with the reciprocal regulation of these pathways documented in some mammalian tissues. Marsh et al. (1989) have examined the regulation of glycolysis more directly by measurement of most glycolytic intermediates, glycogen, and citrate in the pectoralis muscles of summer- and winter-acclimatized American goldfinches. Their findings are discussed in more detail elsewhere in these proceedings (Marsh and Dawson, 1988b). The PFK-catalyzed step in glycolysis is regulated in a manner consistent with previously established patterns of glycogen and glucose use. However, the observed regulation of PFK in American goldfinches does not consistently correlate with the concentration of citrate, contrasting with the situation in some mammalian tissues in which the rise in this metabolite as a result of increased β-oxidation has been proposed to inhibit glycolytic flux (Randle, 1981). In the pectoralis muscles of American goldfinches, either some other metabolic change accompanying augmented β-oxidative capacity inhibits PFK, or the functional explanation of seasonally enhanced HOAD activity lies elsewhere.

Despite the studies of enzymatic activity and metabolite flux completed thus far, we have only very incomplete knowledge of the basis of metabolic acclimatization to winter conditions in birds. The apparent overall correlation between thermoregulatory failure and depletion of muscle glycogen suggested that this compensation involved seasonally improved capacities for conserving carbohydrate reserves (Marsh and Dawson, 1982; Marsh et al., 1984). However, this has not been confirmed with more extensive sampling (Marsh et al., 1989). Winter-acclimatized American goldfinches can sustain high levels of cold-induced thermogenesis even when muscle glycogen levels are low, whereas summer-acclimatized individuals become hypothermic with significant amounts of this storage substance remaining. Furthermore, earlier studies of goldfinches and house finches did not establish a role for hypoglycemia in thermoregulatory failure in the cold (Marsh and Dawson, 1982; Marsh et al., 1984). All these observations persuaded Marsh et al. (1989) that patterns of carbohydrate metabolism in American goldfinches in summer and winter are indicators of metabolic adjustments responsible for the pronounced seasonal differences in thermogenic ability. However, they further concluded that carbohydrate depletion is not necessarily a sine qua non for thermoregulatory failure in the cold.

AVIAN METABOLIC ACCLIMATIZATION TO COLD AND SEASON

It is now apparent that metabolic acclimatization to cold and season in birds involves more than shifts in metabolic level (as indicated by RMR or BMR) or in the level of summit metabolism. Indeed, such changes seem largely absent (general metabolic level) or modest (summit metabolism) in wild birds contending with natural conditions. An accurate characterization of metabolic acclimatization in these animals must recognize the primary role of increased abilities to sustain high rates of thermogenesis in the cold, the importance of winter fattening, and the involvement an array of biochemical changes such as increases in the activity of certain catabolic enzymes, in the adjustment process. This form of acclimatization is readily documented in small birds, whose limited insulative capacities make them especially reliant on chemical thermoregulation in the cold. However, we should also anticipate its occurrence in aquatic birds swimming in cold water and, perhaps, in poorly insulated young birds. Recent information indicates considerable geographical variation in extent of metabolic acclimatization in birds with wide winter distributions. Future tasks in the study of metabolic acclimatization must include both further characterization of underlying mechanisms and analysis of this geographical variation and the factors producing it.

REFERENCES

Ambrose, S. J., and Bradshaw, S. D., 1988, Seasonal changes in standard metabolic rates in the white-browed scrubwren Sericornis frontalis (Acanthizidae) from arid, semi-arid, and mesic environments. Comp. Biochem. Physiol., 89A:79.

Arieli, A., Berman, A., and Meltzer, A., 1979, Cold thermogenesis in the summer-acclimatized and winter-acclimated domestic fowl, Comp. Biochem. Physiol., 63C:7.

Barré, H., 1986, Metabolic and insulative changes in winter- and summer-acclimatized king penguin chicks, J. Comp. Physiol., B 154:317.

Barré, H., Cohen-Adad, F., and Rouanet, J., 1987, Two daily glucagon injections induce nonshivering thermogenesis in Muscovy ducklings. Am. J. Physiol., 252:E616.

Barré, H., and Roussel, B., 1986, Thermal and metabolic adaptation to first cold-water immersion in juvenile penguins, Am. J. Physiol., 251:R456.

Bech, C., 1980, Body temperature, metabolic rate, and insulation in winter and summer acclimatized mute swans (Cygnus olor), J. Comp. Physiol., B 136:61.

Blem, C. R., 1973, Geographic variation in the bioenergetics of the house sparrow, Ornithol. Monogr., 14:96.

Blem, C. R., 1976, Patterns of lipid storage and utilization in birds, Am. Zool., 16:671.

Blem, C. R., and Pagels, J. F., 1984, Mid-winter lipid reserves of the golden-crowned kinglet, Condor, 86:491.

Brackenbury, J., 1984, Physiological responses of birds to flight and running, Biol. Rev., 59:559.

Brody, S., 1945, "Bioenergetics and Growth," Reinhold Publishing Corp., New York.

Calder, W. A., 1984, "Size, Function, and Life History," Harvard Univ. Press, Cambridge, MA.

Callow, M., Morten, A., and Guppy, M., 1986, Marathon fatigue: the role of plasma free fatty acids, Eur. J. Appl. Physiol., 55:654.

Cannon, B., and Nedergaard, J., 1988, Shivering and non-shivering thermogenesis in birds, this volume.

Carey, C., Dawson, W. R., Maxwell, L. C., and Faulkner, J. A., 1978, Seasonal acclimatization to temperature in cardueline finches. II. Changes in body composition and mass in relation to season and acute cold stress, J. Comp. Physiol., B 125:101.

Carey, C., Marsh, R. L., Bekoff, A. C., and Olin, A., 1988, Enzyme activities and muscular patterns of shivering in house finches, this volume.

Clark, J. H., and Conlee, R. K., 1979, Muscle and liver glycogen content: diurnal variation and endurance, J. Appl. Physiol., 47:425.

Dawson, W. R., 1958, Relation of oxygen consumption and evaporative water loss to temperature in the cardinal, Physiol. Zool., 31:37.

Dawson, W. R., Buttemer, W. A., and Carey, C., 1985, A reexamination of the metabolic response of house finches to temperature, Condor, 87:424.

Dawson, W. R., and Carey, C., 1976, Seasonal acclimatization to temperature in cardueline finches. I. Insulative and metabolic adjustments, J. Comp. Physiol., 112:317.

Dawson, W. R., and Marsh, R. L., 1986, Winter fattening in the American goldfinch and the possible role of temperature in its regulation, Physiol. Zool., 59:357.

Dawson, W. R., Marsh, R. L., Buttemer, W. A., and Carey, C., 1983a, Seasonal and geographic variation of cold resistance in house finches Carpodacus mexicanus, Physiol. Zool., 56:353.

Dawson, W. R., Marsh, R. L., and Yacoe, M. E., 1983b, Metabolic adjustments of small passerine birds for migration and cold, Am. J. Physiol., 245: R755.

Dawson, W. R., and Smith, B. K., 1986, Seasonal acclimatization in the American goldfinch (Carduelis tristis), In: "Living in the Cold," H. C. Heller, X. J. Musacchia, and L. C. H. Wang, eds., Elsevier Science Publishing Co., New York.

Dawson, W. R., and Tordoff, H. B., 1959, Relation of oxygen consumption to temperature in the evening grosbeak, Condor, 61:388.

Depocas, F., 1962, Body glucose as fuel in white rats exposed to cold: results with fasted rats, Am. J. Physiol., 202:1015.

Dontcheff, L., and Kayser, C., 1934, Le rythme saisonnier du métabolisme de base chez le pigeon en fonction de la température moyenne du milieu, Ann. Physiol. Physicochim. Biol., 10:285.

Evans, P. R., 1969, Winter fat deposition and overnight survival of yellow buntings (Emberiza citrinella L.), J. Anim. Ecol., 38:415.

Gelineo, S., 1934, Influence du milieu thermique sur la courbe de la thermorégulation, Compt. Rend. Soc. Biol., 117:40.

Gelineo, S., 1955, Température d'adaptation et production de chaleur chez oiseaux de petite taille, Arch. Sci. Physiol., 9:225.

Gelineo, S., 1964, Organ systems in adaptation: the temperature regulating system, In: "Handbook of Physiology, Section 4, Adaptation to Environment," D. B. Dill, ed., American Physiological Society, Washington, D. C.

Gelineo, S., 1969, Heat production in birds in summer and winter, Srpska Akad. Nauka I Umetnosti Belgrad, Bull. Classe Sci. Math. Natur., XXVI, Sci. Natur. (n.s.), no. 12:99

George, J. C., and John, T. M., 1986, Physiological responses to cold exposure in pigeons. In: "Living in the Cold," H. C. Heller, X. J. Musacchia, and L. C. H. Wang, eds., Elsevier Science Publishing Co., New York.

Giaja, J., 1925, Le métabolisme de sommet et le quotient métabolique, Ann. Physiol. Physicochim. Biol., 1:596.

Giaja, J., 1931, Contribution à l'étude de la thermorégulation des oiseaux, Ann. Physiol. Physicochim. Biol., 7:13.

Hart, J. S., 1962, Seasonal acclimatization in four species of small wild birds, Physiol. Zool., 35:224.

Hart, J. S., 1964, Insulative and metabolic adaptations to cold in vertebrates, Soc. Exp. Biol. Symp., 35:31.

Hartman, F. A., 1961, Locomotor mechanisms in birds, Smithsonian Misc. Coll., 143:1.

Harvey, S., Klandorf, H., Foltzer, C., Strosser, M. T., and Phillips, J. G., 1985, Endocrine responses of ducks (Anas platyrhynchos) to treadmill exercise, Gen. Comp. Endocr., 48:415.

Hissa, R., and Palokangas, R., 1970, Thermoregulation in the titmouse (Parus major L.), Comp. Biochem. Physiol., 33:942.

Hohtola, E., 1982, Thermal and electromyographic correlates of shivering thermogenesis in the pigeon, Comp. Biochem. Physiol., 73A:159.

Hohtola, E., and Stevens, E. D., 1986, The relationship of muscle electrical activity, tremor and heat production to shivering thermogenesis in Japanese quail, J. Exp. Biol., 125:119.

Irving, L., Krog, J., and Monson, M., 1955, The metabolism of some Alaskan animals in winter and summer, Physiol. Zool., 28:173.

Johnson, S. R., and McTaggart Cowan, I., 1975, The energy cycle and thermal tolerance of the starlings (Aves, Sturnidae) in North America, Can. J. Zool., 53:55.

Karlsson, J., Nordesjo, L.-O., and Saltin, B., 1974, Muscle glycogen utilization during exercise after physical training, Acta Physiol. Scand., 90:210.

King, J. R., 1972, Adaptive periodic fat storage by birds, Proc. XVth Internat. Ornith. Congr., p. 201.

King, J. R., and Farner, D. S., 1961, Energy metabolism, thermoregulation and body temperature, In: "Biology and Comparative Physiology of Birds," Vol. II, A. J. Marshall, ed., Academic Press, New York.

Kendeigh, S. C., 1944, Effect of air temperature on the rate of energy metabolism of the English sparrow, J. Exp. Zool., 96:1.

Koteja, P, 1986, Maximum cold-induced oxygen consumption in the house sparrow Passer domesticus L., Physiol. Zool., 59:43.

LeClerq, B., 1984, Adipose tissue metabolism and its control in birds, Poultry Sci., 63:2044.

McCumbee, W. D., and Hazelwood, R. L., 1978, Sensitivity of chicken and rat adipocytes and hepatocytes to isologous and heterologous pancreatic hormones, Gen. Comp. Endocr., 34:421.

Marsh, R. L., Carey, C., and Dawson, W. R., 1984, Substrate concentrations and turnover of plasma glucose during cold exposure in seasonally acclimatized house finches, Carpodacus mexicanus, J. Comp. Physiol., B 154:469.

Marsh, R. L., and Dawson, W. R., 1982, Substrate metabolism in seasonally acclimatized American goldfinches, Am. J. Physiol., 242:R563.

Marsh, R. L., and Dawson, W. R., 1988a, Avian adjustments to cold, In: "Animal Adaptation to Cold," L. Wang, ed., Springer-Verlag, Berlin. (In press)

Marsh, R. L., Dawson, W. R., 1988b, Metabolism of energy substrates and seasonal acclimatization, this volume.

Marsh, R. L., Dawson, W. R., Camilliere, J., and Olson, J. M., 1989, Regulation of glycolysis in the pectoralis muscles of seasonally acclimatized American goldfinches exposed to cold, Am. J. Physiol., submitted.

Miller, D. S., 1939, A study of the physiology of the sparrow thyroid, J. Exp. Zool., 80:259.

Minaire, Y., Vincent-Falquet, J.-C., Pernod, A., and Chatonnet, J., 1973, Energy supply in acute cold-exposed dogs, J. Appl. Physiol., 35:51.

Pearce, J., 1977, Some differences between avian and mammalian biochemistry, Internat. J. Biochem., 8:269.

Randle, P. J., Tubbs, P. K., 1979, Carbohydrate and fatty acid metabolism, In: "Handbook of Physiology, Section 2, The Cardiovascular System, Vol. 1, The Heart," R. M. Berne, N. Sperelakis, and S. R. Geiger, eds., American Physiological Society, New York.

Rennie, M. J., Winder. W. W., and Holloszy, J. O., 1976, A sparing effect of plasma fatty acids on muscle and liver glycogen content of the exercising rat, Biochem. J., 156:649.

Riesenfeld, G., Berman, A., and Hurwitz, S., 1979, Glucose kinetics and heat production in normothermic, hypothermic, and hyperthermic fasted chickens, Comp. Biochem. Physiol., 67A:199.

Rogers, C. M., Ketterson, E. D., and Nolan, Jr., V., 1988, Regulation of winter fattening in dark-eyed juncos Junco hyemalis hyemalis: a geographical perspective, unpublished ms.

Rosenmann, M., and Morrison, P., 1974, Maximum oxygen consumption and heat loss facilitation in small homeotherms by $He-O_2$, Am. J. Physiol., 226:490.

Saarela, S., 1988, Thermogenic capacity of greenfinches and siskins in winter and summer, this volume.

Scholander, P. F., Hock, R., Walters, V., Johnson, F., and Irving, L., 1950, Heat regulation in some arctic and tropical mammals and birds, Biol. Bull., 99:237.

Steube, M. M., and Ketterson, E. D., 1982, A study of fasting in tree sparrows (Spizella arborea) and dark-eyed juncos (Junco hyemalis): ecological implications, Auk, 99:299.

Thomas, V. G., and George, J. C., 1975, Changes in plasma, liver, and muscle metabolite levels in Japanese quail exposed to cold, J. Comp. Physiol., 100:297.

Wallgren, H., 1954, Energy metabolism of two species of the genus Emberiza as correlated with distribution and migration, Acta Zool. Fennica, 84:1.

West, G. C., 1965, Shivering and heat production in wild birds, Physiol. Zool., 38:111.

Withers, P. C., 1977, Respiration, metabolism, and heat exchange of euthermic and torpid poorwills and hummingbirds, Physiol. Zool., 50:43.

Yacoe, M. E., and Dawson, W. R., 1983, Seasonal acclimatization in American goldfinches: the role of the pectoralis muscle, Am. J. Physiol., 242:R265.

ENZYME ACTIVITIES IN MUSCLES OF SEASONALLY ACCLIMATIZED HOUSE FINCHES

Cynthia Carey[1], Richard L. Marsh[2], Anne Bekoff[1],
Rebecca M. Johnston[1], and Ann M. Olin[1]

[1]Department of EPO Biology, University of Colorado
Boulder, CO 80309, USA and [2]Department of Biology
Northeastern University, Boston, MA 02115, USA

INTRODUCTION

Members of the avian subfamily Carduelinae have served as subjects of numerous investigations on adaptations of small birds to cold winters (Salt, 1952; Dawson and Tordoff, 1964; West, 1972; Dawson and Carey, 1976; Carey et al. 1978; Weathers et al. 1980; Marsh et al. 1984; Reinertsen, 1986; Yacoe and Dawson, 1983; and others). These birds, including goldfinches, pine siskins, house finches, grosbeaks, crossbills, and redpolls, are particularly well suited for such study because of their principally northern distribution and small body sizes. Their success in cold climates appears to be linked in part to a type of metabolic acclimatization, involving enhanced capabilities for sustained elevated production of heat for prolonged periods (Dawson and Carey, 1976; Dawson et al. 1983a,b; Dawson and Marsh, 1988).

Intensive study of two cardueline finches, American goldfinches (Carduelis tristis) and house finches (Carpodacus mexicanus) has revealed differences between these species in degree of seasonal adjustment to cold. Winter goldfinches exhibit considerable enhancement of endurance during cold stress that is correlated with seasonal accumulation of lipid stores, shifts in activities of β-oxidative and glycolytic enzymes, and increased emphasis on lipid utilization and conservation of glucose during shivering (Dawson and Carey, 1976; Carey et al. 1978; Marsh and Dawson, 1982; Yacoe and Dawson, 1983; Marsh and Dawson 1988a). In contrast, winter house finches exhibit only a modest improvement in thermogenic capacity and a slight increase in levels of stored lipid (Dawson et al. 1983a). Use of carbohydrate during shivering is similar in summer and winter (Marsh et al. 1984).

House finches are plentiful in areas of the western United States in which cold winters are common. They have also recently extended their range throughout the Northeast and northern Midwest following their introduction into New York in 1940 (see Dawson et al. 1983a; Leck, 1987). Their persistence in cold climates is noteworthy considering their apparently limited capacity for seasonal enhancement of thermogenic endurance. In this study we attempted to extend our understanding of the mechanisms by which house finches survive cold winters by addressing the question of whether activities of catabolic enzymes in skeletal muscle vary seasonally.

MATERIALS AND METHODS

Enzyme Activities

House finches were trapped between 0630 and 0900 hrs in December, 1987 (winter), May 1988 (spring), and July 1988 (summer), in Boulder, Boulder County, Colorado (40°01'N, 105°16'W). They were taken immediately to the laboratory where they were weighed to the nearest 0.01 g on a Mettler top-loading balance and then sacrificed by cervical dislocation. The pectoralis muscles and the muscles from both legs were rapidly removed, weighed, and minced on a cold surface. The minced leg and pectoralis muscles were then divided into four groups each. One sample of each tissue was weighed and frozen for analysis of muscle water content by lyophilization. Since water content of house finches carcasses varies seasonally (Dawson et al., 1983a), this measurement was necessary to ensure that seasonal variation in enzyme activity per gram wet mass was not caused by variation in muscle water content. The other three aliquots were weighed and placed in the appropriate buffer for analysis of activities of hexokinase (HK, EC 2.7.1.1.), phosphofructokinase (PFK, EC 2.4.1.1.), β-hydroxyacyl-CoA dehydrogenase, (HOAD, EC 1.1.1.35), and citrate synthase (CS, EC 4.1.3.7). Pectoralis and leg muscles were homogenized in 10 and 20 volumes of buffer, respectively, in hand-held glass-glass homogenizers on ice. The homogenizing buffer for analysis of HK contained 50 mM triethanolamine-HCl, 7.5 mM $MgCl_2$, 1 mM EDTA, 100 mM glucose, and 5 mM mercaptoethanol at pH 7.5. Muscles to be analyzed for PFK, CS and HOAD activity were homogenized, sonicated and stored with methods similar to Marsh (1981).

Activities of the four enzymes in the crude homogenates were measured spectrophotometrically at 340 nm (HK, PFK, HOAD) or 413 nm (CS) with a Gilford Model 250 spectrophotometer in conjunction with an Apple IIe microcomputer. All assays were done at 25°C in a final volume of 1 ml. Hexokinase was assayed by a modification of the method of Crabtree and Newsholme (1972). The assay medium contained 75 mM Tris-HCl, 35 mM glucose, 10 mM $MgCl_2$, 1 mM EDTA, 1.5 mM KCl, 5 mM mercaptoethanol, 0.4 mM $NADP^+$, 2.5 mM ATP, 10 mM creatine phosphate, 10 units creatine phosphokinase, and 1 unit glucose-6-phosphate dehydrogenase at pH 7.5. Activities of the other enzymes were measured according to methods outlined by Marsh (1981) with the exception of the following: 1) the assay medium for HOAD contained 50 mM triethanolamine, 2) the medium for PFK contained 0.3 mM KCN, 1.5 mM $NADH_2$, approximately 0.8 units aldolase, 0.7 units glucose-6-phosphate dehydrogenase, and 7.0 units triose phosphate isomerase, and 3) the pH of the CS assay was 8.2.

Control assays were run for each of the enzymes in the absence of substrate. Duplicate assays were performed for each enzyme and the averaged results are presented as μmoles substrate used (min . g wet mass)$^{-1}$. Mean values for enzyme activity of the two muscle types and other muscle characteristics are presented $\pm$ S.E.M. Comparisons between averages from the three seasons were made with one-way ANOVA. In cases where significant differences were found, the Tukey Honest Sum Difference test was used to identify homogeneous subsets of means that were statistically indistinguishable. Calculation of a regression line by the method of least-squares was used to test for a significant relation between nocturnal low ambient temperature and enzyme activity. These temperatures were provided by the U.S. National Weather Service office in Boulder.

Table 1. Mean (± SEM) body mass and muscle mass from one side (g) of house finches captured in different seasons. Sample sizes are in parentheses. Results of analysis of variance for seasonal differences among means is at the bottom of each column.

Season	Body Mass	Pectoralis	Leg
Winter	21.24 ± 0.39 (11)	1.75 ± .042 (11)	0.25 ± 0.013 (9)
Spring	20.32 ± 0.29 (9)	1.73 ± .032 (9)	0.29 ± 0.013 (9)
Summer	20.48 ± 0.37 (6)	1.75 ± 0.062 (6)	0.30 ± 0.002 (6)
f-ratio	2.04	0.12	3.19
f-probability	.15	.89	.06

RESULTS

Enzyme Activities

Body mass, pectoralis mass and the total mass of the lower leg muscles of house finches did not vary seasonally (Table 1). The water content of the pectoralis and leg muscles averaged 69.9% and 72.0%, respectively in winter. These values did not differ significantly from values in spring (70.8% and 71.5%, respectively) or summer (71.0% and 72.2%, respectively). No significant variation in the activities of HK, PFK, or CS was evident in the pectoralis of winter, spring or summer birds (Table 2), but the activity of HOAD in spring and summer pectoralis was significantly lower than that in winter ($P < 0.001$; Table 2). Activities of HK and PFK in leg muscles did not vary seasonally, but CS of leg muscle in summer was significantly reduced below activity levels in other seasons. HOAD activity in winter leg muscles was significantly higher than in the other two seasons (Table 2).

In each season, HK activity levels were significantly ($P < 0.01$) greater in leg than in pectoralis muscle. Levels of CS, HOAD, and PFK were significantly ($P < 0.01$) greater in pectoralis than in leg in all seasons. No significant relations existed between enzyme activities and the nocturnal low temperature preceding the morning on which a specific bird was captured. However, levels of hexokinase activity in winter leg muscles were inversely correlated with the average low temperature on the three nights previous to the capture of the birds (Fig. 1). The slope of the regression line defining the relation between the average low of the three previous nights and hexokinase activity was significant ($P = 0.008$).

DISCUSSION

Compared to American goldfinches, house finches exhibit relatively modest seasonal increases in the capacity to produce heat under conditions of severe cold (Dawson and Carey, 1976; Dawson et al.

Table 2. Mean (± SEM) enzyme activities [μmol substrate (g wet mass·min)$^{-1}$] of house finch pectoralis muscles in winter (W), spring (Sp), and summer (S). Sample sizes are in parentheses. HK = hexokinase; PFK = phosphofructokinase; HOAD = β-hydroxyacyl-CoA dehydrogenase; CS = citrate synthase. Results from analysis of variance for seasonal differences among the means is at the bottom of each column.

Tissue (Season)	HK	PFK	HOAD	CS
Pectoralis (W)	1.19 ± 0.10 (11)	23.1 ± 1.1 (8)	19.6 ± 1.2 (9)	119.7 ± 4.8 (9)
Pectoralis (Sp)	1.12 ± 0.11 (9)	23.3 ± 1.3 (9)	10.7 ± 0.51 (9)	113.4 ± 6.3 (9)
Pectoralis (S)	0.92 ± 0.07 (6)	22.4 ± 2.7 (6)	12.7 ± 1.0 (6)	100.6 ± 1.8 (6)
f-ratio	1.60	0.08	27.13	2.60
f-probability	.22	.93	.0001	.10

Table 3. Mean (± SEM) enzyme activities of house finch leg muscles in winter (W), spring (Sp), and summer (S). Sample sizes are in parentheses. Units, symbols, f-ratios, and probabilities are the same as in Table 2.

Tissue (Season)	HK	PFK	HOAD	CS
Leg (W)	2.08 ± 0.16 (11)	9.9 ± 1.2 (7)	7.8 ± 0.54 (9)	49.1 ± 2.7 (9)
Leg (Sp)	2.18 ± 0.14 (9)	8.0 ± 1.2 (8)	6.0 ± 0.45 (9)	46.9 ± 1.8 (9)
Leg (S)	1.72 ± 0.16 (6)	9.3 ± 1.8 (6)	5.5 ± 0.19 (6)	36.3 ± 1.3 (6)
f-ratio	1.77	0.64	6.98	8.40
f-probability	0.19	0.54	.005	.002

1983a). The thermogenic response of house finches differs from that of goldfinches primarily in the ability to sustain high levels of heat production, rather than the degree of metabolic expansibility: during exposure to temperatures as low as -70 C, the peak metabolic rates of goldfinches and house finches were 5.5- and 6.4-times standard levels, respectively. Winter goldfinches remain homeothermic for more than 6-8 hr below -60 C, whereas winter house finches can tolerate similar temperatures for no more than 90 min (Dawson and Carey, 1976; Dawson et al. 1983a). The ability to increase endurance in muscular thermogenesis would have obvious survival benefits for small birds in winter, since ambient temperatures are far below thermoneutral ranges for long periods and since capacities for non-shivering thermogenesis are absent or minimal in adult birds (Marsh and Dawson, 1988b).

In our attempt to understand the mechanisms underlying these seasonal shifts in thermogenic endurance, it has been of interest to determine the extent to which biochemical adjustments associated with seasonal increases in thermogenic endurance reflect those made by skeletal muscle in response to endurance exercise training. When mammalian muscle is subjected to aerobic endurance training, overall aerobic capacity of the muscle increases (see Holloszy and Booth, 1976, for review). Underlying these changes are increases in activities of most mitochondrial enzymes (Holloszy and Booth, 1976). Endurance training either doesn't affect or causes small decreases in most glycolytic enzymes. However, the activity of hexokinase, the enzyme responsible for phosphorylation of glucose absorbed from the blood is augmented considerably by this training; HK activity can increase significantly after even a single bout of exercise (Lamb et al. 1969; Barnard and Peter, 1969). Mammalian muscle adapted for endurance exercise derives more of its ATP from fatty acids and less from carbohydrate than untrained muscle (Holloszy and Booth, 1976). Glucose-sparing is thought to increase endurance, since fatigue is correlated with the depletion of muscle glycogen (Newsholme and Leech, 1983). Glucose-sparing in vertebrate muscle has been thought to result from a reduction in the rate of glycolysis due to inhibition of PFK by citrate (Randle, 1981).

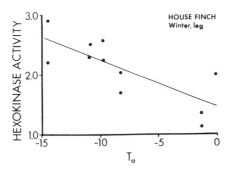

Figure 1. Levels of hexokinase activity [μmoles (g · min)$^{-1}$] of pectoralis of individual house finch as a function of the average nighttime low on the three nights preceeding capture. The thin line is a least square regression line described by the equation HK = 1.46 - 0.076 Ta, where Ta is in °C.

Biochemical changes in avian muscle associated with increased thermogenic endurance share both similarities and differences with this pattern. No seasonal variation in aerobic capacity nor in HK activity occurs in the pectoralis muscle of goldfinches (Carey et al. 1978; Marsh and Dawson, 1982; Yacoe and Dawson, 1983). Apparently the aerobic capacity necessary to support the costs of flight is more than adequate to foster shivering. However, the activity of the β-oxidative enzyme HOAD in the pectoralis of winter birds is increased substantially. Also in winter, goldfinches increase deposits of lipid and muscle glycogen and catabolize less carbohydrate and more lipid during shivering than do summer birds (Carey et al. 1978; Marsh and Dawson, 1982). The regulatory processes which shift reliance from carbohydrate to fatty acids have not been identified (Marsh et al., 1989). Moreover, the precise importance of glucose-sparing is unclear, since some birds exposed to severe cold become hypothermic long after muscle glycogen is depleted (Marsh et al., 1989).

Stable levels of citrate synthase activity in the pectoralis muscle of house finches indicate that the aerobic capacity, for which CS can be used as a marker (see Marsh, 1981), is seasonally invariant (Table 2). Levels of pectoralis CS activity are about 60% of the values for goldfinch muscle at comparable seasons (Marsh and Dawson, 1981). Although the standard metabolic rate of 20-g house finches is 75% of the value for 11 to 14-g goldfinches (Dawson and Carey, 1976; Dawson et al. 1985), the low CS in house finch muscle is not due to effects of body mass as CS activity in the pectoralis of 38-g catbirds (<u>Dumetella carolinensis</u>) is within the range of goldfinch values (Marsh, 1981). The low overall oxidative capacity of house finch pectoralis muscles may limit thermogenesis in this tissue and cause reliance on the leg muscles during cold exposure (see below).

House finches deposit only minimal amounts of neutral lipid in winter: total body fat increases from a low of 6.5% of body mass in summer to 7.5% in winter, one of the smallest increments of fat deposition recorded for wintering birds (Dawson et al. 1983a,b). Even so, stored lipid is quantitatively a much greater potential resource for shivering than is glycogen, and some lipid is catabolized by winter house finches during exposure to severe cold (Dawson et al. 1983a). Rates of glucose-turnover and utilization of muscle glycogen do not vary seasonally in this species (Marsh et al. 1984). Ratios of HOAD to PFK in pectoralis of winter, spring, and summer house finches (.85, .46, and .57, respectively) emphasize the predominance of glycolytic over β-oxidative capacities. These values are substantially below the HOAD/PFK ratio (1.28) observed for winter goldfinch pectoralis (Marsh and Dawson, 1982). The lack of seasonal change in carbohydrate catabolism during shivering prompted Marsh et al. (1984) to predict that HOAD activity levels in the pectoralis of house finches would not change seasonally. However, activities of this enzyme were significantly higher in winter than in spring or summer (Table 2). The higher ratio of HOAD to CS in winter (.16) compared to summer (.09) contrasts with the situation in mammalian muscle after endurance training, in which increases in most mitochondrial enzyme activities roughly parallel the increase in total oxidative capacity (see Marsh and Dawson, 1982). However, despite the increase in HOAD/CS in winter, this index of β-oxidative relative to aerobic capacities remains substantially below the values for goldfinches in either summer or winter, 0.33 and 0.21, respectively (Marsh and Dawson, 1982). Also, the use of plasma glucose as measured by Marsh et al. (1984) may be influenced by the involvement of the legs in shivering (see below).

House finches do not appear to elevate stores of glycogen in the pectoralis in winter (Marsh et al. 1984), but these data were collected on individuals subjected to the stresses of capture and a few hours of captivity. Hormonal responses to these events could have altered glycogen stores from the levels found in free-living individuals and masked seasonal differences. Glycogen content (mg/g) in the pectoralis of house finches was greater than corresponding levels in these muscles

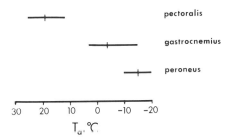

Figure 2. Mean (vertical line) and range (horizontal line) of air temperatures at which the pectoralis, gastrocnemius and peroneus muscle begin shivering, as judged by EMG's. Data were recorded on 12 birds as temperatures were lowered gradually from 30 to −20°C.

of goldfinches held under similar conditions in the laboratory (Marsh and Dawson, 1982; Marsh et al. 1984). Therefore, it is possible that free-living house finches have greater amounts of pectoralis glycogen than do goldfinches. Since levels of HK activity in the house finch pectoralis are significantly below corresponding levels in the leg muscles (Tables 2, 3), it appears that the pectoralis is less reliant on uptake of glucose from the blood than are leg muscles.

The inverse relation between HK activity and the average mean nocturnal low temperature on the three nights preceeding capture (Fig. 1) suggests that the leg muscles are able to take up more glucose from the blood at lower temperatures. This observation raises the possibility that leg muscles participate more extensively in shivering at cold temperatures. We have preliminary data which support this suggestion (Fig. 2). As air temperature decreases below the thermoneutral range, the pectoralis muscle begins shivering immediately and shivering continues in this muscle throughout the range of temperatures studies. In contrast, the gastrocnemius and peroneus, both ankle extensors, do not initiate shivering until temperatures drop to a range of 3 to −15 C, and −10 to −20 C, respectively. Like the response of HK activity in mammalian muscle to endurance training, HK activity in leg muscle may be elevated in response to increased muscle use.

While CS and HOAD activities did not vary seasonally in legs of goldfinches (Marsh and Dawson, 1982), corresponding activities did decrease in summer in house finch leg muscles (Table 3). This reduction is also probably related to the involvement of leg muscles in shivering only at colder temperatures. Nocturnal low temperatures in July, ranging between 9 and 15 C, probably did not elicit much, if any, contribution of leg muscles to thermogenesis. The reduction in metabolic expenditure from the higher levels in spring and winter apparently resulted in a decrease in the activities of these enzymes.

SUMMARY AND CONCLUSIONS

Previous studies have shown that house finches have less thermogenic endurance in winter and a smaller seasonal change in this parameter than American goldfinches (Dawson and Carey, 1976; Dawson et al. 1983a). The present study has revealed that one of the largest biochemical differences between the skeletal muscles of these two species is in their aerobic capacity, as indicated by CS activity. The mass specific activity of this enzyme in the pectoralis and leg muscles of winter house finches is only 59% and 70%, respectively, of the values in the corresponding muscles of goldfinches (Marsh and Dawson, 1982). In this regard, one interesting comparison is between the total aerobic capacity of the pectoralis muscles as indicated by CS activity (muscle mass x mass specific activity) and the total oxygen consumption of birds exposed to severe cold. Winter goldfinches have a total CS activity in the pectoralis muscles of 462 µmol/min (Carey et al., 1978; Marsh and Dawson, 1982) and a total oxygen consumption at −70 °C of approximately 330 ml O_2/h (Dawson and Carey, 1976). In contrast, winter house finches have a lower total activity of CS (419 µmol/min), but have a considerably higher total oxygen consumption during severe cold exposure (400 ml O_2/h; Dawson et al., 1983a). These data raise the possibility that house finches may shiver more at low temperatures with their leg muscles, which have lower mass specific aerobic capacities, as indicated by CS activity and are biased more toward the oxidation of carbohydrate, as indicated by HK/CS and HOAD/CS ratios. Indeed, we report preliminary data that house finches shiver with their leg muscles at temperatures equivalent to the "moderate" cold exposure (−15 °C) used in a previous study of substrate metabolism (Marsh et al., 1984). The responsiveness of the HK activity in the leg muscles to environmental temperature averaged for 3 days prior to capture (Fig. 1) and the winter increases in CS and HOAD activities also point to the possible role of the leg muscles in thermogenesis. No data are available on shivering in goldfinches, but, interestingly, this species does not show the winter increase in CS and HOAD activities in their leg muscles that we have documented in house finches (Marsh and Dawson, 1982). These suggestions must be tempered by our relatively poor understanding of the relation between the aerobic capacity of a muscle and the limits of shivering thermogenesis. Evidence suggests that the muscles of both mammals and birds can consume more oxygen during exercise than during shivering. Thus the full aerobic potential of this tissue apparently cannot be used during shivering (see discussion in Marsh and Dawson, 1988b).

Numerous contrasts exist between the pattern of metabolic adjustment to cold in house finches and goldfinches. The modest improvement in lipid stores, apparently greater emphasis on carbohydrate metabolism during shivering, and lower aerobic capacity of the skeletal muscles would appear to put house finches at a disadvantage in conditions of prolonged cold and food restriction during snowstorms. However, house finches are successful residents in some of the areas of the United States in which cold winters are common and in which goldfinches are also winter residents. The complete explanation for this phenomenon may well require not only more information on biochemical characteristics of house finches, such as gluconeogenic capacities, but also additional aspects of their biology, such as foraging efficiency and utilization of protected microclimates.

ACKNOWLEDGEMENTS

This study was supported in part by grants from the University of Colorado Council on Creative Work to CC, NSF grants BSR 84-07952 and DCB 84-09585 to RLM, and NIH grant NS 20310 to AB.

REFERENCES

Barnard, R. J., and Peter, J. B., 1969, Effect of training and exhaustion on hexokinase activity of skeletal muscle, J. Appl. Physiol., 27:691.

Carey, C., Dawson, W. R., Maxwell, L. C., and Faulkner, J. A., 1978, Seasonal acclimatization to temperature in cardueline finches. II. Changes in body composition and mass in relation to season and acute cold stress, J. Comp. Physiol. B, 125:101.

Crabtree, B. and Newsholme, E. A., 1972, The activities of phosphorylase, hexokinase, phosphofructokinase, lactate dehydrogenase, and the glycerol-3-phosphate dehydrogenases in muscles from vertebrates and invertebrates, Biochem. J. 126:49.

Dawson, W. R., Buttemer, W. A., and Carey, C., 1985, A reexamination of the metabolic response of house finches to temperature, Condor, 87:424.

Dawson, W. R. and Carey, C., 1976, Seasonal acclimatization to temperature in cardueline finches. I. Insulative and metabolic adjustments, J. Comp. Physiol., 112:317.

Dawson, W. R. and Marsh, R. L., 1988. Metabolic acclimatization to cold and season in birds, this volume.

Dawson, W. R., Marsh, R. L., Buttemer, W. A., and Carey, C., 1983a, Seasonal and geographic variation of cold resistance in house finches Carpodacus mexicanus, Physiol. Zool., 56:353.

Dawson, W. R., Marsh, R. L., and Yacoe, M. E., 1983b, Metabolic adjustments of small passerine birds for migration and cold. Am. J. Physiol., 245:R755.

Dawson, W. R., and Tordoff, H. B., 1964, Relation of oxygen consumption to temperature in the red and white-winged crossbills, Auk, 81:26.

Holloszy, J. R., and Booth, F. W., 1976, Biochemical adaptations to endurance exercise in muscle. Ann. Rev. Physiol., 38:273.

Lamb, D. R., Peter, J. B., Jeffress, R. N., and Wallace, H. A., 1969, Glycogen, hexokinase, and glycogen synthetase adaptations to exercise. Am. J. Physiol., 217:1628.

Leck, C. F., 1987, Update on house finch range. Rec. New Jersey Birds, 13:18-19.

Marsh, R. L., 1981, Catabolic enzyme activities in relation to premigratory fattening and muscle hypertrophy in the gray catbird Dumetella carolinensis. J. Comp. Physiol., 141:417.

Marsh, R. L., Carey, C., and Dawson, W. R., 1984, Substrate concentration and turnover of plasma glucose during cold exposure in seasonally acclimatized house finches, Carpodacus mexicanus. J. Comp. Physiol. B, 154:469.

Marsh, R. L. and Dawson, W. R., 1982, Substrate metabolism in seasonally acclimatized American goldfinches. Am. J. Physiol., 242:R563.

Marsh, R. L. and Dawson, W. R., 1988a, Metabolism of energy substrates and seasonal acclimatization, this volume.

Marsh, R. L. and Dawson, W. R., 1988b, Avian adjustments to cold, In: "Animal Adaptation to Cold," L. Wang, ed., Springer-Verlag, Berlin. (In press).

Marsh, R. L., Dawson, W. R., Camilliere, J. J., and Olson, J. M., 1989, Regulation of glycolysis in the pectoralis muscles of seasonally acclimatized American goldfinches exposed to cold, Am. J. Physiol., submitted.

Newsholme, E. A. and Leech, A. R., 1983, "Biochemistry for the Medical Sciences", John Wiley and Sons, New York.

Randle, P. J., 1981, Molecular mechanisms regulating fuel selection in muscle, in: "Biochemistry of Exercise", J. Poortmans and G. Niset, eds, University Park, Baltimore.

Reinertsen, R. E., 1986, Hypothermia in northern passerine birds, in: "Living in the Cold: Physiological and Biochemical Adaptations;" H. C. Heller, X. J. Musaccia and L. D. H. Wang, eds., Elsevier, New York.

Salt, G. W., 1952, The relation of metabolism to climate and distribution in three finches of the genus Carpodacus, Ecol. Monogr., 22:121.

Weathers, W. W., Shapiro, C. J., and Astheimer, L. B., 1980, Metabolic responses of Cassin's finches (Carpodacus cassinii) to temperature, Comp. Biochem. Physiol., 65A:235.

West, G. C., 1972, The effect of acclimation and acclimatization on the resting metabolic rate of the common redpoll, Comp. Biochem. Physiol., 43A:293.

Winder, W. W., Baldwin, K. M., and Holloszy, J. O., 1974, Enzymes involved in ketone utilization in different types of muscle: adaptation to exercise. Eur. J. Biochem., 47:461.

Yacoe, M. E. and Dawson, W. R., 1983, Seasonal acclimatization in American goldfinches: the role of the pectoralis muscle. Am. J. Physiol., 245:R265.

ENERGY SUBSTRATES AND METABOLIC ACCLIMATIZATION IN SMALL BIRDS*

Richard L. Marsh and William R. Dawson

Department of Biology
Northeastern University
Boston, MA 02115, U.S.A.

Museum of Zoology and Department of Biology
The University of Michigan
Ann Arbor, MI 48109, U.S.A.

INTRODUCTION

Many small birds wintering at middle or high latitudes encounter temperatures that require elevated thermogenic rates for months at a time. Although selection of favorable microclimates (Marsh and Dawson, 1988) and shallow nocturnal hypothermia (Reinertsen, 1983) may reduce metabolic demand, these energy saving mechanisms only partially offset the need for regulatory thermogenesis. In this context, survival depends critically on the ability to sustain high rates of thermogenesis, i.e., *thermogenic endurance*. Thermogenic endurance of small birds varies seasonally (Dawson and Marsh, 1988), and we regard this variation as one of the definitive characteristics of metabolic acclimatization to cold. Improved cold tolerance by winter birds is demonstrable both in acute studies (e.g., Dawson and Carey, 1976) and in longer-term exposures (e.g., Blem, 1973). This short review examines a series of studies that attempt to elucidate the role of metabolism of energy substrates in the process of metabolic acclimatization.

POSSIBLE MECHANISMS OF METABOLIC ACCLIMATIZATION IN SMALL BIRDS

Endurance during periods of elevated aerobic metabolism has often been linked to the maximum attainable metabolic rate, which in cold exposure is referred to as *summit metabolism*. A general inverse relation between endurance and the fraction of maximum metabolic rate that is maintained forms the basis for this linkage. Thus, if summit metabolism increases, the endurance at any given absolute metabolic rate should rise because this rate is a smaller fraction of the summit. Such a prospect could be important in small mammals, which commonly show 50% elevations of summit metabolism during winter acclimatization (Marsh and Dawson, 1988). However, most birds show smaller seasonal changes in this vari-

---

* Research of R. L. Marsh, W. R. Dawson, and coworkers supported by U.S.A. National Science Foundation grants BSR 84-07952 and DCB 84-09585

able. For example, the American goldfinch (*Carduelis tristis*) in Michigan shows an increase in summit metabolism of only 16% between summer and winter (Dawson and Smith, 1986; see also Dawson and Marsh, 1988). Thus the dramatic seasonal alteration in endurance during cold exposure in this and other species of birds involves an increased ability to sustain a high fraction of summit metabolism.

Regulatory thermogenesis in small birds appears linked primarily with aerobic energy metabolism in skeletal muscle. Most current evidence favors shivering as the major form of regulatory heat production in birds (Marsh and Dawson, 1988; Cannon and Nedergaard, 1988). Electromyograms of passerine birds indicate shivering at all temperatures below thermoneutrality (e.g., West, 1965; Carey et al., 1988). Moreover, an alternative highly aerobic tissue such as brown adipose tissue has not been found in birds (Barré et al., 1986; Olson et al., 1988). Therefore, we assume that most muscular heat production results from coupling mitochondrial ATP production to ATP hydrolysis during contraction. However, none of the interpretations presented below would be affected significantly if mitochondrial respiration were uncoupled from ATP production as long as skeletal muscle is the site of thermogenesis. This tissue has been suggested as the site of nonshivering thermogenesis in cold-acclimated ducklings (see, Barré et al., 1988).

The flight muscles, particularly the large pectoralis muscles, probably provide a major portion of the regulatory thermogenesis in small passerines. These muscles are large (15% or more of body mass) and have very high mass-specific aerobic capacities, as measured by marker enzymes for the TCA cycle or electron transport and by *in vitro* oxygen consumption (Carey et al., 1978; Marsh and Dawson, 1982; Yacoe and Dawson, 1983). Sustained shivering by these muscles might be limited by any of the interrelated systems that support aerobic metabolism in skeletal muscle (Fig. 1). The limitation could involve transport of respiratory gases, cellular aerobic capacity (TCA cycle and electron transport), mobilization and delivery of energy substrates, and the ability to catabolize the substrates in the muscle. Summit metabolism changes little with season and flight metabolism exceeds the maximum metabolism elicited by cold (see Marsh and Dawson, 1988). Consequently, the transport of respiratory gases probably does not limit thermogenic

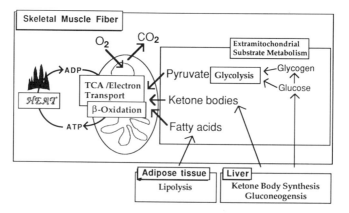

Fig. 1. Thermogenesis in skeletal muscle showing the possible energy substrates.

endurance. Increases in cellular aerobic capacity are likely involved in cold adaptation of mammals (e.g., Wickler, 1981), and a change in this factor could possibly influence endurance and not maximum metabolism (Davies et al., 1984). In the American goldfinch, however, the aerobic capacities of the pectoralis and leg muscles vary little over the year (Carey et al., 1978; Marsh and Dawson, 1982; Yacoe and Dawson, 1983). House finches (*Carpodacus mexicanus*) also show seasonally stable aerobic capacity in their pectoralis muscles; however, the citrate synthase (CS) activity in their leg muscles increases during winter suggesting some adaptive change. Electromyograms indicate that house finches use their leg musculature for shivering at low temperatures (Carey et al., 1988). The relative constancy of aerobic capacity at the organismal and cellular level has led us to focus on the role of energy substrates in the process of metabolic acclimatization.

SYNTHESIS, STORAGE, AND MOBILIZATION OF SUBSTRATES

One of the necessary foundations of thermogenic endurance is simply having enough substrate on hand to sustain aerobic metabolism. Because of the well known limits on the storage of fuel as carbohydrate in vertebrates, triacylglycerol (fat) is expected to be the major form of stored energy, as indeed is the case for birds. Even so, the small amounts of glycogen stored in the liver of winter-acclimatized cardueline finches are somewhat surprising (Marsh and Dawson, 1982; Marsh et al., 1984). Apparently, little or no liver glycogen is mobilized during cold exposure in these birds. Thus, muscle glycogen is the only carbohydrate store called upon (see below) and its quantitative significance is limited. The reliance on fat storage makes the phenomenon of winter fattening an essential aspect of seasonal acclimatization in birds (Dawson and Marsh, 1988). However, we have concluded that aspects of substrate metabolism other than simply "adding fuel to the tank" may be involved in seasonal acclimatization (Dawson and Marsh, 1988).

The capacity to mobilize fatty acids from adipose tissue could limit the availability of this fuel to the muscle, but current evidence does not favor this possibility. The factors determining the rate of uptake of fatty acids into skeletal muscle have not been investigated in birds, but this rate in mammals depends on plasma concentration and metabolic rate (Shaw et al., 1975). Plasma concentrations of fatty acids and metabolic rate are seasonally stable in goldfinches and house finches exposed to cold (Marsh and Dawson, 1982; Marsh et al., 1984).

Plasma glucose is continually catabolized during cold exposure, either in the thermogenic tissues or in crucial nonthermogenic tissues, such as brain. In small birds that do not accumulate significant stores of liver glycogen, the glucose requirements of fasting animals must be met predominantly from gluconeogenesis. The irreversible removal of plasma glucose is probably balanced mainly by synthesis from amino acids and glycerol. (Lactate can be gluconeogenic, but this primarily represents glucose recycling, for most lactate is produced from glycolysis.) Gluconeogenesis has not been systematically investigated in birds in different acclimatization states, but data on liver glycogen concentrations in American goldfinches and house finches raise the possiblity of seasonal changes in this process. Glycogen normally should be mobilized from the liver during times of increased metabolic demand (Hers, 1976). However, winter-acclimatized goldfinches and house finches actually accumulate glycogen in the liver under conditions of moderate cold exposure (Marsh and Dawson, 1982; Marsh et al., 1984). Perhaps the capacity for gluconeogenesis from glycerol is altered seasonally. Glycerol is released from adipose tissue when fatty acids are mobilized for use as thermogenic substrates.

SUBSTRATE USE BY THE THERMOGENIC TISSUES

Evidence from material balance studies and respiratory exchange ratios indicates that fatty acids provide the major substrate for shivering thermogenesis in birds (see Marsh and Dawson, 1988), which is perhaps not surprising given that fat provides the major form of stored energy in these animals. Nevertheless, other substrates may contribute significantly to heat production and alterations of their metabolism could contribute to metabolic acclimatization. Two alternative substrates are indicated in Fig. 1, carbohydrate and ketone bodies. Glucose has the advantage over fatty acids of being a readily soluble and easily diffusible fuel. Glycolysis from glucose or glycogen produces pyruvate that can be converted to acetyl-CoA for use in the TCA cycle, and also can be used in anaplerotic reactions that replenish TCA cycle intermediates (e.g., Lee and Davis, 1979). This latter role may be essential for continuous flux through the cycle. Ketone bodies are produced mainly in the liver as a result of β-oxidation of fatty acids (Robinson and Williamson, 1980). They represent soluble easily diffusible substrates that can substitute for fatty acids in the production of acetyl-CoA for catabolism in the TCA cycle (Fig. 3). Additionally, ketone bodies may serve as signals that control other metabolic pathways (Robinson and Williamson, 1980; Cherel and Le Maho, 1985).

Carbohydrate Metabolism

In mammalian muscle carbohydrate may be an essential substrate for the TCA cycle. When energy expenditure increases in mammals due to muscle contraction either during exercise or cold stress, carbohydrate use increases, even under conditions that favor the oxidation of fatty acids (Depocas, 1962; Minaire et al., 1973; Ahlborg and Felig, 1976). Limited data suggest a similar situation in some birds (Riesenfeld et al., 1979; Marsh et al., 1984). The use of carbohydrate and its depletion has also been linked to fatigue during periods of high energy expenditure in mammals. Numerous studies have focused on glycogen depletion in working muscles (e.g., Karlsson et al., 1974), but others point to a correlation between hypoglycemia and fatigue (e.g., Callow et al., 1986). In human runners, high energy output can continue with depleted muscle glycogen if supplemental glucose feeding prevents hypoglycemia (Coyle et al., 1986).

Marsh and Dawson (1982) were prompted to examine the use of carbohydrate during cold exposure by seasonally acclimatized American goldfinches, because of the substantial seasonal variation in thermogenic endurance noted in these birds (Dawson and Carey, 1976). During cold exposure, winter-acclimatized goldfinches use less glycogen and have lower turnover of plasma glucose than do summer animals. During moderate cold exposure, resulting in more than a doubling of metabolic rate over thermoneutral values, winter-acclimatized birds show no significant increase in turnover of plasma glucose or breakdown of muscle glycogen. This result is surprising in view of the mammalian data cited above, which include information on cold-acclimated rats (Depocas, 1962). The possible link between carbohydrate sparing in winter goldfinches and thermogenic endurance is strengthened by the results of a comparative study on house finches (Marsh et al., 1984). Compared to American goldfinches, house finches vary less in thermogenic endurance between summer and winter than do goldfinches, and, interestingly, do not show significant seasonal changes in glucose turnover or glycogen depletion during cold exposure (Marsh et al., 1984).

Carbohydrate sparing occurs in cold-exposed winter goldfinches despite a small (14-28%) increase in the activity of phosphofructokinase (PFK) in their pectoralis muscles (Marsh and Dawson, 1982; Yacoe and Dawson, 1983). This enzyme catalyzes a key regulated reaction in glycolysis (Marsh et al., 1989). Marsh and Dawson (1982) suggested a possible link between the sparing of carbohydrate and a substantial increase (50-100%) in the activity of the β-oxidative enzyme, β-hydroxyacyl-CoA dehydrogenase (HOAD) (see also, Yacoe and Dawson, 1983). This winter increase in HOAD activity may not produce an increase in the potential for energy release from fatty acids, for, judging from the data of Yacoe and Dawson (1983), the rate of oxidation of the acetyl-CoA provided by β-oxidation is limited by the overall aerobic capacity of the mitochondria and does not change seasonally. However, reciprocal regulation of β-oxidation and glycolysis occurs in some mammalian tissues (Randle and Tubbs, 1979). Consequently, an increase in the availability of acetyl-CoA from β-oxidation in winter birds could produce a greater inhibition of glycolysis. This regulatory role would provide an adaptive explanation for the increase in HOAD activity during winter.

Marsh et al. (1989) examined the regulation of glycolysis in summer- and winter-acclimatized American goldfinches. They measured the concentrations of most glycolytic intermediates as well as glycogen and citrate in the pectoralis muscles of birds exposed to thermoneutral temperatures (30 °C), moderate cold (-15 °C), and severe cold (0 °C in an atmosphere of 21% oxygen and 79% helium). A portion of these data is summarized in Fig. 2. Activation of an enzyme in a metabolic pathway can be identified if concentrations of the substrates for the enzyme decrease under conditions when the flux through the pathway increases (see Newsholme and Start, 1973). As predicted from work on other tissues, PFK appears to be the main regulated step in glycolysis. A drop in glucose-6-phosphate and fructose-6-phosphate concentrations under conditions known to mobilize glycogen and increase turnover of plasma glucose (Marsh and Dawson, 1982) provides the principle evidence for this regulation. Glycolysis remains inhibited when winter-acclimatized goldfinches are exposed to moderate cold (oxygen consumption ≈ 2.5 x thermoneutral values), which contrasts markedly with the activation of PFK under these conditions in summer animals (Fig. 2). Pyruvate levels are generally elevated in winter animals, suggesting that the pyruvate dehydrogenase complex (PDH) may also be inhibited to a greater extent in the pectoralis muscles of winter goldfinches. The possible coordinate regulation of PFK and PDH *in vivo* is interesting as the data of Yacoe and Dawson (1983) revealed no seasonal change in the *in vitro* regulation of PDH.

The major current hypothesis for glycolytic regulation in vertebrate muscle under aerobic conditions involves an inhibitory role for citrate, the first metabolite in the TCA cycle. According to this model, increasing availability of fatty acids or ketone bodies for oxidation leads to an increase in citrate concentrations in the mitochondria and subsequently the sarcoplasm. Inhibition of PFK by citrate has been quantified *in vitro* (Newsholme et al., 1977). However, in goldfinch muscle the concentration of citrate in the pectoralis muscle does not correlate consistently with the observed regulation of PFK (Fig. 2). Marsh et al. (1989) concluded that glycolysis is inhibited in the pectoralis muscles of winter birds, but factors other than citrate appear to be involved in this regulation. Perhaps some other metabolic change accompanying the augmented β-oxidative capacity inhibits PFK, or, alternatively, the functional explanation for the increased HOAD activity lies elsewhere.

Earlier investigations of acclimatization in cardueline finches seemed consistent with the idea that depletion of muscle glycogen is involved in thermogenic fatigue and consequent hypothermia in small birds (Marsh and Dawson, 1982; Marsh et al., 1984). However, later data (Marsh et al., 1989) cast some doubt on this correlation. Summer birds become hypothermic during a severe cold challenge with apparently large glycogen reserves remaining in the pectoralis muscles. Additionally, winter-acclimatized birds deplete their muscle glycogen during a 20-min exposure to severe cold, but these birds can remain normothermic for 2 (or more) hours. The correlation between glycogen depletion and fatigue has been documented in mixed fiber type muscles of man and other mammals. Perhaps the

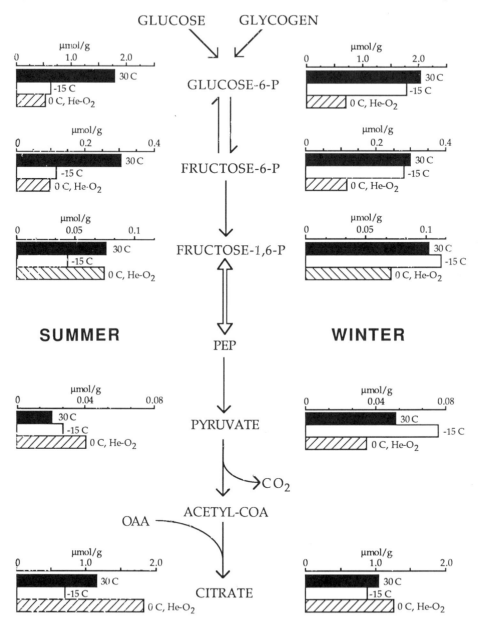

Fig. 2. Aerobic glycolysis and the initial step of the TCA cycle. Metabolites measured in the pectoralis muscles of goldfinches are shown with concentrations indicated by bar graphs beside the name of each metabolite. Birds were exposed for 2 h to the temperatures indicated beside each bar. (PEP, phosphoenolpyruvate; OAA, oxaloacetic acid) Based on Marsh et al. (1989).

ability of American goldfinches to shiver with depleted glycogen reserves is related to the purely oxidative character of their pectoralis muscles (Carey et al., 1978). Previous studies of goldfinches and house finches (Marsh and Dawson, 1982; Marsh et al., 1984), have not noted significant reductions in plasma glucose associated with hypothermia. These data and the more recent information on muscle glycogen suggest that fatigue and hypothermia are not direct consequences of carbohydrate depletion. However, this conclusion should be considered tentative as the plasma concentrations of substrates should be interpreted with caution. As pointed out by Marsh et al. (1984), the hypothermic birds in these studies are not metabolically stable. Plasma levels of glucose are maintained by a balance between production in the liver and peripheral uptake. We know that summer goldfinches are using more plasma glucose during shivering than their winter counterparts (Marsh and Dawson, 1982), and evidence of Marsh et al. (1989) indicates that some of this glucose is probably being used in the pectoralis muscles. During the development of hypothermia, concentrations of plasma glucose will be determined by the relative effects of the lowered temperatures on glucose output in the liver and uptake in the peripheral tissues, including the muscles. Because of this complex situation, the role of carbohydrate depletion in the onset of hypothermia is ambiguous; the available evidence neither establishes or excludes such a role. Establishing a single cause for hypothermia in these animals may be difficult in view of the well known problems of identifying a single cause for a systemic failure in any complex physiological system. These problems are compounded by the small size of American goldfinches, which restricts the use of such techniques as serial blood sampling with implanted cannulae.

We believe that the lack of precise coupling between depletion of carbohydrate and hypothermia does not lessen the potential importance of sparing carbohydrate in the metabolic economy of winter-acclimatized birds. Such animals routinely face long nights without food and periodically are exposed to longer periods of restricted feeding due to inclement weather. Sparing carbohydrate under these conditions preserves gluconeogenic precursors such as amino acids, and protein sparing has been shown to be important to survival during a fast (e.g., Cherel and Le Maho, 1985).

Metabolism of Ketone Bodies

Ketone bodies, acetoacetate (AcAc) and β-hydroxybutyrate (β-HB), are produced mostly in the liver from acetyl-CoA derived from β-oxidation of fatty acids (Zammit, 1981). These substrates are released into the blood and can be oxidized by various peripheral tissues. The plasma concentrations of these substrates increase in fasting mammals and birds, and in mammals they have been implicated prominently as substitutes for glucose in metabolism of the brain (Robinson and Williamson, 1980). However, other mammalian tissues, including heart and some types of skeletal muscle, have the capacity to oxidize ketone bodies. In addition to their roles as energy yielding substrates, these compounds may also serve as signals that control other aspects of energy metabolism. Elevated plasma concentrations of ketone bodies may inhibit release of free fatty acids from adipose tissue (a form of negative feedback on plasma free fatty acid concentrations) and their uptake by the heart is also known to reduce oxidation of glucose, presumably by inhibiting glycolysis (Robinson and Williamson, 1980). Much less is known of the roles of these compounds in birds, although a correlation between elevated plasma concentrations of ketone bodies and protein sparing during fasting has been noted in geese and penguins (Le Maho et al., 1981; Cherel and Le Maho, 1985).

In collaboration with J.O. Olson and T. van Hoft, we have recently examined aspects of ketone body metabolism in American goldfinches during summer and winter (unpublished observations). We measured circulating levels of ketone bodies and the activity of several enzymes involved in the synthesis and degradation of these substrates (Fig. 3). As in other birds, β-HB is the major circulating ketone body with plasma concentrations of 1-1.5 μmol/ml in fed birds in both summer and winter; AcAc concentrations are only about 5% of these values. Concentrations of β-HB increase during fasting at thermoneutral temperatures and the levels reached after 4-5 h are significantly greater in winter birds than in summer birds (approximately 5.5 versus 2.2 μmol/ml). A 2 h exposure to -15 °C causes β-HB levels to fall to near the levels in fed birds, eliminating the seasonal difference in concentration. No further fall in these levels occurs with exposure to more severe cold in winter. The similarity in plasma β-HB levels in goldfinches exposed to cold in summer and winter argues against the possibility that the availability of this substrate contributes to the inhibition of glycolysis discussed above. The actual quantitative contribution of β-HB to the energy expenditure in the cold cannot be judged by the data on plasma concentrations alone. In fact, the drop in plasma levels during cold exposure could be due to a decrease in synthesis as well as an increase in peripheral uptake.

An examination of the activity of several enzymes involved in the synthesis and degradation of ketone bodies confirms that the liver is the major site of synthesis of these substrates in goldfinches. Activities of HMG-CoA synthase are 20- or 30-fold higher in liver than in heart or skeletal muscle. The activity of β-hydroxybutyrate dehydrogenase increases in the liver of winter birds, which may help account for the greater accumulation of β-HB during a fast. Both heart and pectoral muscle have the enzymatic capacity for oxidation of ketone bodies, with the heart having considerably higher activities of oxoacid-CoA transferase.

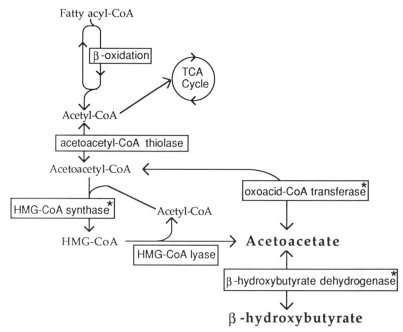

Fig. 3. Pathways for synthesis and degradation of ketone bodies. The HMG pathway is used for synthesis and oxoacid-CoA transferase is used in degradation. Enzymes measured by Olson et al. (unpublished) are indicated by *.

SUMMARY

Metabolic acclimatization in small birds, including cardueline finches, imparts an impressive capacity for thermogenic endurance in winter. The mechanistic basis for this acclimatization appears quite different from that found in small mammals, which rely on increases in summit metabolism and cellular aerobic capacity. Most of the major seasonal changes documented in cardueline finches relate to the metabolism of energy substrates. The changes between summer and winter include fattening, carbohydrate sparing, increased β-oxidative capacity in the pectoralis muscles, inhibition of glycolysis in these muscles, and increased accumulation of ketone bodies during fasting. Further comparative work should help clarify the interrelation among these changes and their relative importance to thermogenic endurance.

REFERENCES

Ahlborg, G., and Felig, P., 1976, Influence of glucose ingestion on fuel-hormone response during prolonged exercise, *J. Appl. Physiol.*, 41:683.

Barré, H., Cohen-Adad, F., Duchamp, C., and Rouanet, J. -L., 1986, Multilocular adipocytes from muscovy ducklings differentiated in response to cold acclimation, *J. Physiol., Lond.*, 375:27.

Barré, H., Duchamp, C., Rouanet, J. -L., Dittmar, A., and Delhomme, G., 1988, Muscular nonshivering thermogenesis in cold-acclimated ducklings, this volume.

Blem, C. R., 1973, Geographic variation in the bioenergetics of the house sparrow, *Ornith. Monogr.*, 14:96.

Callow, M., Morton, A., and Guppy, M., 1986, Marathon fatigue: the role of plasma fatty acids, muscle glycogen and blood glucose, *Eur. J. Appl. Physiol.*, 55:654.

Cannon, B., and Nedergaard, J., 1988, Shivering and nonshivering thermogenesis in birds, this volume.

Carey, C., Dawson, W. R., Maxwell, L. C., and Faulkner, J. A., 1978, Seasonal acclimatization to temperature in cardueline finches. II. Changes in body composition and mass in relation to season and acute cold stress, *J. Comp. Physiol.*, 125:101.

Carey, C., Marsh, R. L., Bekoff, A., Johnston, R., and Olin, A. M., 1988, Enzyme activities in seasonally acclimatized house finches, this volume.

Cherel, Y., and Le Maho, Y., 1985, Five months of fasting in king penguin chicks: body mass loss and fuel metabolism, *Am. J. Physiol.*, 249:R387.

Coyle, E. F., Coggan, A. R., Hemmert, M. K., and Ivy, J. L., 1986, Muscle glycogen utilization during prolonged strenuous exercise when fed carbohydrate, *J. Appl. Physiol.*, 61:165.

Davies, K. J. A., Donovan, C. M., Refino, C. J., Brooks, G. A., Packer, L., and Dallman, P. R., 1984, Distinguishing effects of anemia and muscle iron deficiency on exercise bioenergetics in the rat, *Am. J. Physiol.*, 246:E535.

Dawson, W. R., and Carey, C., 1976, Seasonal acclimatization to temperature in cardueline finches. I. Insulative and metabolic adjustments, *J. Comp. Physiol.*, 112:317.

Dawson, W. R., and Marsh, R. L., 1988, Metabolic acclimatization to cold and season in birds, this volume.

Dawson, W. R., and Smith, B. K., 1986, Metabolic acclimatization in the American goldfinch (*Carduelis tristis*), in: "Living in the Cold," H. C. Heller, X. J. Musacchia, and L. C. H. Wang, eds., Elsevier Science Publishing Co., New York.

Depocas, F., 1962, Body glucose as fuel in white rats exposed to cold: results with fasted rats, *Am. J. Physiol.*, 202:1015.

Hers, H. G., 1976, Control of glycogen metabolism in the liver, *Annu. Rev. Biochem.*, 45:167.

Karlsson, J., Nordesjo, L. -O., and Saltin, B., 1974, Muscle glycogen utilization during and after exercise, *Acta Physiol. Scand.*, 90:210.

Lee, S. -H., and Davis, E. J., 1979, Carboxylation and decarboxylation reactions. Anaplerotic flux and removal of citrate cycle intermediates in skeletal muscle, *J. Biol. Chem.*, 254:420.

Le Maho, Y., Vu Van Kha, H., Koubi, H., Dewasmes, G., Girard, J., Ferre, P., and Gagnard, M., 1981, Body composition, energy expenditure, and plasma metabolites in long-term fasting geese, *Am. J. Physiol.*, 241:E342.

Marsh, R. L., and Dawson, W. R., 1982, Substrate metabolism in seasonally acclimatized American goldfinches, *Am. J. Physiol.*, 242:R563.

Marsh, R. L., and Dawson, W. R., 1988, Avian adjustments to cold, in: "Animal Adaptation to Cold," L. Wang, ed., Springer-Verlag, Berlin. (In Press)

Marsh, R. L., Carey, C., and Dawson, W. R., 1984, Substrate concentrations and turnover of plasma glucose during cold exposure in seasonally acclimatized house finches, *Carpodacus mexicanus, J. Comp. Physiol.*, B, 154:469.

Marsh, R. L., Dawson, W. R., Camilliere, J. J., and Olson, J. O., 1989, Regulation of glycolysis in the pectoralis muscles of seasonally acclimatized American goldfinches exposed to cold, *Am. J. Physiol.* (Submitted)

Minaire,Y., Vincent-Falquet, J. -C., Pernod, A., and Chatonnet, J., 1973, Energy supply in acute cold-exposed dogs., *J. Appl. Physiol.*, 35:51.

Newsholme, E. A., and Start, C., 1973, "Regulation in Metabolism," John Wiley and Sons, New York.

Newsholme, E. A., Sugden, P. H., and Williams, T., 1977, Effect of citrate on the activities of 6-phosphofructokinase from nervous and muscle tissues from different animals and its relationship to the regulation of glycolysis, *Biochem. J.*, 166:123.

Olson, J. M., Dawson, W. R., and Camilliere, J. J., 1988, Fat from black-capped chickadees: avian brown-adipose tissue?, *Condor.* (In Press)

Randle, P. J., and Tubbs, P. K., 1979, Carbohydrate and fatty acid metabolism, in: "Handbook of Physiology, Section 2, The Cardiovascular System, Vol. 1, The Heart," R. M. Berne, N. Sperelakis, and S. R. Geiger, eds., American Physiological Society, New York.

Reinertsen, R. E., 1983, Nocturnal hypothermia and its energetic significance for small birds living in the arctic and subarctic regions. A review, *Polar Res.*, 1:269.

Riesenfeld, G., Berman, A., and Hurwitz, S., 1979, Glucose kinetics and heat production in normothermic, hypothermic, and hyperthermic fasted chickens, *Comp. Biochem. Physiol.*, 67A:199.

Robinson, A. M., and Williamson, D. H., 1980, Physiological roles of ketone bodies as substrates and signals in mammalian tissues., *Physiol. Rev.*, 60:143.

Shaw, W. A. S., Issekutz, T. B., and Issekutz, B., 1975, Interrelationship FFA and glycerol turnovers in resting and exercising dogs, *J. Appl. Physiol.*, 39:30.

West, G. C., 1965, Shivering and heat production in wild birds, *Physiol. Zool.*, 38: 111.

Wickler, S. J., 1981, Seasonal changes in enzymes of aerobic heat production in the white-footed mouse, *Am. J. Physiol.*, 240:R289.

Yacoe, M. E., and Dawson, W. R., 1983, Seasonal acclimatization in American goldfinches: the role of the pectoralis muscle, *Am. J. Physiol.*, 245:R265.

Zammit, V. A., 1981, Intrahepatic regulation of ketogenesis, *Trends Biochem. Sci.*, 6:46.

THERMOGENIC CAPACITY OF GREENFINCHES AND SISKINS IN WINTER AND SUMMER

Seppo Saarela, Bernt Klapper* and
Gerhard Heldmaier*

Department of Zoology, University of Oulu
SF-90570 Oulu, Finland
*Fachbereich Biologie, Philipps-Universität
D-3550 Marburg, Federal Republic of Germany

INTRODUCTION

The maintenance of thermal balance and high body temperature (T_b) during cold exposure of birds requires the improvement of thermal insulation and the increase of heat production (HP). Strategies for responding to low ambient temperature (T_a) are dependent on the body size. Large birds like the black grouse (Lyrurus tetrix, body mass about 1 kg) and the capercaillie (Tetrao urogallus, body mass 2-5 kg) compensate for increased cold load mainly by improvement of thermal insulation (Rintamäki et al.,1983; Hissa et al., 1983). The small body size of finches minimizes the significance of thermal insulation. The maintenance of high T_b in the cold is thus mainly dependent on the efficiency of HP.

The metabolic adaptations to cold climates of American Passerine birds have been studied in the laboratory of Dawson (Dawson and Carey, 1976, Carey et al., 1978, Dawson et al., 1983a,b, 1985). Greenfinches (Carduelis chloris) and siskins (Carduelis spinus) used in this study are small European finches, with 26 g and 13 g of body mass, respectively. They are distributed up to the latitudes of 65-66°N in Scandinavia. The thermogenic capacity of these finches must be sufficient to endure the lowest T_a (-40°C) which may occur in winter. The seasonal variation in the thermogenic capacity of finches is not well known. To elucidate this topic we measured maximum heat production (HP_{max}) of greenfinches and siskins in winter and again in summer. In the larger sized greenfinches we additionally recorded T_b telemetrically.

MATERIAL AND METHODS

Birds and housing conditions

Adult birds of both sexes of greenfinch (Carduelis chloris) and siskin (Carduelis spinus) were captured in the Marburg area (51°N, West Germany) from their respective habitats during winter (December-March). The body mass of the birds was: the greenfinch 26.6±0.68 g, the siskin 13.2±0.22 g. The birds were kept in outdoor aviaries (volume 16 m3) constructed from wire netting. Water and food were supplied in the cages ad libitum. During winter the water was prevented from freezing by means of a thermostat. A commercial premix of seeds (sunflower, millet etc.) for forest small birds was given. Shrub or tree vegetation, flower and leaf buds of bushes, were also available from spring to autumn.

Experimental procedure

Birds were measured in winter (Dec-Mar) and again in summer (July-Sept). The procedure of tests was as follow. The birds were captured from the aviary and weighed and placed into the metabolic chambers (small transparent plastic boxes, volume 3 l). The boxes were then transferred into a dark climatic chamber (Weiss 500 SD/500-60 DU). The birds were left first at the thermoneutral zone to obtain their basal metabolic rate (BMR) for about 2 h. Air was sucked through the boxes into gas analyzers (a paramagnetic O2-analyzer Oxytest-S, Hartman and Braun, and an infrared CO2-analyzer Uras 2T, Hartman and Braun, or an Applied Electrochemistry Oxygen Analyzer, Model S-3A). The gas analysis was performed continuously at 1-minute-intervals. After stabilization of oxygen consumption (VO2) the temperature of the metabolic chamber was reduced gradually from 30 to -75°C (in temperature steps of 10-15°C) for the subsequent 8 hours. The Ta was measured continuously from the top of the metabolic chamber with a thermoelement. The cooling of the metabolic chamber was continued as long as HP increased in response to the increasing cold load. When HP remained constant or was even reduced following a decrease in Ta, the experiment was terminated.

The Tb of greenfinches was measured with temperature transmitters (Mini Mitter, Model X, weight 1.2 g) allowing the continuous recording during the experiment. The transmitters were implanted into the abdominal cavity two weeks prior to the experiments. The siskins were too small to carry Mini Mitters.

The analog output of the gas analyzers and the temperature recordings were processed on-line by a computer (Tektronix 4051).

Calculations

For the calculation of HP the following equation was used:

HP = (4.44 + 1.43 * RQ) * VO2

where HP = heat production (mW/g), RQ = respiratory quotient and VO2 = oxygen consumption (ml/(g*h)) (Heldmaier, 1975).

The regression lines for increasing VO2 against the cooling of Ta were calculated in each experiment using the three lowest values of each temperature step (resting metabolic rate, RMR). Regression lines were used to calculate the cold limit (CL) and lower critical temperature (Tlc) of each bird (see Fig. 1).

Thermal conductance of the greenfinch was calculated according to the formula C = VO2/(Tb-Ta) giving the values C as mW/(g*h*°C) if VO2 is indicated as mW/g (see Calder and King, 1974).

Statistical significance was tested by Student's t-test.

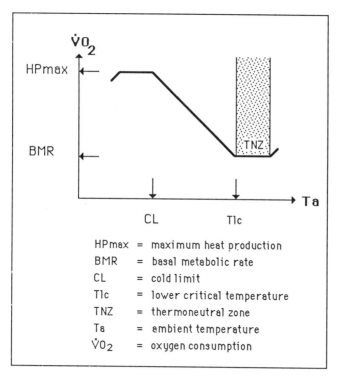

Figure 1. A schematic picture showing the parameters determined in this study.

RESULTS

Basal Metabolic Rate (BMR)

BMR of the greenfinches and the siskins did not differ significantly between winter and summer (Table 1). This was observed for weight specific as well as for total BMR. BMR values were compared with corresponding predicted values calculated on the basis of body mass of birds (Aschoff and Pohl, 1970; Table 1). The winter BMR values of the greenfinches and siskins did not differ from their predicted values. The summer BMR of the greenfinches was 20.8 % higher than predicted.

Table 1. Means (±S.E.M.) of body weight and basal metabolic rate (BMR, mW/g, RQ = 0.8) of greenfinches and siskins in winter and summer. Observed BMR values are compared with corresponding predicted values calculated according to the equation of Aschoff and Pohl (1970).

Bird/ Season	Number of birds	Weight (g)	BMR (mW/g)	BMR (mW/bird)	
				Observed	Predicted*
Greenfinch					
winter	16	26.6±0.68	21.2±0.95	560	531
summer	14	27.1±0.63	24.0±1.01	650	538
Siskin					
winter	8	13.2±0.22	25.1±1.84	331	324
summer	3	12.8±0.37	26.3±0.78	337	317

*$M = 140.9 * W^{0.704}$ (Aschoff and Pohl, 1970), where M = kcal/(24h*bird) and W = body weight (kg), kcal/24h were converted to mW by multiplication with 48.4.
BMR of summer greenfinches $P < 0.001$ between observed and predicted values

Maximum Heat Production (HPmax)

Figure 2 shows that there was no difference in HPmax values between winter and summer birds. The HPmax values were: 65.4±4.98 mW/g in the winter greenfinch, 76.6±3.08 mW/g in the summer greenfinch, 115.0±4.52 mW/g in the winter siskin and 109.4±2.57 in the summer siskin.

The RMR of summer birds was higher at each temperature step than the RMR of winter birds. Winter greenfinches and siskins were able to maintain, however, a high metabolic rate down to -60 - -75°C although HPmax was reached at higher Ta's of 15-20°C. Summer greenfinches and siskins did not endure Ta's lower than the temperature of the CL. A sudden decrease of VO2 (and Tb in the greenfinch) was found when the metabolic chamber was further cooled.

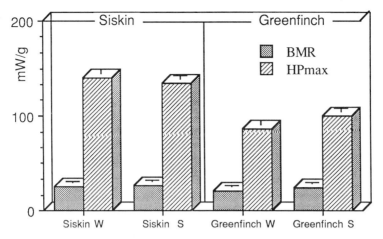

Figure 2. Means (±S.E.M.) of basal metabolic rate (BMR, mW/g) and maximum heat production (HPmax, mW/g) of greenfinches and siskins in winter (W) and summer (S). Number of birds is indicated in Table 1.

Thermogenic Capacity

Thermogenic capacity (HPmax/BMR) of greenfinches was in winter 3.5±0.20 and 3.3±0.16 in summer (N.S.). Corresponding values in siskins were 4.7±0.27 in winter and 4.4±0.04 in summer (N.S.).

Lower Critical Temperature (Tlc) and Cold Limit (CL)

Figure 3 describes Tlc and CL values in winter and summer of greenfinches and siskins. In winter Tlc of the greenfinch was not lower than in summer (23.3±1.32°C and 24.9±0.65°C, respectively). In the siskin Tlc was significantly lower in winter than in summer (28.0±0.76 and 31.6±0.55°C, respectively, $P < 0.05$).

The cold limit (CL) is the value describing the cold resistance of the bird. Figure 3 shows that winter acclimatization of the siskin improved cold resistance significantly (14.3°C, $P < 0.01$; -61.2±2.17°C in winter and -46.9±3.30°C in summer). Corresponding values for the greenfinch were -41.3±2.29°C in winter and -32.6±3.93°C in summer.

Body Temperature and Thermal Conductance of the Greenfinch

Greenfinches were able to maintain a higher Tb in winter than in summer (Table 2). At -20°C Tb of winter greenfinches was 0.8°C higher than Tb of summer greenfinches ($P < 0.01$). At -40°C Tb was 3.2°C higher in winter than in summer ($P < 0.001$). The homeothermy of winter greenfinches endured down to -40°C, while summer greenfinches became hypothermic already at 0°C.

The values of conductance show that the thermal insulation of greenfinches was significantly better in winter than in summer (Table 2).

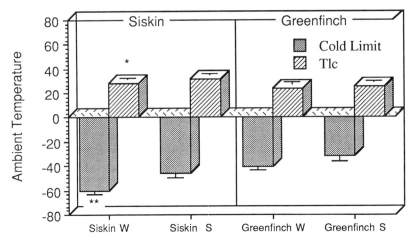

Figure 3. Means (±S.E.M.) of cold limit (CL, °C) and lower critical temperature (Tlc, °C) of greenfinches and siskins in winter (W) and summer (S). Number of birds is indicated in Table 1. * = $P < 0.05$, ** = $P < 0.01$ W vs. S.

Table 2. Means (±S.E.M.) of body temperature (°C, recorded telemetrically) and thermal conductance (mW/(g*h*°C)) of greenfinches in winter and summer. Number of birds was 5 in winter and 7 in summer.

T_a (°C)	Body Temperature (°C)		Thermal Conductance (mW/(g*h*°C))	
	Winter	Summer	Winter	Summer
30	41.2±0.24	41.1±0.31	1.8±0.06	2.2±0.08
25	40.7±0.16	40.6±0.22	1.1±0.05	1.5±0.06
0	41.0±0.11	40.7±0.18	0.9±0.01	1.1±0.06
-20	40.9±0.15	40.1±0.18	0.9±0.03	1.1±0.08
-40	41.2±0.24	38.0±0.27	0.9±0.02	1.1±0.06
-60	35.6±0.67		0.7±0.03	

T_b:
  winter vs. summer $P < 0.01$ at $T_a = -20°C$
  $P < 0.001$ at $T_a = -40°C$
Conductance:
  winter vs. summer $P < 0.01$ at $T_a = 30, 25, 0°C$
  $P < 0.05$ at $T_a = -40°C$

DISCUSSION

Winter acclimatization improved the cold tolerance of greenfinches and siskins. This is in accordance with seasonal improvements of cold tolerance observed in other species of birds. Dawson and Carey (1976) found that winter-acclimatized American goldfinches (Carduelis tristis) endure exposure to severe cold (Ta = -70°C) for 6-8 h, while summer-acclimatized goldfinches become hypothermic within an hour. Improved cold tolerance has also been observed in evening grosbeaks (Hesperiphona vespertina) (Hart, 1962), house sparrows (Passer domesticus) (Barnett, 1970) and starlings (Sturnus vulgaris) (Lustick and Adams, 1977) in winter. However, house finches (Carpodacus mexicanus) are only slightly more resistant to cold in winter than in summer (Dawson et al., 1983b). The physiological mechanisms responsible for improvement of cold tolerance during seasonal acclimatization are unclear.

Our data show that there is no significant seasonal difference in basal metabolic rate (BMR) and none in maximum capacity for thermoregulatory heat production (HPmax) in greenfinches nor in siskins. However, both species tolerate greater cold load in winter. In greenfinches we were able to measure body temperature simultaneously with heat production, which allowed calculation of thermal conductance. These measurements demonstrate that seasonal improvements of cold tolerance were due to improvements of thermal conductance (20 %). Furthermore, under severe cold stress winter-acclimatized greenfinches could tolerate hypothermia with Tb of 35.6°C while still maintaining HPmax, whereas in summer-acclimatized greenfinches even a slight hypothermia with Tb>40°C caused a rapid drop in thermoregulatory heat production. In siskins we did not measure body temperature and thermal conductance directly, but the lower RMR at any given cold load strongly suggests that they acclimatize also by improved thermal insulation in winter to a similar extent as observed in greenfinches. Whether they also develop a hypothermia under severe cold load has to remain unanswered, since no Tb records of undisturbed, freely moving birds could be taken due to their small size. Readings of cloacal temperature with a hand held device are of no use since handling strongly affected body temperature in these small birds.

The improved endurance of cold stress in winter has been discussed by Dawson et al. (1983b). Mobilization of lipids and oxidation of fatty acids are enhanced in winter. Recently has been shown that there is a seasonal shift in the metabolism of energy substrates in birds (Marsh, 1988). This metabolic acclimatization to winter due to inhibition of glycolytic enzymes associated with release of β-oxidative enzymes in the pectoral muscles.

In summary better cold tolerance of greenfinches and siskins in winter was not dependent on their thermogenic capacity. The thermal insulation of greenfinches was better in winter than in summer.

Acknowledgement

The support of the Alexander von Humboldt Foundation for this study is gratefully acknowledged.

REFERENCES

Aschoff, J., and Pohl, H., 1970, Rhythmic variations in energy metabolism, Fed. Proc., 29:1541-1552.
Barnett, L. B., 1970, Seasonal changes in temperature acclimatization of the house sparrow, Passer domesticus, Comp. Biochem. Physiol., 33:559-578.
Calder, A., and King, J. R., 1974, Thermal and caloric relations of birds, in: "Avian Biology," D. S. Farner, and J. R. King, eds., Academic Press, New York, London.
Carey, C., Dawson, W. R., Maxwell, L. C., and Faulkner, J. A., 1978, Seasonal acclimatization to temperature in Cardueline finches. II. Changes in body composition and mass in relation to season and acute cold stress, J. Comp. Physiol., 125:101-113.
Dawson, W. R., Buttemer, W. A., and Carey, C., 1985, A reexamination of the metabolic response of house finches to temperature, Condor, 87:424-427.
Dawson, W. R., and Carey, C., 1976, Seasonal acclimatization to temperature in Cardueline finches. I. Insulative and metabolic adjustments, J. Comp. Physiol., 112:317-333.
Dawson, W. R., Marsh, R. L., Buttemer, W. A., and Carey, C., 1983a, Seasonal and geographic variation of cold resistance in house finches Carpodacus mexicanus, Physiol. Zool., 56(3):353-369.
Dawson, W. R., Marsh, R. L., and Yacoe, M. E., 1983b, Metabolic adjustment of small passerine birds for migration and cold, Am. J. Physiol., 245:R755-R767.
Hart, J. S., 1962, Seasonal acclimatization in four species of small wild birds, Physiol. Zool., 35:224-236.
Heldmaier, G., 1975, Metabolic and thermoregulatory responses to heat and cold in the Djungarian hamster, Phodopus sungorus, J. Comp. Physiol., 102:115-122.
Hissa, R., Saarela, S., Rintamäki, H., Lindén, H., and Hohtola, E., 1983, Energetics and development of temperature regulation in capercaillie Tetrao urogallus, Physiol. Zool., 56(2):142-151.
Lustick, S., and Adams, J., 1977, Seasonal variation in the effect of wetting on the energetics and survival of starlings (Sturnus vulgaris), Comp. Biochem. Physiol., 56A:173-177.
Marsh, R. L., 1988, Metabolism of energy substrates and seasonal acclimatization, in: "Physiology of Cold Adaptation in Birds", R. E. Reinertsen, and C. Bech, eds., Plenum, New York.
Rintamäki, H., Saarela, S., Marjakangas, A., and Hissa, R., 1983, Summer and winter temperature regulation in the black grouse Lyrurus tetrix, Physiol. Zool., 56(2):152-159.

HEAT INCREMENT OF FEEDING IN THE KESTREL, Falco tinnunculus, AND ITS NATURAL SEASONAL VARIATION

Dirkjan Masman[1,2], Serge Daan[1] and Maurine Dietz[1]

1) Zoological Laboratory, University of Groningen, P.O.Box 14, 9750 AA Haren, The Netherlands

2) Centre for Isotope Research, University of Groningen, Westersingel 34, 9718 CM Groningen, The Netherlands

ABSTRACT

We measured the Heat Increment of Feeding (H) as the difference in oxygen consumption between the fed and fasting state in the kestrel at different ambient temperatures. Total H is a constant fraction of metabolizable energy intake and amounts to 16.6% of the energy assimilated. The effect of a meal was detected as elevation in metabolic rate lasting until 20 hours after the meal, independently of meal size. Below thermoneutral temperatures only part of H compensates for the cost of thermoregulation (T) up to a reallocation of 50% of H to T. Even at low temperatures (-12 °C) metabolic rate of fed kestrels is thus higher than that of fasting birds.

Using field data on daily metabolizable energy intake and meal-timing we conclude that average daily energy expenditure for H, after compensation for thermoregulation costs, varies from 7 % of daily energy expenditure in April to 15 % in August. Due to the habit of selecting sheltered roost sites, the nocturnal operative temperature is higher than that in daytime. Thereby the typical pattern of evening feeding, as observed in the kestrel during winter, can not be functionally explained by the reduction of total daily costs for thermoregulation.

INTRODUCTION

Food intake induces an increase in metabolic rate, a phenomenon described as Heat Increment of Feeding (HIF) or Specific Dynamic Action (SDA; Rubner 1910). This SDA is probably associated with a short term increase in rates of protein synthesis and turnover following feeding (Kirkwood

1981; Jobling 1983). Considerable knowledge on SDA has been assembled since a long time in domestic animals, livestock and humans (Brody 1945; Kleiber 1961; Webster et al. 1975; Kirkwood 1981). However, little attention has been paid to the role of SDA in freeliving wild animals (Calder and King 1974; King 1974; Kendeigh et al 1977; Mugaas and King 1981). The heat associated to the digestion of food may be a considerable part of the daily energy requirements. Further, it may (Rubner 1910; Kleiber 1961; Berman and Snapir 1965; Costa and Kooyman 1984; Biebach 1984) or may not (Ricklefs 1974; Klaassen 1989) be used to substitute for other costs like thermogenesis and incubation. To evaluate the importance of the heat increment of feeding (here indicated by H, in Watt) in the energy budget of free-living organisms we have to establish the relation between H and food intake, the time over which H is produced and the possible substitution of H for other components of energy expenditure.

The typical time course of the difference in metabolic rate between fed and fasting birds has rarely been described. The time course is of particular importance for application to the field situation, and especially for such predators where feeding is discontinuous and meals occur sporadically. The question of the compensation of heat produced as H for thermoregenesis has received attention since a long time but only a few studies demonstrated the ecological importance. Rubner (1910) originally demonstrated that SDA substitutes for thermoregulatory costs below the thermal neutral zone in mammals. Rubner (1910) and Kleiber (1961) argue that substitution of SDA is complete if thermostatic requirements are sufficiently high, and that substitution would be particularly substantial in animals on a protein rich diet (Simek 1976). For sea otters (Enhydra lutris) on a herring diet, Costa and Kooyman (1984) suggested that this marine mammal utilizes heat produced from SDA to replace heat otherwise generated from activity. In the starling (Sturnus vulgaris) SDA appears to replace shivering for thermoregulation during incubation, especially at low ambient temperatures (Biebach 1984).

The role of H in free-living energy requirements was important in our study on the behavioural energetics of the kestrel, in which we tried to evaluate energy expenditure and its different components on a year-round basis (Rijnsdorp et al. 1981; Daan and Aschoff 1982; Masman 1986; Masman et al. 1986, 1987, 1988a, 1988b; Dijkstra 1988; Meijer 1988; Daan et al. 1989). In particular, the hypothesis had been advanced that the daily pattern of food intake would contribute to a reduction of daily energy expenditure (DEE) by exploiting the supposed compensatory effect. Kestrels in winter have the habit of caching some of the prey they capture in the course of the day and retrieve these caches just before darkness. Thereby they concentrate meals especially in the last hour of the day (Fig.1). Rijnsdorp et al (1981) have suggested that the evening H might substitute for thermogenesis costs during the following night.

To quantify heat increment of feeding in the kestrel, we studied gas exchange in birds following a standard meal, at different temperatures in the laboratory. We established the time course of H as well as partial compensation for thermogenesis at low ambient temperatures. However, in contrast to the above hypothesis, the habit of evening-feeding in winter does not appear to contribute to a reduction in DEE.

METHODS

Kestrels (*Falco tinnunculus*) were caught by bal-chatri (Cavo' 1960) near our study site (Lauwersmeer 53°20' N, 6°12' E) and trained to sit quietly, tethered on a perch (Glasier 1978). The birds were housed in the laboratory under the actual seasonal LD cycle at an ambient temperature ranging from 15 to 20°C. They were fed ad libitum once per day (ca. 4 pm) with laboratory mice

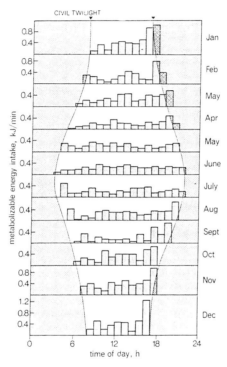

Fig.1. Variation in metabolizable energy intake (kJ/min) with time of day in free-living kestrels. Bars indicate the average energy intake in each hour interval for all observation days per month (653 days in total). Shaded area indicates nocturnal rest. (From Masman et al. 1986).

6 days per week and day-old cockerels once per week, both supplemented with a calcium-vitamin mixture (Carnicon R). Kestrels were weighed (to 0.1 g) each day just prior to feeding.

Metabolic heat production of perched birds was calculated from oxygen consumption and CO_2 production in an open-flow system (Daan et al 1989). We used a conversion factor of 19.5 kJ.L O_2^{-1} based on the average measured RQ (0.769; n=397 nights; SE= 0.001) and the composition of kestrel food (Masman and Klaassen 1987). Metabolizable energy intake was quantified by weighing the food offered (to 0.1 g) using an energy content of 7.8 kJ.g^{-1} fresh, and an assimilation quotient of 0.78 (Masman et al 1986).

To quantify variations with sex, body mass, season and time of day in BMR and heat increment of feeding (H) in the thermoneutral zone (17.5 - 35 °C) we ran an experiment (series I) for a year (LD cycle adjusted to natural daylength. On alternate weeks we recorded oxygen consumption by two birds and CO_2 production by one of these birds, continuously for 3 consecutive night and days. A preweighed mass of laboratory mice was offered inside the chamber at the beginning of the experiment and after 48 hours. We operationally defined H, associated with a standard meal, as the difference in oxygen consumption between a day following this meal (day 1 and 3) and a postabsorptive day (day 2).

Compensation of H for thermostatic requirements was evaluated from two series of experiments. We measured the oxygen consumption of one male and one female at various ambient temperatures during the night (on a 48 h feeding schedule) under fasting and fed conditions over the course of a year, once a month for 6 successive nights at two temperatures per night (Series II). In addition we measured the oxygen consumption by two males and two females in a 4 day run, again on a 48 h feeding schedule: two days at 25°C and two days at 5°C (Series III). The compensation effect was evaluated by comparison of the fasting and fed metabolic rates at different ambient temperatures.

RESULTS

Difference between digesting and postabsorptive birds

The heat increment of feeding (H) is here operationally defined as the difference between oxygen consumption in the fasting, postabsorptive state and that following a standard meal. The typical time course for VO_2 in series I experiments (in thermoneutrality) is shown in Fig.2, where the average VO_2 values are plotted for each hour of all experiments performed in one bird. Meals were presented at 16:00 on day 1 and day 3 (arrows in Fig.2). We calculated the difference between each hour of night 1 and the corresponding hour of night 2 (replotted as open dots in Fig.2), and similarly between night 3 and night 2, day 2 and day 3, and day 4 and day 3. This difference was obtained for 57 meals of three size classes and averages were plotted as a function of time (Fig.3).
The linear regressions fitted to these data differ significantly from 0 in both intercept and slope (Table 1). The results suggest that the total time during which VO_2 is increased due to feeding, is independent of meal size, whereas the size of H is roughly proportional to the size of the preceding meal for any of the 20 hours following the meal (Table 1). Average nocturnal H-values, which estimate the heat increment of feeding 10 hours after the meal, were obtained for each of 62 experimental nights. These were positively correlated with meal size (r= 0.336; p<0.005). Extrapolated to the whole 20-h period this gives an integrated estimate of 54.6 ml O_2.g^{-1}, or 0.166 kJ per kJ metabolizable energy intake. The estimates of H for the three meal size classes are presented in Table 1.

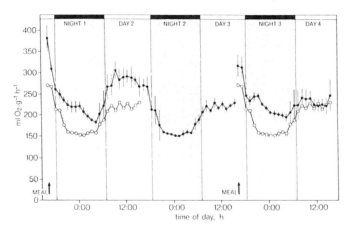

Fig.2. Average oxygen consumption of male J during 6 3-day experiments (Series I, winter 1984-85). Indicated are the hourly means for VO_2 ($\pm$ SD). To estimate the Heat Increment of Feeding ($\overline{H}$) the post-absorptive oxygen consumption (night 2 and day 3) was replotted (open symbols) in the first and last 24-hours where food was degisted (arrows indicate the meals).

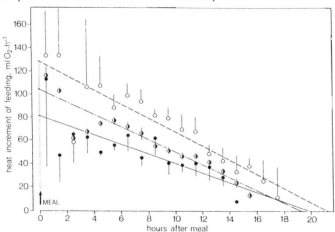

Fig.3. Heat Increment of Feeding (H) following meals of different size, plotted as a function of the time since food intake (arrow). Indicated are hourly means ($\pm$ SE) for meals of 30.2 g (open dots), 18.5 g (half-open dots) and 11.6 g (solid dots). See Table 1 for details.

TABLE 1. Heat Increment of Feeding (H in ml $O_2 \cdot h^{-1}$) as a function of time (t) elapsed since the meal (experiments series I). Indicated are the number of trials (n), meal size (mean $\pm$SD in g), Metabolizable energy intake (M in kJ), total Heat increment produced (H in kJ). In all cases $p < 0.001$.

n	meal size	M	H as function of t	r	Total H
3	11.6$\pm$ 3.6	71.5	81.3 - 4.1 t	- 0.786	15.8
49	18.5$\pm$ 1.0	114.0	104.9 - 5.4 t	- 0.866	19.8
5	30.2$\pm$ 7.2	186.1	128.3 - 6.0 t	- 0.904	26.6

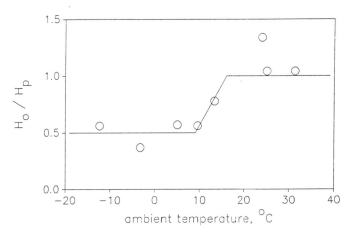

Fig.4. Substitution of Heat increment of feeding (H) for thermoregulation. Indicated is the fraction of H predicted (H_p; based on Table 1., assuming no substitution) and H observed (H_o; the increment above Fasting Metabolic Rate), as a function of ambient temperature in experiments of series II and III (see Table 2). Line: see text.

Subtitution of H for thermostatic requirements

While under thermoneutral conditions the heat increment of feeding is clearly defined and operationally measurable, it remains to be discussed to what extent the extra heat lost following a meal can substitute for thermoregulatory expenditure below the lower critical temperature (17.5 °C; Masman 1986). To address the question of substitution we have compared VO_2 values in the fasting state and following a meal in the nights of experimental series II, at ambient temperatures of 31.3 °C, 23.9 °C, 13.3 °C, 9.5 °C, -3.3 °C and -12.3 °C (Table 2). The resulting H-values were compared with predictions for the corresponding part of the night based on H-measurements in thermoneutrality in series I. H-values obtained within the thermoneutral zone (31.3 °C, 23.9 °C) are close to or above their predicted values. However, at lower temperatures the observed H falls below prediction. Our experiments in series III, where we compared the metabolic rate over complete nights fasting or following a meal at 25 °C and 5 °C, show the same results (Table 2). The observed H is reduced relative to predicted H at the lower temperatures. The average observed H below thermoneutrality amounts to 57% (SD=15;n=5 temperatures sampled) of the predicted H. The four values of the ratio $H_{observed}$ / $H_{predicted}$ for temperatures below 10 °C do not vary systematically with ambient temperature and converge on an average of ca 0.5. The value at 13.3 °C (0.78) is intermediate between this level and 1.0 for the thermoneutral zone. This suggest that there is a narrow transition zone between 10 °C and 17.5 °C over which the fraction of H which may substitute for costs of thermogenesis drops rapidly with increasing temperature. This interpretation of the data has been indicated by the solid line in Fig.4.

TABLE 2. Substitution of Heat Increment of Feeding (H) for energy expenditure for Thermoregulation (T). Data from birds in experiment series II after a meal of 34.6 g (SD=3.6), equivalent to 210.5 kJ metabolizable energy and data from birds in series III after a meal of variable size but at least 20 g or 121.7 kJ. Indicated are ambient temperature (t_a in °C), Fasting Metabolic Rate (FMR in ml $O_2.h^{-1}$±SD (n)), H (ml $O_2.h^{-1}$±SD (n)) as observed and H (ml $O_2.h^{-1}$) as predicted under the assumption that no substitution takes place, calculated using the relationships as presented in Table 1. Notes: 1) measurements of series II in the first part of the night (5 to 7 hours after feeding); 2) measurements of series II in the second part of the night (9 to 11 hours after feeding); 3) measurements of series III over the total night.

t_a	FMR	H obs.	H pred.	$H_{obs.}/H_{pred.}$
31.3[1]	175.2±17.4(5)	111.8±13.6(5)	107.5	1.04
25.0[3]	180.0±31.2(3)	79.5±58.1(3)	76.3	1.04
23.9[2]	164.5±17.5(4)	103.3±28.0(4)	77.3	1.34
13.3[1]	227.1±33.4(15)	80.9±28.0(15)	103.1	0.78
9.5[2]	253.7±46.0(15)	39.6±27.6(15)	70.5	0.56
5.0[3]	296.9±50.1(5)	39.0±37.8(5)	68.2	0.57
- 3.3[2]	368.1±66.2(17)	27.1±33.4(17)	74.1	0.37
-12.3[1]	467.4±63.6(14)	62.6±32.6(14)	112.7	0.56

DISCUSSION

Heat increment of feeding as a fraction of energy assimilated

In the kestrel we observed a total H production of 16.6 % of the metabolizable energy intake, independend of the size of the meal. This is identical to the value of 17 % found by Kirkwood (1981) in the same species also on a mouse diet. In Harp Seals, fed with herring, Gallivan and Ronald (1981) found comparable H values of at least 16% of gross energy intake. It is well established that the quality and quantity of the diet and the type of animal influences the magnitude of the extra heat produced after a meal (Calder and King 1974; Ricklefs 1974; Webster et al. 1975). In birds SDA ranges from 31% of assimilated energy in protein to only 6% in sucrose as based on theoretical considerations (Ricklefs 1974). Therefore it is suggested that in birds as well as mammals the magnitude of H will primarily be determined by the rate of protein synthesis and turnover (Kirkwood 1981; Jobling 1983).

Time course of H produced, following a meal

The effect of a meal on the metabolic rate of kestrel extended over 20 h, independent of meal size. This again is in exact agreement with the observations of Kirkwood (1981) in the kestrel. The duration of H in birds has been reported to vary between a few hours and a few days (Ricklefs 1974). Gallivan and Ronald (1981) observed a duration of H in the Harp seals of at least 7 hours, a minimal estimate since the experiments were terminated before postabsorbtive levels were reached. For the Largemouth Bass, Beamish (1974) reported the duration to increase from one to three days with ration size in the largest fish (200 g), in contrast with our observations.

The operational defenition of the heat increment of feeding obviously does not specify the precise biochemical source of the difference in energy expenditure involved between fasting and fed animals. Apart from the energy initially required for tearing the prey apart and the Specific Dynamic Action of food digested, a difference in basal metabolic rate may be involved. We know

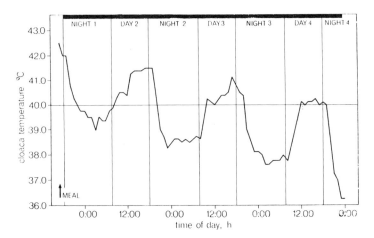

Fig.5. Body temperature (thermistor inserted in cloaca) of a male kestrel, fasting for 79 hours, in complete darkness at -2 °C (data obtained by J.de Vries).

that in the kestrel prolonged fasting results in a gradual decrease in body temperature.(Fig.5; Daan et al. 1989), similar to findings in other birds (Biebach 1974; Chaplin et al. 1984). In the American kestrel (Falco sparverius) fasting results in a gradual decrease in nocturnal metabolic rate (Shapiro and Weathers 1981). These effects can be seen as an adaptive energy saving response to the fasting condition (Daan et al. 1989). It implies that, especially in the later part of the response, the differences in metabolic rate between the fed and fasting situation are perhaps no longer primarily determined by SDA, but by a reduction in body temperature and thereby in thermoregulation costs by fasting birds.

Compensation of H for thermogenesis

Apparently, in the kestrel H substitutes for thermoregulatory costs to a certain extent, which conforms to Rubner's substitutions principle (1910). However, H remains detectable even when at -12 °C thermostatic requirements greatly surpass the predicted H (Fig.4). Rubner's principle was formulated to account for the 'Specific Dynamic Action' (SDA) of food, i.e., the costs of biochemical degradation of food in the process of digestion. As stated above, the operational definition of H involves two factors besides SDA: Firstly the mechanical action of tearing the mice apart entails significant muscular activity which contributes to the initially very high values of H during a meal (Fig.2); Secondly, a gradual decrease in body temperature in prolonged fasting. These contributions to H may give rise to incomplete substitution for thermostatic requirements.

So far data obtained in some studies have been interpreted as complete absence of thermostatic substitution (Ricklefs 1974; Klaassen 1989). Other studies have shown partly or complete substitution of H for T (Berman and Snapir 1965; Simek 1975, 1976). The present study, however, clearly shows partially substitution in the kestrel of about 50 % of H at temperatures below 10 °C. Below we evaluate its consequences for the energy balance of free-living kestrels.

Annual and daily variation in H and T in free-living kestrels

The energy expenditure for thermoregulation by free-living kestrels is determined not only by ambient temperature (t_a) but also by wind and radiation. The combined effects of these meteorological conditions in the field were quantified by the use of heated taxidermic mounts in the field and laboratory studies on living kestrels. These effects are expressed in the operative temperature (t_{op}; Bakken et al. 1981; Buttemer 1985; Buttemer et al. 1986; Masman et al. 1988b). The average operative temperature depends both on measured temperature, wind and radiation and on the behavioural position of the birds as assesed in 366 continuous dawn to dusk observations of focal birds. The annual variations in t_a and t_{op} during the day and night are displayed in Fig 6A and 6B. Kestrel behaviour leads to higher values for T during the day (Fig 6C). During the night kestrels shelter against the wind in their night roost (mainly nestboxes). During the day, in winter, when they mainly forage from perches (Masman et al. 1988a) they are exposed to relatively strong winds.

The questions that arise now are: 1) how much H is produced per day, and 2) what fraction of H substitutes for the energy expenses for T. We quantified the average daily metabolizable energy intake (M) in our kestrel population which ranges in males from 152 kJ/day in September to 414 kJ/day in June (Masman et al. 1986, 1988b). From these data we calculated total H produced (H = 0.166 x M; Fig.6D). We calculated the amount of H that is used to cover the expenses for T from data on meal timing (Fig.1; Masman et al. 1986), the data on intensity and duration of H (Fig.2; Table 1.) and substution of H for T (Fig.4) at different operative temperatures. By subtraction of this fraction of H, reallocated to T, from the total H produced, we derived the actual heat loss due to H in the field (Fig. 6D). During winter about 25% of the produced H is estimated to substitute for thermoregulation. This fraction reduces during summer to become negligable in August.

We are now in the position to evaluate the hypothesis, postulated by Rijnsdorp et al. (1981) and Daan and Aschoff (1982), that the tendency of evening feeding, as performed by kestrels during winter (Fig.1), reduces total daily energy expenditure by the concentration of digestion in the night and thus to exploit H-T substitution to cover energy expenses in nocturnal thermogenesis. Since the expenses for thermoregulation are higher during the day than during the night (Fig.6C) overall values for T are not likely to be reduced by evening feeding. In fact we computed total costs of thermogenesis throughout the year assuming two purely hypothetical distributions of meals: one where all food was taken as a single meal at sunset, the other where food was distributed evenly over the daily light period in n meals, where n equals the average number of meals observed. The resulting thermoregulatory costs were essentially equal for these two extreme cases. Therefore, we conclude that temporal

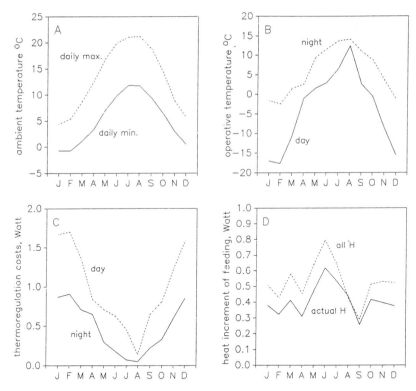

Fig.6. Annual variation in climatic conditions, energy expenditure for heat increment of feeding and thermogenesis in free-living male kestrels. A. Diurnal maxima (day) and minima (night) in ambient temperature (oC); B. Estimation of the operative temperature (°C) during the night and day as experienced by kestrels in the field, based on taxidermic mount measurements and full day behavioural observations; C. Estimation of energy expenditure for thermoregulation (Watt) during the day and night, based on data in Fig.6B and laboratory measurements; D. Estimation of daily heat increment of feeding produced (H-produced, calculated as 0.166 times daily metabolizable energy, Watt) and the residue after compensation for thermogenesis (H-lost, Watt). See for details Masman et al. 1988b.

patterning of meals does not contribute to a reduction in energy turnover through H-T substitution.

Since a meal induces an increase in the mass a bird has to carry around during the day, evening-feeding reduces the costs for flight. Flight costs are directly proportional to body mass (Masman and Klaassen 1987). The average time spent flying during winter is 1.5 h/day (Masman et al. 1988a). By an extra load of 15 g (average prey mass during winter; Masman et al.1986) and an average body mass of 207 g (average male winter mass; Dijkstra 1988) total daily energy expenditure would increase by 2%. On days with extreme high activity levels (up to 8 h/day of flight; Masman et al.1988a) this could increase up to 10%. This indicates that in situations, worse than average, the observed meal patterning may be of energetic advantage, as postulated also by Rijnsdorp et al. (1981) and Daan and Aschoff (1982).

Furthermore, kestrels may occasionally encounter periods of bad weather which prevent them from foraging and eventually force them to migrate. The habit of caching food enables temporal separation of foraging from feeding, a mechanism that can provide energy when it is urgently needed. Evening-feeding, thereby, may be seen as an expression of a strategy of management of the available energy stores, but not primarily as providing an internal heat source during nocturnal cold.

ACKNOWLEDGMENTS

This study was supported financially by grant 14.10.25 from the Netherlands Foundation for Biological Research (BION) to S.Daan and J.J.Videler. We thank Hans Beldhuis, Cor Dijkstra, Marijke Gordijn, Marcel Klaassen and Jan de Vries for their participation in this project.

REFERENCES

Bakken,G.S., W.A.Buttemer, W.R.Dawson and D.M.Gates. 1981. Heated taxidermic mounts: a means of measuring the standard operative temperature affecting small animals. Ecology 62: 311-318.
Beamish, F.W.H. 1974. Apparent Spesific Dynamic Action of Large mouth Bass, Micropterus salmoides. J.Fish.Res.Board Can. 31: 1763-1769.
Berman, A. and N.Snapir. 1965. The relation of fasting and resting metabolic rates to heat tolerance in the domestic fowl. Brit. Poultry Science 6: 207-216.
Biebach, H. 1977. Reduction des Energiestofwechsels und der Korpertemperatur hungernder Amseln (Turdus merdula). J. fur Ornithologie 118 (3): 294-300.
Biebach, H. 1984. Effect of clutch size and time of day on the energy expenditure of incubating Starlings (Sturnus vulgaris). Physiological Zoology 57(1):26-31.
Brody, S. 1945. Bioenergetics and growth. Reinold, New York.
Buttemer, W.A. 1985. Energy relations of winter roost-site utilization by American Goldfinches (Carduelis tristis). Oecologia 68: 126-132.
Buttemer, W.A., A.H.Hayworth, W.W.Weathers and K.A.Nagy. 1986. Time-budget estimates of avian energy expenditure: Physiological and meteorological considerations. Physiol.Zool. 59: 131-149.
Cave, A.J. 1968. The breeding of the kestrel, Falco tinnunculus, in the reclaimed area of Oostelijk Flevoland. Neth. J.Zool. 18: 313-407.
Chaplin, S.B., D.A.Diesel and J.A.Kasparie. 1984. Body temperature regulation in Red-tailed Hawks and Great-horned Owls: responses to air temperature and food deprivation. The Condor 86(2): 175-182.
Costa, D.P. & G.L.Kooyman. 1984. Contribution of Specific Dynamic Action to heat balance and thermoregulation in the Sea otter, Enhydra lutris. Physiological Zoology 57(2):199-203.
Daan, S. and J.Aschoff. 1982. Circadian contributions to survival. In: J.Aschoff, S.Daan and G.A.Groos (eds.). Vertebrate circadian systems. Springer Verlag, 305-321.
Daan, S., D.Masman, A.Strijkstra and S.Verhulst. 1989. Intraspecific allometry of Basal Metabolic Rate: Relations

with body size, temperature, composition and circadian phase in the kestrel, Falco tinnunculus. J. Biol. Rhythms (in press).

Dijkstra, C. 1988. Reproductive tactics in the kestrel, Falco tinnunculus, a study in evolutionary biology. Ph.D.Thesis University of Groningen, The Netherlands.

Gallivan, G.J. and K.Ronald. 1981. Aparent Specific Dynamic Action in the Harp Seal (Phoca groenlandica). Comp. Biochem. Physiol. 69A: 579-581.

Glasier,Ph. 1978. Falconry and Hawking. Bt. Batsford Ltd., London.

Jobling, M. 1983. Towards an explanation of Specific Dynamic Action (SDA). J. Fish. Biol. 23: 549-555.

Kendeigh, S.C., V.R.Dolnik and V.M.Gavrilov. 1977. Avian energetics. In: J.Pinowski and S.C.Kendeigh (eds.). Granivorous birds in ecosystems. Cambridge University Press pp.127-377.

King, J. 1974. Seasonal allocation of time and energy resources in birds. In: R.A.Paynter (ed.). Avian energetics. Nuttall Orn. Club, Cambridge, Mass. pp. 4-85.

Kirkwood, J.K. 1981. Bioenergetics and growth in the kestrel (Falco tinnunculus). Ph.D. thesis, University of Bristol, UK.

Klaassen, M., C.Bech and G.Slagsvold. 1989. Basal metabolic rate and thermal conductance in Arctic Tern chicks (Sterna paradisaea) and the effect of heat increment of feeding on thermoregulatory expenses. (in press).

Kleiber, M. 1961. The fire of life, an introduction to animal energetics. R.E.Krieger Publ.Comp. Huntington, New York.

Masman, D. 1986. The annual cycle of the kestrel, Falco tinnunculus. A study in behavioural energetics. PhD Thesis University of Groningen, The Netherlands.

Masman, D.,M.Gordijn, S.Daan and C.Dijkstra. 1986. Ecological energetics of the European Kestrel: Field estimates of energy intake throughout the year. Ardea 74:24-39.

Masman, D. and M.Klaassen. 1987. Energy expenditure for free flight in trained and wild Kestrels, Falco tinnunculus. Auk 104: 603-616.

Masman, D., S.Daan and C.Dijkstra. 1988a. Time allocation in the kestrel, Falco tinnunculus, and the principle of energy minimization. J.Anim.Ecol. 57: 411-432.

Masman, D., S.Daan and H.J.A.Beldhuis. 1988b. Ecological energetics of the Kestrel: Daily energy expenditure throughout the year based on Time-Energy Budget, Food Intake and Doubly Labeled Water methods. Ardea 76: 64-81.

Meijer, T. 1988. Reproductive decisions in the kestrel, Falco tinnunculus: a study in physiological ecology. PhD. Thesis University of Groningen, The Netherlands.

Mugaas, J.N. and J.R.King.1981. Annual variation of Daily Energy Expenditure by the Black-billed Magpie: A study of thermal and behavioural energetics. Studies in Avian Biology No.5, The Cooper Ornithological Society. Allan Press, Inc., Lawrence, Kansas.

Ricklefs, R.E. 1974. Energetics of reproduction in birds. in: Avian Energetics. R.A.Paynter (ed.). Publ.Nuttall Ornith. Club no.15. Cambridge, Massachusetts.

Rubner, M. 1910. Uber Kompensation und Summation von functionellen Leistungen des Korpers. Sitzungsberichte der Koniglichen Preussischen Akademie der Wissenschaften 1910: 316-324.

Rijnsdorp, A., S.Daan and C.Dijkstra. 1981. Hunting in the

Kestrel, <u>Falco</u> <u>tinnunculus</u>, and the adaptive significance of daily habits. Oecologia 50: 391-406.

Shapiro, C.J. and W.W. Weathers. 1981. Metabolic and behavioral responses of American Kestrels to food deprivation. Comp.Biochem.Physiol. 68A: 111-114.

Simek, V. 1975. Specific dynamic action of a high protein diet and its significance for thermoregulation in Golden Hamsters. Physiologia Bohemoslovaca 24: 421-424.

Simek, V. 1976. Influence of single administration of different diets on the energy metabolism at temperatures of 10, 20 and 30°C in the Golden Hamster. Physiologia Bohemoslovaca 25: 251-253.

Webster, A.J.F., P.O.Osuji, F.White and J.F.Ingram. 1975. The influence of food intake on portal blood flow and heat production in the digestive tract of sheep. Br.J.Nutr.34: 125- 139.

THERMOREGULATION AND ENERGETICS OF ARCTIC SEABIRDS

Geir W. Gabrielsen and Fridtjof Mehlum

Department of Biology
Norwegian Polar Research Institute
1330 Oslo Lufthavn, Norway

INTRODUCTION

Svalbard and the Barents Sea are inhabited by one of the largest concentrations of seabirds in the world, comprising several million birds (Løvenskiold 1964, Norderhaug et al. 1977). These birds constitute a major component of the marine ecosystem and form an important link between the terrestrial and marine ecosystem by transporting organic material and nutrients from sea to land.

Seabirds such as the Fulmars (Fulmarus glacialis), Glaucous Gulls (Larus hyperboreus), Ivory Gulls (Pagophila eburnea) and the Brünnich's Guillemots (Uria lomvia) stay in the Svalbard area all the year round, frequenting the flow edge throughout the winter months, whereas Kittiwakes (Rissa tridactyla), Black Guillemots (Cepphus grylle) and Little Auks (Alle alle), after spending the winter further south, arrive in early spring when the sea-ice starts to break up and the ambient temperature is averaging $-10\ ^0C$. Our studies were performed during the breeding period (June-August), regarded as the most energy demanding period for seabirds (Ricklefs 1983).

Resting metabolic rate (RMR), thermal conductance (TC) and daily energy expenditure (DEE) in birds have been estimated from various allometric equations (Aschoff and Pohl 1970, Lasiewski and Dawson 1967, Herreid and Kessel 1967, Kendeigh et al. 1977, Aschoff 1981, Walsberg 1983). Such estimates are frequently used in models of seabird energetics in which RMR, TC and DEE are unknown.

This paper reviews our studies on thermoregulation and energetics for five species of arctic seabirds (Gabrielsen et al. 1987, 1988 and 1989). Data of two more species, not previously reported on, are also included. We have determined RMR and cost of thermoregulation at different ambient temperatures (T_A) in seven of the above mentioned seabird species. Field metabolic rate (FMR), using the doubly-labeled water (DLW) method, was used to determine DEE and food consumption in single individuals and populations of seabirds.

In this paper we first deal with TC, RMR and FMR measurements from arctic seabirds. Based on these measurements we discuss the reliability of different allometric equations. At the end we summarize our metabolic data and show how it can be used to indicate the impact of seabird populations on the marine and terrestrial ecosystem in the Svalbard area.

MATERIAL AND METHODS

Laboratory studies were performed at the Research Station of the Norwegian Polar Research Institute at Ny-Ålesund, Svalbard (79 °N). Field studies were carried out in Kongsfjorden (79 °N), and Hornsund (77 °N) areas and at the island of Hopen (76 °N). Measurements were done from June to August when the average ambient temperature was +2 to +5 °C.

In the laboratory we determined resting metabolic rate (RMR) and thermal conductance (TC) in Kittiwakes, Ivory Gulls, Glaucous Gulls, Fulmars, Black Guillemots, Brünnich's Guillemots and Little Auks. Birds were either trapped on the nest, using a bamboo pole with a nylon snare or they were caught in flight with a net connected to two poles or with a net-gun (Coda Enterprises). The birds were kept in individual outdoor cages for 1-2 days. They were not fed but were given water ad libitum. The first metabolic measurement was done 10-12 h after capture.

Metabolic measurements were carried out as described by Gabrielsen et al.(1988). Briefly, a metabolic chamber, made of plexiglass which allowed surveillance, was placed inside a climatic chamber where the ambient temperature (T_A) could be controlled within ± 1 °C from -25 to +25 °C. Temperature in the climatic/metabolic chamber was measured by thermocouples connected to a thermometer. Body temperature (T_B) was measured by means of a small thermocouple inserted about 2-3 cm into the colon. Based on values for gas flow and composition of oxygen and carbon dioxide, when a stable resting period was obtained, we were able to calculate O_2 consumption (ml O_2/g·h), CO_2 production (ml CO_2/g·h), respiratory quotient (RQ) and energy expenditure (KJ/day). All results are given at STPD. Thermal conductance (TC) was calculated by dividing the oxygen consumption by $T_B - T_A$ at a T_A below thermoneutrality. TC is expressed as "wet" conductance since evaporative heat loss from respiration is not excluded. For calculating energy expenditure from oxygen consumption, the conversion coefficient of 4.8 kcal per liter oxygen was used (Schmidt-Nielsen 1975; 1 kcal=4.184).

Metabolic rate (CO_2 production) and water flux rate was measured in free ranging seabirds using the doubly labeled water (DLW) method (Lifson and McClintock 1966, Nagy 1980, and Nagy and Costa 1980). The method was used on one or both members of a breeding pair of Kittiwakes, Black Guillemots and Little Auks during the chick rearing period. After the DLW injection the experimental animals were recaptured within 2-3 days. The experimental procedure is described by Gabrielsen et al. (1987).

RESULTS

The mean value of body weight from each species was used when calculating the specific resting metabolic rate (RMR)(Table 1). The mean RQ values were 0.73, 0.72, 0.73, 0.73, 0.72, 0.72 and 0.75 in Kittiwake, Ivory Gull, Glaucous Gull, Fulmar, Black Guillemot, Brünnich's Guillemot and Little Auk, respectively. The Glaucous Gull had the lowest and the Little Auk the highest specific RMR (Tab. 1).

The RMR values obtained in this study, at thermoneutrality, are all higher than those predicted by Lasiewski and Dawson (1967) and Aschoff and Pohl (1970) (Table 2). High RMR values were most pronounced in the Ivory Gull, being 220 % and 190 % of predicted values, respectively. It has to be emphasized, however, that the data on the Ivory Gull is from one specimen only. The least deviation was in the Fulmar with 131 % and 113 % of predicted values, respectively.

Tab.1. Mean body weight (BW), specific resting metabolic rate (RMR), thermal conductance (TC), and lower critical temperature (T_{LC}) in seven species of arctic seabirds. N is number of birds.

Species	N	BW (gram)	RMR (ml O_2/g·h)	TC (ml O_2/g·h·$°C$)	T_{LC} ($°C$)	Source
Kittiwake	16	365	1.64	0.0466	+4.5	*
Ivory Gull	1	508	1.81	0.0488	+0.5	***
Glaucous G.	9	1326	0.88	0.0248	+2.0	***
Fulmar	16	651	1.00	0.0336	+9.0	*
Black G.	13	342	1.59	0.0475	+7.0	*
Br. Guillem.	11	819	1.11	0.0282	+2.0	*
Little Auk	23	153	2.42	0.0630	+4.5	**

From: * Gabrielsen, Mehlum and Karlsen (1988)
     ** Gabrielsen, Taylor, Konarzewski and Mehlum (1989)
    *** Gabrielsen and Mehlum, unpublished

Tab. 2. Resting metabolic rate (RMR) and thermal conductance (TC) in seven species of arctic seabirds compared to literature data.

Species	BW (g)	Mean RMR			TC			Source
		This study (KJ/day)	L&D (1967)	A&P (1970)	This Study (ml O_2/g·h·$C°$)	H&K (1967)	A (1981)	
Kittiwake	365	289	183	158	0.0466	115	86	*
Ivory Gull	508	443	220	190	0.0448	132	97	***
Glaucous Gull	1326	562	140	120	0.0248	122	85	***
Fulmar	651	314	131	113	0.0336	113	82	*
Black Guillem.	342	262	174	151	0.0475	113	85	*
Brünnich's G.	819	438	154	133	0.0282	107	77	*
Little Auk	153	178	212	184	0.0630	98	76	**

Data from this study are compared (as %) to values predicted according to the equations given by Lasiewski and Dawson (1967), Aschoff and Pohl (active phase)(1970), Herreid and Kessel (1967) and Aschoff (1981). Sources as in Tab.1.

The mean body temperature (T_B) was stable over the whole range of ambient temperatures (T_A) (between -25 to +20 °C). T_B was 40.2, 40.3, 39.6, 38.7, 39.9, 39.6 and 40.1 °C in Kittiwakes, Ivory Gulls, Glaucous Gulls, Fulmars, Black Guillemots, Brünnich's Guillemots and Little Auks, respectively. Ivory Gull which had the second highest RMR had the highest T_B. Fulmars had the lowest T_B, though the second lowest RMR.

Thermal conductance (TC) and the lower critical temperature (T_{LC}) values are presented in Table 1. Our TC values are expressed as ml O_2/g·h·°C in order to compare our results with allometric equations. TC values were highest in the Little Auk and lowest in the Brünnich's Guillemot and the Glaucous Gull. TC values obtained in all species were close to values obtained by equations made by Herreid and Kessel (1967) and Aschoff (1981) (Table 2).

Foraging Kittiwakes had a field metabolic rate (CO_2 production) (FMR) of 4.04 ml CO_2/g·h which is equivalent to 992 KJ/day (RQ=0.77) or 3.2 times RMR (Table 3). RMR was estimated for birds of the same mass as those used in the FMR experiments, using the mass specific values given in Table 1. Non-foraging (brooding) Kittiwakes had a significantly lower ($p<0.01$) metabolic rate, averaging 2.43 ml CO_2/g·h or 581 KJ/day (1.9 times RMR). Foraging Black Guillemots had a FMR of 3.63 ml CO_2/g·h which is equivalent to 866 KJ/day or 3.0 times RMR (Table 3). FMR in foraging Little Auks averaged 7.14 ml CO_2/g·h which is equivalent to 733 KJ/day or 3.8 times RMR.

Tab.3. *Mean body weight (BW), resting metabolic rate (RMR), field metabolic rate (FMR) and FRM/RMR relationship in three species of seabirds. N is number of birds.*

Species	N	BW (g)	RMR (KJ/day)	FMR (KJ/day)	FMR/RMR	Source
Kittiwake	13	392	310	992	3.2	*
Black Guillem.	11	381	292	866	3.0	**
Little Auk	14	164	191	733	3.8	***

From: * Gabrielsen, Mehlum and Nagy (1987)
       ** Mehlum, Gabrielsen and Nagy, unpublished
      *** Gabrielsen, Taylor, Konarzewski and Mehlum (1989)

DISCUSSION

Except for the Fulmar the specific resting metabolic rate (RMR) in the other six species were considerably higher than expected. The Fulmar belongs to the Procellariiformes and its RMR value was similar to values obtained by Adams and Brown (1984) of ten species of Sub-Antarctic Procellariiformes. Compared to the other species studied, the Fulmar has a low body temperature (T_B), a high lower critical temperature (T_{LC}) and good insulation due to a low RMR. Together with its ability to store oils in the proventriculus (Rosenheim and Webster 1927) this may be adequate adaptation for survival through long winter nights in the Arctic with extended periods of fasting.

Our RMR values from the other six species, were 120-220 % of predictions based on the allometric equations of Lasiewski and Dawson

(1967) and Aschoff and Pohl (1970). While Lasiewski and Dawson (1967) based their allometric equation for basal metabolic rate (BMR) in non-passerines only using body mass (BMR = $327.8 \, m^{0.723}$), Aschoff and Pohl (1970) also accounted for the influence of diurnal rhythm on BMR (BMR, active phase = $381.0 \, m^{0.729}$ and BMR, resting phase = $307.7 \, m^{0.734}$, where BMR is in KJ/day and m is mass in kg). These equations are based primarily on metabolic data of birds from temperate and tropical areas. Weathers (1979) proposed that BMR in birds was a function of breeding latitude, so that species breeding at high latitudes have higher BMR than those in temperate and tropical areas. Later Ellis (1984) presented data from 16 species of seabirds which supported this. Even so, when our RMR measurements were applied to Ellis' (1984) modified equation (BMR = $381.8 \, m^{0.721}$) we obtained values which were 120-189 % of his predictions. If the RMR values from the present and other arctic studies (Johnson and West 1975, Roby and Ricklefs 1986) are representative, then the equations proposed by previous authors seriously underestimate metabolic rates of arctic seabirds. More data are obviously needed, especially studies of the same species at different latitudes, in order to show the effect of climate on RMR in seabirds.

What is the reason for a high RMR in arctic seabirds compared to birds from tropical or temperate areas? Climate and the temperature of the water, in which they are foraging, may account for their high RMR. The average ambient temperature in the study area is close to or lower than the birds' lower critical temperature (T_{LC}). This indicates that the metabolic rate in brooding birds is higher than RMR. Our doubly labeled water (DLW) studies of brooding Kittiwakes has shown that the daily energy expenditure (DEE) is 1.9 times RMR. Despite that there is some energy demand for activity and specific dynamic action these data may indicate that arctic brooding seabirds are in negative heat balance. Though solar radiation may put the birds into a positive heat balance this has not been investigated at present. When foraging at sea, water may compress feathers thereby increasing TC. Despite that the ambient temperature may be lower, the higher conductivity of water may therefore cause more severe cold stress than when exposed to air. Studies of Kittiwakes and Brünnich's Guillemots swimming on cold water under laboratory conditions show an energy expenditure which is 2-2.5 times RMR (Gabrielsen and Mehlum, unpublished).

We have priviously hypothesised (Gabrielsen et al. 1988) that the activity level in seabirds has an effect on RMR. Among the seven species studied there is a great variation in activity level. Some species such as the Fulmars and Glaucous Gulls are gliders, while Kittiwakes and Ivory Gulls must work hard to stay in the air. Masman and Klaassen (1987) have made an equation for calculating flight cost in birds. According to their equation Fulmars and Glaucous Gulls have lower flight cost than Kittiwakes and Ivory Gulls. We therefore propose that this low cost of flight allow the retention of a low RMR as shown for the Glaucous Gulls and the Fulmars.

Scholander et al. (1950) described two main adjustments for animals living in the cold. One was to lower the heat loss by increasing the insulation, the other was to increase the heat production during cold stress by raising the metabolism. While the Fulmars, Glaucous Gulls and Brünnichs Guillemots seem to use the former mechanism, the latter principle seem to be used by the Kittiwakes, Ivory Gulls, Black Guillemots and the Little Auks. Below the T_{LC} the metabolism increased in the usual homeothermic fashion. Projection along the metabolic line intercepted the abscissa close to T_B. The lowest thermal conductance (TC) values were found in Brunnich's Guillemots, Glaucous Gull and Fulmars, which are also the largest of the species studied. TC in all species were close to allometric values calculated for birds of similar size. By using the Herreid and Kessel (1967) equation for dead birds (dry conductance) we got values 98-132 % of predicted while using Aschoff's (1981) equation (active phase) we got 76-97 % of the predicted values. When compensating for respiratory heat loss, which by Drent and Daan (1971) is estimated at 12 % of total heat loss, we got an even better agreement for the Herreid and Kessel (1967) equation than that of Aschoff (1981). According to Bligh and Johnson (1973) TC should be expressed as $W/m^2 \cdot {}^0C$. Based on Meeh's (1879) equation $A = k \cdot m^{0.667}$ (where m is mass in kg and k is a constant of 10 according to Benedict (1934) we calculated that of the seven species studied the lowest conductance or the best insulation was found in the Brünnich's Guillemot and the Glaucous Gull, while the Ivory Gull had the highest conductance or had poorest insulation.

When comparing TC values of seabirds further south we found considerable interspecific variability. TC values of Brünnich's Guillemots measured in the high arctic (79 ^{0}N) were 54 % lower than measurements of the same species (<u>Uria</u> spp.) further south (65 ^{0}N) (Johnson and West 1975). Gulls however, of similar body size, measured in temperate areas (Herreid and Kessel 1967, Drent and Stonehouse 1971, Lustic et al. 1978, Gavrilov 1985, Hüppop 1988), have lower conductance or better insulation than arctic gulls. But when compared to other arctic birds, of similar body weight, such as ptarmigans (<u>Lagopus</u> spp.) (West 1972, Herreid and Kessel 1967, Mortensen and Blix 1985) they have lower TC values or better insulation than arctic seabirds.

Scholander et al. (1950) reported a RMR 47 % higher than ours in the Glaucous Gulls measured at Point Barrow (70 ^{0}N). Scholander did his measurement in April, we in July-August, indicating a possible seasonal difference in RMR. Based on measurements of Kittiwakes, also measured in April, and Gavrilov's (1985) study of 16 species of non-passerine species, which show little or no seasonal difference in RMR, we are at the present unable to explain the high resting metabolic rate in Glaucous Gulls measured in Alaska.

In models hypothesised for estimating food consumption of single individuals and populations of seabirds, we often find daily energy expenditure budgets using different allometric equations (Wiens and Scott 1975, Furness 1978, Furness and Barrett 1985). Kendeigh et al. (1977) and Walsberg (1983) equations are most often used in model studies explaining the interactions of seabird and fish populations. The Kendeigh et. al (1977) equation (DEE (KJ/d) at 0 ^{0}C $= 33.73\ m(g)^{0.50}$) and the Walsberg (1983) equation (DEE $= 13.05\ m^{0.605}$) is based on studies of birds from tropical/temperate areas. When these equations were applied on foraging seabirds during the chick rearing period our DEE values were 35-156 % higher. This suggest that great caution should be taken, at least for arctic breeding birds, when modelling seabird energetics using only allometric equations.

Our DLW studies show that DEE in foraging seabirds varies between 3 and 4 times RMR. This is consistent with Drent and Daan (1980), who proposed a "maximum sustained working level" of 4 times RMR during the chick rearing period. DLW studies of birds of similar body mass, performed further south (Roby and Ricklefs 1986) also show the 3-4 times FMR/RMR relationship. Arctic birds show the same relationship but at a higher "background" (resting) level. This means that birds in the Arctic spend more energy on heat production and therefore have less energy for foraging compared to the same species further south. It may therefore be more "stressful" to raise chicks in the Arctic compared to tropical/temperate areas. Access to good nesting sites and rich food sources are probably important factors allowing them to stay in these areas.

Based on studies on RMR, thermoregulation, food and energy requirements of seabirds in the Svalbard area we have been able to calculate an estimate of food consumption for the breeding seabird community at Bear Island ($74°N$), during the chick rearing period (30 days). The total number of birds, adults and chicks, was estimated in 1986 at 1.582 mill. birds (Bakken and Mehlum 1988). The community consists of the following species; Kittiwake, Fulmar, Glaucous Gull, Little Auk, Common and Brünnich's Guillemot. During the breeding season these seabirds feed mainly on capelin (Mallotus villosus). They consume an estimated 357 tons of capelin per day. During one month this colony therefore consume 10700 tons or 2.4 % of the total fisheries catch of capelin in Norway during 1985. At a dry matter digestibility of about 75 % (Brekke, Gabrielsen and Mehlum, unpublished) an estimated 667 tons (dry matter) of nitrogen-rich guano may be added to the Bear Island terrestrial and marine ecosystems during the chick rearing period.

CONCLUSION

The RMR values obtained in our studies are much higher than expected in seabirds, while the TC values were close to expected. Factors such as climate, activity, size and lack of diurnal phases may account for the high RMR values in arctic seabirds. DEE values of foraging seabirds were 3-4 times RMR, and also higher than expected. Our studies show that great caution should be taken when modelling seabird energetics based on allometric equations. The impact of a seabird community on a fish population was calculated for the Bear Island ecosystem. A community of 1.58 mill. birds consume 10700 tons capelin per month. Not only will these consumption contribute to the enrichment of the terrestrial and marine ecosystem via guano but it may also have ramification for the fishery in the area.

ACKNOWLEDGEMENTS

Our seabird project "The ecological function of seabirds in ice-filled waters around Svalbard" is part of the Norwegian Research Program for Marine Arctic Ecology (PRO MARE). The project was supported by the Norwegian Research Council for Sciences and the Humanities and the Norwegian Polar Research Institute. We are grateful to Drs. J.B. Steen and C.Cyler for valuable suggestion regarding the manuscript.

REFERENCES

Adams, N.J., and Brown, C.R. 1984. Metabolic rates of sub-Antarctic Procellariiformes: a comparative study. Comp Biochem Physiol, 77 A: 169-173.
Aschoff, J., 1981. Thermal conductance in mammals and birds; Its dependence on body size and circadian phase. Comp Biochem Physiol, 69A: 611-619.
Aschoff, J., and Pohl, H., 1970. Der Ruheumsatz von Vögeln als Funktion der Tageszeit und der Körpergrösse. J. Orn, 111: 38-47
Bakken, V., and Mehlum, F., 1988. AKUP - Sluttrapport sjøfuglundersøkelser nord for N $74°$ / Bjørnøya. Norsk Polarinstitutt Rapportserie Nr 44.
Benedict, F.G., 1934. Die Oberflächenbestimmung verschiedener Tiergattung. Ergeb. Physiol, 36: 300-346.
Drent, R.H., and Stonehouse, B., 1971. Thermoregulatory responses of the Peruvian penguin Sphensicus humboldti. Comp Biochem. Physiol, 40A: 689-710.
Drent, R.H., and Daan, S., 1980. The prudent parent energetic adjustments in avian breeding. Ardea, 68: 225-252.

Ellis, H.I. 1984. Energetics of free ranging seabirds. In: Whittow G.C., and Rahn, H. eds. Seabird energetics. Plenum Press, New York, London, pp 203-234.

Furness, R.W. 1978 Energy requirements of seabird communities: A bio-energetics model. J Anim Ecol, 47: 39-53.

Furness, R.W., and Barrett, R.T., 1985, The food requirements and ecological relationships of a seabird community in North Norway. Ornis Scand, 16: 305-313.

Gabrielsen, G.W., Mehlum, F., and Nagy K.A. 1987. Daily energy expediture and energy utilization of free ranging Black legged Kittiwakes *(Rissa tridactyla)*. Condor, 89: 126-132.

Gabrielsen, G.W., Mehlum, F. and Karlsen, H.E., 1988. Thermoregulation in four species of arctic seabirds. J Comp Physiol B, 157: 703-708.

Gabrielsen, G.W., Taylor, I., Konarzewski, M., and Mehlum, F., 1989. Thermoregulation, daily energy expediture and energy utilization in Little Auks *(Alle alle)*. J. Comp. Physiol. B. (submitted).

Gavrilov, V.M.,1985. Seasonal and circadian changes of thermoregulation in passerine and non-passerine birds; Which is more important? In: "Acta, XVIII Congressus Int. Ornithol", V.D. Ilyichev, and V.M. Gavrilov, eds. vol.2., Moscow Nauka, Moscow.

Herreid, C.F. and Kessel, B., 1967. Thermal conductance in birds and mammals. Comp Biochem Physiol 21: 405-414.

Hüppop, O., 1987. Der Einfluss von Wachstum Thermoregulation und Verhalten auf den Energihaushalt der Silbermöwe *(Larus argentatus,* PONTOPPIDAN, 1763). PhD thesis, Hamburg Universität.

Johnson, S.R., and West, G.C., 1975. Growth and development of heat regulation in nestling and metabolism of adult common and Thick-billed murres. Ornis Scand. 6: 109-119.

Kendeigh, S.C., Dol'nik, V.R., and Gavrilov, V.M., 1977. Aivan energetics. In: J. Pinowski and S.C. Kendeigh eds. Granivorous birds in ecosystems. Cambridge University Press, Cambridge.

Lasiewski, R.C. and Dawson, W.R., 1967. A re-examination of the relation between standard metabolic rate and body weight in birds. Condor 69:13-23.

Lifson, N., and McClintock, R., 1966. Theory of use of the turnover rates of body water for measuring energy and material balance. J Theoret Biol, 12:46-74.

Lustick, P.B.S., Battersby, B., and Kelty, M., 1978. Behavioral thermoregulation: orientation toward the sun in herring gull, Science, 200:881-83.

Løvenskiold, H.L. 1964. Avifauna Svalbardensis. Norsk Polarinstitutts Skrifter 129.

Masman, D. and Klaassen, M. 1987. Energy expenditure during free flight in trained and free living Eurasian Kestrels (Falco tinnunculus). The Ank, 104:603-616.

Meeh, K., 1879. Oberflächen messungen des menschlichen körpers. Z. Biol. 15: 426-458.

Mortensen, A., and Blix, A.S., 1985. Seasonal changes in resting metabolic rate and mass - specific conductance in Svalbard Ptarmigan, Norwegian Rock Ptarmigan and Norwegian Willow Ptarmigan. Ornis Scand, 17: 8-13.

Nagy,, K.A., 1980. CO_2 production in animals: analysis of potential errors in the doubly labeled water method. Am J Physiol 238 (Reg Int Comp Physiol, 7): R 466- R 473.

Nagy, K.A., and Costa, D.P., 1980. Water flux in animals: analysis of potential errors in the tritiated water method. Am J Physiol 238 (Reg Int Comp Physiol 7): R 454-465.

Norderhaug, M., Brun, E., and Uleberg Møller, G., 1977. Barentshavets sjøfuglressurser. Norsk Polarinstitutt Meddelelser Nr 104.

Ricklefs, R.E., 1983. Some considerations on the reproductive energetics of pelagic seabirds. Stud Avian Biol 8:84-94.

Roby, D.D., and Ricklefs, R.E., 1986. Energy expenditure in adult least auklets and diving petrels during the chick-rearing period. Physiol Zool 59: 661-678.

Rosenheim, O., and Webster, T.A. 1927. The stomach oil of the fulmar petrel *(Fulmarus glacialis)*. Biochem J 21: 11-118.

Schmidt-Nielsen, K. 1975. In: Animal physiology: Adaption and environment. Cambridge University Press.

Scholander, P.F., Hock, R., Walters, V., and Irving, L. 1950. Adaptation
to cold in Arctic and tropical mammals and birds in relation to body temperature. Insulation and basal metabolic rate. Biol Bull, 99: 259-271.

Walsberg, G.E., 1983. Avian ecological energetics. in: D.S. Farner, J.R. King and K.C. Parks, eds. Avian Biology, Vol. 7, Academic Press, New York.

Weathers, W.W., 1979. Climate adaptions in avian standard metabolic rate. Oecologia 42: 81-89.

West, G.C., 1972. Seasonal differences in resting metabolic rate of Alaskan ptarmigan. Comp. Biochem Physiol. 42A: 867-876.

Wiens, J.A, Scott, J.M., 1975. Model estimation of energy flow in Oregon coastal seabird populations. Condor 77: 439-452.

THERMOREGULATORY ADAPTATIONS TO COLD IN WINTER-ACCLIMATIZED LONG-TAILED DUCKS (CLANGULA HYEMALIS)

Bjørn Munro Jenssen and Morten Ekker

Department of Zoology
University of Trondheim
N-7055 Dragvoll, Norway

INTRODUCTION

The extreme thermal conditions that prevail during wintertime in maritim arctic and subarctic regions are probably the reason why so few homeotherms choose to winter there. In addition to low air temperatures, aquatic animals are exposed to water temperatures as low as -1.8 °C. In order to remain homeothermic in water, some animal groups, such as the pinnipeds and cetaceans, have evolved a thick insulating layer of blubber (Aschoff, 1981). Other groups, such as otters and birds are dependent on a water-repelling pelt or plumage that creates an insulating air layer around the animal (Costa and Kooyman, 1982, Kooyman et al., 1976). The only birds that spend most of their time on the sea surface during winter are found among loons, auks and seaducks. Few studies have been made to investigate the particular adaptations that these species, which may winter as far north as 71 °N, have evolved in order to live in such harsh conditions. In a previous study (Jenssen et al., 1989) we studied the thermoregulation of Common eiders (Somateria mollissima) acclimatized to winter conditions. The study showed that the Common eider, which is the largest of the seaducks, has evolved a highly insulative plumage. This species, in addition, depends upon extensive use of peripheral vasoconstriction to minimize its heat loss. The basal metabolic heat production of the Common eider is not particurlarly high, as is found to be the case among other seabirds from high lattitudes (Ellis, 1984).

The Long-tailed duck (Clangula hyemalis) inhibits approximately the same distributional area in winter as the Common eider (Cramp and Simmons, 1977). The Long-tailed duck, however, is the smallest seaduck species. Because it has a higher surface to volume ratio, the Long-tailed duck is potentially exposed to a greater heat loss than the Common eider. The aim of the present study was to study the thermoregulatory adaptations to a marine life in Long-tailed ducks acclimatized to winter conditions.

MATERIAL AND METHODS

In this study we used 6 winter acclimatized Long-tailed ducks (body weight = 490 ± 26 g) from a population wintering in the Trondheimsfjord (63 °N). During the experimental period the birds were kept indoors, with access to a freshwater pool (1m x 1m x 3.5m) with air and water temperatures between 5 and 7 °C and light conditions corresponding to the natural conditions.

The oxygen consumption of resting post-absorptive (>12 hours) Long-tailed ducks was measured in an open-circuit system (Depocas and Hart, 1957) during daytime (in the birds activity-phase). The birds were placed, one at a time, in a darkened respiration chamber (20 L in the air experiments, 20L water/20L air in the water experiments) in a temperature controlled cabinet. In the water experiments the water and air temperatures (T_W and T_A, respectively) were the same. The air from the respiration chamber (4-5 L min^{-1}) was dried with Silica-gel, and the oxygen tension determined by using an oxygen analyzer (Servomex Series 1100). The oxygen consumption of resting birds was calculated according to Depocas and Hart (1957) and corrected to STP-conditions. Heat production (H) was calculated using an energetic equivalent of 5.501 W kg^{-1} pr.ml O$_2$ g^{-1} h^{-1} assuming an respiratory exchange ratio of 0.71. Simultaneous measurements of body temperature (T_B) were made by means of thermosensitive radio-transmitters (T-PM, Mini-Mitter Company, Oregon, USA).

The thermal insulation values (I, °C m^2 W^{-1}) was calculated as the inverse value of the total thermal conductance. Thermal conductance (C_T, W °C^{-1} m^{-2}) was calculated according to the equation:

$$C_T = H / (T_B - T_{A/W}) \times A \qquad \text{I}$$

where A is the surface area of the plumage (m^2). Thus, C_T represents the overall ("wet") thermal conductance according to Aschoff (1981). The surface area of the plumage was calculated according to Meeh's equation, which expresses the relationship of body mass (M, kg) to surface area.

$$A = k \times M^{2/3} \qquad \text{II}$$

The plumage-surface of the Long-tailed ducks was calculated using a mass-coefficient (k) of 0.1.

RESULTS

The metabolic heat production of the Long-tailed ducks at different air and water temperatures is presented in Fig. 1. The mean basal metabolic heat production (H_{BMR}) in air and water respectively, was 5.60 W kg^{-1} (SD=0.32, n=5) and 5.59 W kg^{-1} (SD=0.56, n=5). The H_{BMR} in air and in water did not differ significantly (t=0.83, N.S., df=8). The lower critical temperature in air (T_{LCA}) was about 18 °C, and the heat production increased at lower air temperatures, as indicated by the regression line H=9.06-0.19T_A (r=0.89, n=17, p<0.001). The lower critical temperature in water (T_{LCW}) was about 12 °C, and the heat production increased at lower temperatures, as shown by the regression line $H=11.87-0.50T_W$ ($r=0.88$, $n=13$, $p<0.001$).

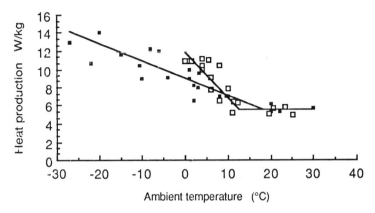

Fig. 1. Metabolic heat production of winter acclimatized Long-tailed ducks (Clangula hyemalis) at exposure to different air (solid symbols) and water (open symbols) temperatures.

The body temperatures of the Long-tailed ducks at different air and water temperatures are presented in Fig. 2. In air the mean T_B was 40.8 °C (SD=0.7, 21 measurements), while the mean T_B was 40.6 °C (SD=1.1, 12 measurements) during exposure to water. The results show that the body temperature increased at exposure to low water temperatures.

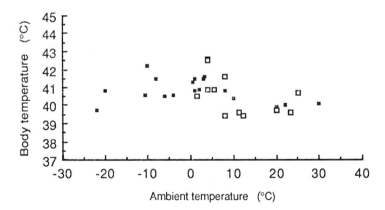

Fig. 2. Body temperature in winter acclimatized Long-tailed ducks (Clangula hyemalis) at exposure to different air (solid symbols) and water (open symbols) temperatures.

The total thermal insulation of Long-tailed ducks exposed to various air and water temperatures are presented in Fig. 3. In air, the thermal insulation increased with decreasing temperatures down to 5 °C, and at exposure to lower air temperatures the insulation was 0.6 W^{-1} °C m^2. When the ducks were exposed to water, the insulation increased with decreasing water temperatures down to about 10 °C. However, at lower water temperatures the insulation decreased, so that the thermal insulation averaged 0.45 W^{-1} m^2 °C at water temperatures below 10 °C.

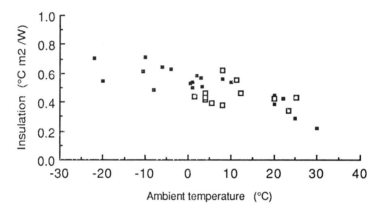

Fig. 3. Thermal insulation in winter acclimatized Long-tailed ducks (<u>Clangula hyemalis</u>) at exposure to different air (solid symbols) and water (open symbols) temperatures.

DISCUSSION

The body temperature of the Long-tailed ducks lay within the range previously found for other species of ducks (Dawson and Hudson, 1970, Jenssen et al., 1989.), and is close to that predicted (40.4 °C) from allometric equations (McNab, 1966). The body temperature increased with decreasing water temperature. This could be caused by an increased peripheral vasoconstriction. It has previously been demonstrated that during cold water exposure, Common eiders reduce their skin temperature, presumably through peripheral vasoconstriction, as a mechanism for decreasing their heat loss (Jenssen et al., 1989). Restriction of the blood supply to peripheral tissues will reduce the heat loss and could also lead to an increase of the core temperature. The observed increase in T_B of Long-tailed ducks at exposure to low water temperatures could also have been due to shivering in muscles close to the colon where we measured body temperature, or even simply due to the increase in heat production observed at the low water temperatures.

The heat production of the Long-tailed ducks predicted from allometric equations (Aschoff and Pohl, 1970), is 5.34 W kg^{-1}. The observed heat production was only 5% higher than the predicted values. The conclusion therefore is that the Long-tailed duck has not evolved a high metabolic rate in order to cope with the low ambient temperatures prevailing in its winter habitat. It has previously been demonstrated that the H_{BMR} in birds living in cold climates tend to be higher than those predicted from the birds' body mass (Kendeigh et al., 1977, Weathers, 1979, Hails, 1983, Ellis, 1984). Weathers (1979) showed that the H_{BMR} of 10 species occurring at latitudes exceeding 50° was on average 47% higher than that predicted from the birds' body mass. Our results indicate that ducks which winter in northern marine environments has not evolved a high metabolic heat production to cope with low ambient temperatures.

In air the lower limit of the thermoneutral zone (lower critical temperature, T_{LC}) for Long-tailed ducks lay at about 18 °C, compared to about 12 °C in water. Previous studies on seabirds have shown that their T_{LC} is higher in water than in air (Kooyman et al., 1976, Stahel and Nicol, 1982, Jenssen et al., 1989). In contrast, the Long-tailed duck has a T_{LC} which is lower in water than in air. Their thermal insulation thus seem to be higher in water than in air at temperatures between 10 °C and 12 °C (Fig. 3). This difference could be caused by a more extensive use of peripheral vasoconstriction producing an increase in the thermal insulation of the tissues.

At temperatures below the T_{LC}, the metabolic heat production increased more rapidly in water than in air. Thus, below about 10 °C, heat production was higher in water than in air. The regression line for the relationship between heat production and water temperature below the T_{LC}, intersects with zero metabolism at 23.7 °C, a value which is considerably below the body temperature of the birds. This phenomenon could be due to a reduction in thermal insulation at water temperatures below 10°C (Fig. 3). This decrease in insulation at lower water temperatures is probably due to the increase in body temperature found at low water temperatures (Fig. 1), but could also be due to dilatation of the peripheral vessels at low water temperatures. Due to the heat requirement of the body tissues at low water temperatures, cyclic vasodilatation (hunting-responses, Lewis, 1930) may be responsible for the observed decrease in insulation at low water temperatures.

Aschoff (1981) presents an allometric equation which relates the minimal total thermal conductance of birds to their body mass. According to this equation, the minimal conductance of a 490 g bird is 3.32 W m^{-2} °C^{-1}. The observed minimal conductance of the water exposed Long-tailed ducks was only 2.22 W m^{-2} °C^{-1}, 67% of the predicted value. The minimal conductance of air exposed ducks was even lower; 1.66 W m^{-2} °C^{-1}, or 50% of the predicted value. The minimal thermal conductance of Common eiders exposed to air and water respectively, is 0.58 W m^{-2} °C^{-1} and 0.91 W m^{-2} °C^{-1}, which is 55% and 108% of the predicted values (Jenssen, et al., 1989) This demonstrates that even though the Long-tailed duck (because of its lower body mass) has a higher thermal conductance than Common eiders, the species has managed to reduce reduced its potentially higher heat loss by evolving a highly insulative plumage. The low conductance also allows the Long-tailed duck to spend relatively little energy on heat production for thermoregulation. As previously suggested by Jenssen, et al. (1989), low energy requirements is probably favorable for species dwelling in the cold and dark winter environments of arctic and subarctic regions.

ACKNOWLEDGEMENTS

The authors wish to thank Dr. Claus Bech and Prof. Karl Erik Zachariassen for their support. This study was carried out as a part of the BECTOS (Biological Effects of Chemical Treatment of Oilspills at Sea) project, which was a research programme supported by the Fina Oil Company.

REFERENCES

Aschoff, J., 1981, Minireview: Thermal conductance in mammals and birds: its dependence on body size and circadian phase. Comp. Biochem. Physiol. 69A: 611.

Aschoff, J., and Pohl, H., 1970, Rhythmic variations in energy metabolism. Fed. Proc. 29: 1541.

Cramp, S., and Simmons, K.E.L., 1977, Handbook of the birds of Europe, the Middle East and North Africa: The birds of the Western Palearctic. Vol. 1: Ostriches-ducks. Oxford University Press, London. 722p.

Costa, D.P., and Kooyman, G.L., 1982, Oxygen consumption, thermoregulation, and the effects of fur oiling and washing on the sea otter, Enhydra lutris. Can. J. Zool. 60: 2761.

Dawson, W.R., and Hudson, J.W., 1970, Birds. In "Comparative physiology of thermoregulation, Vol. I, Invertebrates and non-mammalian vertebrates" (G.C. Whittow, Ed.). p. 223. Academic Press, New York.

Depocas, F.A., and Hart, J.S., 1957, Use of the Pauling oxygen analyzer for measurement of oxygen consumption of animals in open circuit system and in a short-lag, closed-circuit apparatus. J. Appl. Physiol. 10: 388.

Ellis, H.I., 1984, Energetics of free-ranging seabirds. In "Seabird energetics". (G.C. Whittow & H. Rahn, Eds.). p. 203. Plenum Press, New York/London.

Hails, C.J., 1983, The metabolic rate of tropical birds. Condor 85: 61.

Jenssen, B.M., Ekker, M., and Bech, C., 1989, Thermoregulation in winter acclimatized Common eiders, Somateria mollissima, in air and water. Can. J. Zool. (in press).

Kendeigh, S.C., Dol'nik, V.R., and Gavrilov, V.M., 1977, Avian energetics. In "Granivorous birds in ecosystems" (J. Pinowski & S.C. Kendeigh, Eds.). p. 127. Cambridge University Press, Cambridge.

Kooyman, G.L., Gentry, R.L., Bergman, W.P., and Hammel, H.T., 1976, Heat loss in penguins during immersion and compression. Comp. Biochem. Physiol. 54A: 75.

Lewis, T, 1930, Observations upon the reactions of the vessels of the human skin to cold. Heart 15: 177.

McNab, B.K., 1966, An analysis of the body temperature of birds. Condor 68: 47.

Stahel, C.D., and Nicol, S.C., 1982, Temperature regulation in the Little penguin, Eudyptula minor, in air and water. J. Comp. Physiol. 148: 93.

Weathers, W.W., 1979, Climatic adaptation in avian standard metabolic rate. Oecologia (Berlin) 42: 81.

RESPIRATION AND GAS EXCHANGE IN BIRDS

Johannes Piiper and Peter Scheid

Abteilung Physiologie, Max-Planck-Institut für
experimentelle Medizin, Göttingen, F.R.G. and
Institut für Physiologie, Ruhr-Universität
Bochum, F.R.G.

INTRODUCTION

Lungs of birds are homologous to lungs of mammals, and both are phylogenetically derived from those of their reptilian ancestors. There exist, however, fundamental differences between avian and mammalian lungs in anatomical design, leading to differences in respiratory gas flow pattern and in gas exchange function. The subject has been recently reviewed by Fedde (1976), Scheid (1979, 1982), Scheid and Piiper (1987; in press), Powell and Scheid (in press).

A. ANATOMY: LUNGS AND AIR SACS

The respiratory airways (Figure 1) comprise (1) the bronchial system (trachea, primary or main bronchi, and secondary bronchi); (2) the lung (parabronchi and air capillaries) which serves gas exchange; and (3) the air sacs, which allow the volume changes required for tidal ventilation without significant gas exchange across their walls. Unlike the mammalian lung, the structural elements subserving ventilation and gas exchange are thus separated.

(1) The trachea divides into the right and left main bronchi. After entering the lung, each main bronchus gives off two series of secondary bronchi: The medioventral secondary bronchi or ventrobronchi occupy with their extensive ramification the ventral surface of the lung, while the mediodorsal secondary bronchi or dorsobronchi and their branches form the dorsolateral surface of the lung. In addition, there are lateroventral and laterodorsal secondary bronchi. The second lateroventral bronchus generally provides the direct connection to the caudal thoracic air sac.

(2) The two main sets of secondary bronchi, the ventrobronchi and dorsobronchi, are connected by the intrapulmonary airways, the tertiary bronchi or parabronchi which are long, narrow tubes. They extend across the lungs and display only a mild degree of anastomosing. From the parabronchial lumen radiates a network of very fine (some μm in diameter) tubular air capillaries, which is interposed into a similar network of blood capillaries. Air capillaries and blood capillaries make up most of the periparabronchial tissue, which may be viewed as a gas exchange mantle

around the parabronchial tubes. In some orders of birds, the air capillaries are functionally blind-ending tubes. In others, the air capillaries of neighboring parabronchi interconnect, but significant convective gas movement through these communicating air capillaries is not expected since there probably is little pressure difference between adjacent parabronchi. The avian lungs are compact, small structures which occupy the dorsal portion of the thoracic cavity and are limited ventrally by a membrane, the horizontal septum, composed mainly of connective tissue. The horizontal septum, however, is neither homologous nor functionally equivalent to the mammalian diaphragm.

(3) In most birds there exist nine air sacs, four paired and one unpaired (interclavicular). The air sacs are usually divided into two groups according to their bronchial connections. The cranial group (cervical, clavicular, and cranial thoracic air sacs) connects to the ventrobronchi, while the sacs of the caudal group (caudal thoracic and abdominal air sacs) have direct communication with the main bronchus. The laterobronchus to the caudal thoracic air sac departs from the main bronchus, opposite to the origins of the large, cranial dorsobronchi, while the main bronchus itself opens caudal to the lung into the abdominal sac.

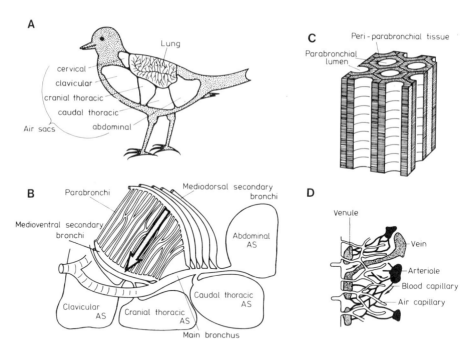

Fig. 1. Scheme of respiratory system in birds. It comprises the lung and a number of air sacs (A and B). The lung is composed of long narrow tubes, parabronchi (B and C), through which gas flows in the same direction during both inspiration and expiration: from mediodorsal (dorsobronchi) to medioventral secondary bronchi (ventrobronchi as shown by the arrow in B). The periparabronchial tissue, a dense network of air capillaries and blood capillaries (D), constitutes the site for respiratory gas exchange. (Modified from Scheid, 1979).

The features of the respiratory system so far described, and schematically represented in Figure 1, have been found to be present in all birds investigated (Duncker, 1971; 1972; 1974). Duncker (1971) has proposed the term "paleopulmo" for this basic arrangement of the lung with parabronchi extending exclusively between ventrobronchi and dorsobronchi. It is the only arrangement found in penguins and emus. In all other birds (studied till now) an additional network of parabronchi, the neopulmo, is developed, to varied degrees. The neopulmo extends from the main bronchus and dorsobronchi to the caudal air sacs, entering particularly into their bronchial connections. In song birds and fowl-like birds, the neopulmo is particularly well developed. Unlike the situation in the paleopulmo, the parabronchi in the neopulmo form a meshwork. Aside from that, the histologic structure of both types of parabronchi appears to be identical.

B. VENTILATION: UNIDIRECTIONAL AIR FLOW

Inspired air moves into the respiratory system as a result of the expansion of the thoraco-abdominal cavity, which is executed by the inspiratory muscles; and during expiration, air is expelled by the action of expiratory muscles. To this extent, the situation is not basically different from that in mammals. But the parabronchial lung appears to undergo only small volume changes in the respiratory cycle. Although the parabronchial tissue has been shown to be distensible (Macklem et al., 1979) such changes appear to be opposed by the phasic activity of the costopulmonary muscles which connect the ribs with the horizontal septum (Fedde et al., 1964). In contrast to the views of earlier authors, all air sacs appear to be effectively ventilated, and in the duck the ratio of ventilation to volume is similar for all air sacs in the duck (Scheid et al., 1974b).

Air reaches the alveoli of mammalian lungs after passing through a number of branching bronchiolar tubes, the airways thus forming a dead-end, to-and-fro system for air flow. The flow of air in the avian lung, on the other hand, cannot easily be predicted from structural considerations. Since the parabronchial lung is open at both ends, air can flow through it in either direction, and the path taken by the air between trachea and the air sac is likewise unpredictable. In fact, this problem has long been a matter of controversy and imaginative speculation (reviewed by Bretz and Schmidt-Nielsen, 1971; Scheid, 1979). More recently, several investigators attempted to directly measure airflow direction in the avian respiratory tract (Bretz and Schmidt-Nielsen, 1971; Brackenbury, 1971; Scheid and Piiper, 1971). All these authors observed air to flow in the dorsobronchi during both inspiration and expiration, flow direction being the same in these bronchi in both respiratory phases, viz. from the main bronchus into the dorsobronchi. The anatomy of the bronchial connections implies that air flow through the parabronchi of the paleopulmo is also in the same direction during inspiration and expiration, from dorsobronchi towards ventrobronchi (Figure 2). Support for the unidirectional flow thesis was also obtained by observation of test gases at various sites in the respiratory system after bolus injections of the gases at key sites (Bouverot and Dejours, 1971; Bretz and Schmidt-Nielsen, 1972).

Although the unidirectional flow through the (paleopulmonic) parabronchi was thus demonstrated beyond reasonable doubt, there remained controversies about the pattern of flow in other parts of the respiratory system. In particular, direction of inspiratory flow across the ventrobronchial orifices into the main bronchus was much debated. Mass spectrometric measurements of the partial pressure profiles for CO_2 and O_2 in the main bronchus and the ventrobronchi of the spontaneously breathing duck led to the conclusion that there was no flow across the ventrobronchial orifices during inspiration (Powell et al., 1981). Debate has also been about whether on expiration all flow from caudal air sacs is directed into the (paleopulmonic) lung or whether some gas passes the direct way out through the main bronchus. The best evidence to date suggests that this flow bypassing the lungs is small if not negligible (Bretz and Schmidt-Nielsen, 1971; Powell et al., 1981).

Air flow through the (paleopulmonic) airways of the resting bird, according to the results outlined above, is schematically represented in Figure 2.

The simplest explanation for the unidirectional air flow would be the existence of anatomic valves opened and closed in phase with breathing. Although claimed by a number of authors in the past, the existence of valves or similar anatomical elements that would cause unidirectional airflow could not be demonstrated by anatomical techniques. Sequential filling and emptying of the air sacs would be another potential mechanism. Small pressure differences appear to exist between the cranial and caudal air sacs and they may play a role in the rectification of flow (Brackenbury, 1972 a, b). Scheid et al. (1972) have shown that the unidirectional dorsobronchial flow persists in the paralyzed, pump-ventilated, and even in the dead animal. Even isolated parts of the bronchial system, excised from an animal that was fixed by glutaraldehyde, showed partial flow rectification or direction-dependent airflow resistance. They inferred that some structure-specific aerodynamic mechanisms were involved. Similarly, Brackenbury (1972b) concluded from his experiments that aerodynamic mechanisms were involved in the rectification of respiratory flow. Recently, Banzett and co-workers have found experimental evidence as well as theoretical support for the importance of inertial forces in producing aerodynamic valving during inspiration, i.e. air flow via dorsobronchi-parabronchi-ventrobronchi instead of direct shunting from the main bronchi via ventrobronchi into the cranial air sacs (Banzett et al., 1987; Butler et al., 1988). Kuethe (1988) has derived evidence for the role of both inertial forces and pressure differences in producing unidirectional parabronchial flow from experiments on fixed lungs and physical airway models.

It is tempting to assume that the unidirectional flow in the parabronchi is of benefit for gas exchange with blood. This would in fact be the case if the parabronchial lung operated as a counter-current system as was proposed earlier (Schmidt-Nielsen, 1971). The actual arrangement, however, of blood and gas flow in the parabronchi suggests a different model for gas exchange, the cross-current, the efficiency of which does

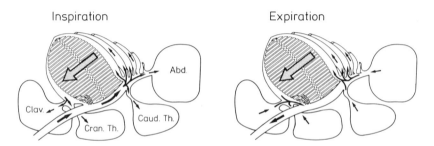

Fig. 2. Representation of air flow in the respiratory tract of birds during both phases of ventilation. (Modified from Scheid, 1979).

not depend on the direction of flow (see below). Thus, rectification of parabronchial gas flow is not required for the functioning of the gas exchange system. The following factors may constitute some advantages of unidirectional flow.

(1) The rates of drop in Po_2 and rise in Pco_2 during the flow stop produced by flow reversal could be intolerably high due to the small parabronchial gas volume. Direct evidence showed, however, that gas exchange with intermittent ventilatory flow was only slightly impaired compared with gas exchange during continuous ventilatory flow, probably because parabronchial gas was mixed during breath-holding with that in the adjoining airways by way of convection and diffusion (Scheid et al., 1977; Scheid, 1978b).

(2) During panting, air flow in the dorsobronchi, and thus in the lung, is still unidirectional (Bretz and Schmidt-Nielsen, 1971; Scheid and Piiper, 1971). In fact, the oscillations in air flow within a respiratory cycle appear to be greatly attenuated at high breathing rates. Reversal of flow at the high respiratory rates during panting could indeed be detrimental.

(3) Intrapulmonary CO_2 receptors seem to be located at the caudal (inflow) end of the parabronchus (Scheid et al., 1974a). At this location, the receptors are exposed to a rather low Pco_2 where their sensitivity curve is steep. Reversal of flow, on the other hand, would raise the Pco_2 in their microenvironment and would possibly shift the receptors into an unfavorable sensitivity range.

It is important to consider that all the above hypotheses and experimental evidence concerned the major, paleopulmonic, compartment of the avian lungs. Anatomical constraints dictate airflow in the neopulmonic parabronchi (see above) to reverse direction from inspiration to expiration In the duck, in which the neopulmo is only moderately developed, the neopulmonic gas exchange has been estimated at about 15% of the total pulmonary gas exchange (Piiper, 1978). The increased development of the neopulmo in the progressive songbirds seems to indicate that the unidirectionality of airflow in the parabronchi is not essential for achieving efficient gas exchange.

C. GAS EXCHANGE: CROSS-CURRENT MODEL

The unidirectional air flow in avian lungs suggested presence of a counter-current gas exchange system (Schmidt-Nielsen, 1971). This hypothesis was experimentally tested by measurement of gas exchange upon experimental reversal of parabronchial airflow in the duck (Scheid and Piiper, 1972). The striking result was that these partial pressures remained virtually unaffected when airflow was reversed, a result unexpected if the counter-current model was operative. Moreover, overlap between partial pressures in gas and blood was observed, particularly for CO_2, during both air flow directions. These results could be explained on the basis of an alternative model, the serial-multicapillary or cross-current model (Scheid and Piiper, 1972; Scheid, 1979) which had already been considered by Zeuthen (1942). In contrast to the counter-current hypothesis, this model is in agreement with the anatomical arrangement of blood capillaries (Duncker, 1974).

A schema of the cross-current model is shown in Figure 3. When air traverses the parabronchial tube, it is continually depleted of O_2 and enriched in CO_2 as it contacts the blood capillaries in serial order. The blood in the blood capillaries, in turn, will be arterialized to varying degrees, depending on their location along the parabronchial tube; the arterialization will be best at the gas inflow end of the parabronchus, and worst at its outflow end. Systemic arterial blood thus constitutes a mixtures of blood from all capillaries, and it is easy to conceive that an overlap in gas and blood partial pressures can occur, whereby, e.g., P_{O_2} in this arterial blood exceeds P_{O_2} in end-parabronchial gas. Hence, the cross-current model can, like the counter-current system, explain the experimentally observed fact of the gas/blood overlap. The cross-current system is flow direction-independent. Suppose that airflow is reversed in the system of Figure 3. As a result, the capillary that had been at the inflow end before would now be at the outflow end. This reversal of sequence of capillaries would affect the degree of arterialization of blood in individual capillaries. For the (mixed) arterial blood, however, no change would occur.

It is important to point out that frequently the blood-gas overlap is not observed, particularly not for O_2. This is attributable to factors known to produce inefficiency of pulmonary gas exchange in mammalian lungs: unequal distribution of ventilation to perfusion, shunt and diffusion limitation. Unequal distribution of ventilation to blood flow has

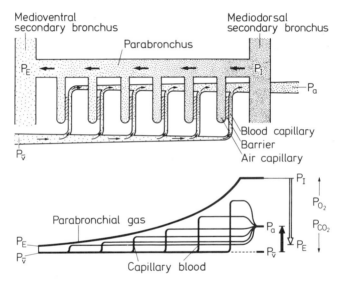

Fig. 3. Cross-current model for avian parabronchial gas exchange. Above, schema of parabronchus with radially departing air capillaries, and with blood capillaries running from the periphery towards the lumen of the parabronchus and contacting air capillaries. Below, partial pressure profile in the gas phase, from initial-parabronchial (P_I) to end-parabronchial values (P_E); and in blood of capillaries, the arterial value (P_a) resulting from mixing of blood from all capillaries. Arrows to the right show the overlap in gas (open arrow) and blood (closed arrow) partial pressures.

been quantified in duck lungs on the basis of analysis of O_2 and CO_2 exchange (Burger et al., 1979) and using the multiple inert gas elimination technique (Powell and Wagner, 1982a,b). Pulmonary diffusing capacity in duck lungs has been estimated by Burger et al. (1979) and by Geiser et al. (1984). Calculations on diffusion limitation in air capillaries have been performed by Scheid (1978a). Thus, although potentially the cross-current model allows a higher efficiency of gas exchange, whereby higher O_2 extraction and arterial P_{O_2} can be achieved for the same O_2 uptake as compared to mammalian lungs, this potential may not always be exploited. In any case, the high O_2 extractions observed in some birds exposed to cold (Brent et al., 1983; 1984) may be made possible by the cross-current system.

D. CO_2 TRANSFER: ENHANCEMENT BY HALDANE EFFECT IN CROSS-CURRENT MODEL

In alveolar lungs, the interaction among H^+ ions and O_2 in binding to hemoglobin, expressed by the Bohr-Haldane effect of blood, results in an increased gas exchange efficiency. This effect operates in the avian as it does in the mammalian lung. However, the Haldane effect may exert a further particular enhancement of CO_2 exchange in the parabronchial lung due to its peculiar structural arrangement.

The mechanisms are illustrated in Fig. 4. Due to its higher diffusivity in tissue and higher solubility in blood, CO_2 transfer in the initial (inflow) segments of the parabronchus is higher than O_2 transfer, i.e. the gas exchange ratio R is high. At the downstream end of the parabronchus the CO_2 transfer rate is much reduced, whereas the O_2 transfer is still high (due to the shape of the O_2 dissociation curve of the blood), i.e. the R value is very low. When R drops below about 0.3, the increase of P_{CO_2} due to the Haldane effect outweighs the decrease of P_{CO_2} due to CO_2 transfer, thus leading to an increase of blood P_{CO_2} compared to mixed venous P_{CO_2} (with R = 0, i.e. in absence of CO_2 transfer, a still higher increase in blood P_{CO_2} is reached). The end-parabronchial blood P_{CO_2} is essentially equal to end-parabronchial gas P_{CO_2}, i.e. to end-expired gas P_{CO_2}. Thus the resulting order of P_{CO_2} is: inspired < arterial < mixed venous < end-expired.

This sequence of P_{CO_2} was observed in chicken by Davies and Dutton (1975) and confirmed by Meyer et al. (1976). Davies and Dutton invoked the blood/gas P_{CO_2} difference assumed to occur as a result of the charged membrane effect, which had been postulated by Gurtner et al. (1969) to be present in buffer-perfused capillaries in mammalian lungs but questioned by others (reviewed by Piiper, 1986). Because the P_{CO_2} difference was abolished in rebreathing equilibrium, Meyer et al. disclaimed this hypothesis and proposed the above-described combined action of Haldane effect and cross-current system, which had been qualitatively predicted by Zeuthen (1942). This effect may be significant in allowing normal acid-base balance during reduced ventilation or in hypercapnia due to other factors. In many cases the observed P_{CO_2} sequence is inspired < arterial < end-expired < mixed-venous or even inspired < end-expired < arterial < mixed-venous. This can be attributed to distribution and diffusion limitations in lungs as discussed above for O_2 exchange.

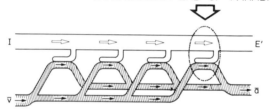
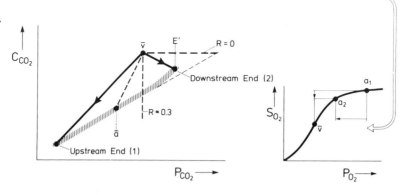

I. Gas exchange at low R

  a) CO_2 transfer much reduced (because close to completed upstream)
  b) Oxygenation little reduced (due to shape of blood O_2 dissociation curve)

II. End-parabronchial ≈ expired (E') P_{CO_2}

Fig. 4. Action of the Haldane effect in the cross-current model. Above, cross-current model, corresponding to Fig. 3. Lower left, CO_2 dissociation curves of blood. By oxygenation, blood CO_2 values are displaced from the mixed venous value (v̄) to end-capillary values lying on the shaded band from (1) to (2); the (mixed) arterial value (ā) is indicated. The downstream and P_{CO_2} ($P_{E'}$) is higher than $P_{\bar{v}}$ when the local gas exchange ratio (R) is below 0.3. Lower right, O_2 dissociation curve of blood, to show minor decrease of O_2 uptake along the parabronchus (S_{O_2} for a_2 being only slightly less than for a_1). (Modified from Scheid, 1979).

REFERENCES

Banzett, R.B., Butler, J.P., Nations, C.S., Barnas, G.M., Lehr, J.L., and Jones, J.H., 1987, Inspiratory aerodynamic valving in goose lungs depends on gas density and velocity, Respir. Physiol., 70: 287-300.

Bouverot, P. and Dejours, P., 1971, Pathway of respired gas in the air sacs-lung apparatus of fowl and ducks, Respir. Physiol., 13: 330-342.

Brackenbury, J.H., 1971, Airflow dynamics in the avian lung as determined by direct and indirect methods, Respir. Physiol., 13: 319-329.

Brackenbury, J.H., 1972a, Lung-air sac anatomy and respiratory pressures in the birds, J. Exp. Biol., 57: 543-550.

Brackenbury, J.H., 1972b, Physical determinants of air flow pattern within the avian lung, Respir. Physiol., 15: 384-397.

Brent, R., Rasmussen, J.G., Bech, C., and Martini, S., 1983, Temperature dependence of ventilation and O_2 extraction in the Kittiwake (Rissa tridactyla), Experientia, 39: 1092-1093.

Brent, R., Pedersen, P.F., Bech, C., and Johansen, K., 1984, Lung ventilation and temperature in the European coot Fulica atra, Physiol. Zool., 57: 19-25.

Bretz, W.L. and Schmidt-Nielsen, K., 1971, Bird respiration: Flow patterns in the duck lung, J. Exp. Biol., 54: 103-118.

Bretz, W.L. and Schmidt-Nielsen, K., 1972, The movement of gas in the respiratory system of the duck, J. Exp. Biol., 56: 57-65.

Burger, R.E., Meyer, M., Graf, W., and Scheid, P., 1979, Gas exchange in the parabronchial lung of birds: Experiments in unidirectionally ventilated ducks. Respir. Physiol., 36: 19-37.

Butler, J.P., Banzett, R.B., and Fredberg, J.J., 1988, Inspiratory valving in avian bronchi: aerodynamic considerations, Respir. Physiol., 72: 241-256.

Davies, D.G. and Dutton, R.E., 1975, Gas-blood Pco_2 gradients during avian gas exchange, J. Appl. Physiol., 39: 405-410.

Duncker, H.R., 1971, The lung air sac system of birds, Ergeb. Anat. Entwicklungsgesch., 45: Heft 6.

Duncker, H.R., 1972, Structure of avian lungs, Respir. Physiol., 14: 44-63.

Duncker, H.R., 1974, Structure of the avian respiratory tract, Respir. Physiol., 22: 1-19.

Fedde, M.R., 1976, Respiration, in: "Avian Physiology", P.D. Sturkie, ed., pp. 122-145, Springer, Berlin, Heidelberg, New York.

Fedde, M.R., Burger, R.E., and Kitchell, R.L., 1964, Electromyographic studies of the effects of bodily position and anesthesia on the activity of the respiratory muscles of the domestic cock, Poultry Sci., 43: 839-846.

Geiser, J., Gratz, R.K., Hiramoto, T., and Scheid, P., 1984, Effects of increasing metabolism by 2,4-dinitrophenol on respiration and pulmonary gas exchange in the duck, Respir. Physiol., 57: 1-14.

Gurtner, G.H., Song, S.H., and Farhi, L.E., 1969, Alveolar to mixed venous Pco_2 difference under conditions of no gas exchange, Respir. Physiol., 7: 173-187.

Kuethe, D.O., 1988, Fluid mechanical valving of air flow in bird lungs, J. Exp. Biol., 136: 1-12.

Macklem, P.T., Bouverot, P., and Scheid, P., 1979, Measurement of the distensibility of the parabronchi in duck lungs, Respir. Physiol., 38: 23-35.

Meyer, M., Worth, H., and Scheid, P., 1976, Gas-blood CO_2 equilibration in parabronchial lungs of birds, J. Appl. Physiol., 41: 302-309.

Piiper, J., 1978, Origin of carbon dioxide in caudal air sacs of birds, in: "Respiratory Function in Birds, Adult and Embryonic", J. Piiper, ed., pp. 148-153, Springer, Heidelberg, New York.

Piiper, J., 1986, Blood-gas equilibrium of carbon dioxide in lungs: a continuing controversy, J. Appl. Physiol., 60: 1-8.

Powell, F.L. and Scheid, P., in press, Physiology of gas exchange in the avian respiratory system, in: "Form and Function in Birds", A.S. King and J. McLelland, eds, Vol. 3, Academic Press, London.

Powell, F.L. and Wagner, P.D., 1982a, Measurement of continuous distributions of ventilation-perfusion in non-alveolar lungs, Respir. Physiol., 48: 219-232.

Powell, F.L. and Wagner, P.D., 1982b, Ventilation-perfusion inequality in avian lungs, Respir. Physiol., 48: 233-241.

Powell, F.L., Geiser, J., Gratz, R.K., and Scheid, P., 1981, Airflow in the avian respiratory tract: variations of O_2 and CO_2 concentrations in the bronchi of the duck, Respir. Physiol., 44: 195-213.

Scheid, P., 1978a, Analysis of gas exchange between air capillaries and blood capillaries in avian lungs, <u>Respir. Physiol.</u>, 32: 27-49.

Scheid, P., 1978b, Estimation of effective parabronchial gas volume during intermittent ventilatory flow: theory and application in the duck, <u>Respir. Physiol.</u>, 32: 1-14.

Scheid, P., 1979, Mechanisms of gas exchange in bird lungs, <u>Rev. Physiol. Biochem. Pharmacol.</u>, 86: 137-186.

Scheid, P., 1982, Respiration and control of breathing, <u>in</u>: "Avian Biology", D.S. Farner and J.R. King, eds., Vol. VI, pp. 405-453, Academic Press, New York, San Francisco, London.

Scheid, P. and Piiper, J., 1971, Direct measurement of the pathway of respired gas in duck lungs, <u>Respir. Physiol.</u>, 11: 308-314.

Scheid, P. and Piiper, J., 1972, Cross-current gas exchange in avian lungs: effects of reversed parabronchial air flow in ducks. <u>Respir. Physiol.</u>, 16: 304-312.

Scheid, P. and Piiper, J., 1987, Gas exchange and transport, <u>in</u>: "Bird Respiration", T.J. Seller, ed., Vol. I, pp. 97-129, CRC Press, Boca Raton FL.

Scheid, P. and Piiper, J., in press, Respiratory mechanics and air flow in birds, <u>in</u>: "Form and Function in Birds", A.S. King and J. McLelland, eds., Vol. 3, Academic Press, London.

Scheid, P., Slama, H., and Piiper, J., 1972, Mechanisms of unidirectional flow in parabronchi of avian lungs: measurements in duck lung preparations, <u>Respir. Physiol.</u>, 14: 83-95.

Scheid, P., Slama, H., Gatz, R.N., and Fedde, M.R., 1974a, Intrapulmonary CO_2 receptors in the duck. III. Functional localization. <u>Respir. Physiol.</u>, 22: 123-136.

Scheid, P., Slama, H., and Willmer, H., 1974b, Volume and ventilation of air sacs in ducks studied by inert gas wash-out. <u>Respir. Physiol.</u>, 21: 19-36.

Scheid, P., Worth, H., Holle, J.P., and Meyer, M., 1977, Effects of oscillating and intermittent ventilatory flow on efficacy of pulmonary O_2 transfer in the duck, <u>Respir. Physiol.</u>, 31: 251-258.

Schmidt-Nielsen, K., 1971, How birds breathe, <u>Sci. Am.</u>, No. 6, 225: 72-79.

Zeuthen, E., 1942, The ventilation of the respiratory tract in birds, <u>Kgl. Dans. Vidensk. Selsk. Biol. Med.</u>, 17: 1-50.

GAS EXCHANGE DURING COLD EXPOSURE IN

PEKIN DUCKS (ANAS PLATYRHYNCHOS)

Claus Bech and Hege Johannesen

Department of Zoology
University of Trondheim
N-7055 Dragvoll, Norway

INTRODUCTION

In homeotherms, the inhaled air is warmed during inspiration, while during expiration heat and water, to a variable degree, is lost via the exhaled air. Ventilation will consequently always exert an influence on body temperature regulation. Despite the importance of ventilation for both gas exchange and body temperature regulation, only little is known so far about the function of the gas exchange system at ambient temperatures which present a thermoregulatory challenge, i.e. temperatures beyond the thermoneutral range. Several studies have dealt with the respiratory responses to high ambient temperatures, e.g. panting, but surprisingly few studies have been concerned with respiration at low ambient temperatures.

Three factors determine the respiratory heat loss of homeotherms. These are: 1) the relative humidity of the expired air, 2) the temperature of the expired air, and 3) the volume of air concerned in ventilation. These three variables thus also represent the three potential parameters available to birds for active regulation of the heat and water losses during respiration. The most well known strategy for reducing the respiratory heat loss at low temperatures is a reduction in the temperature of the exhaled air. This drastically reduces the amount of heat and water lost via the respiratory tract (Schmidt-Nielsen et al., 1970).

The third variable, which could theoretically be used to reduce the respiratory losses of heat and water, is a reduction in the ventilatory requirement, i.e. an increase in the oxygen extraction rate (Eo_2). This would imply that the bird could ventilate less for any given amount of oxygen taken up. Fig 1 illustrates this potential by showing the relationship found between ventilation and oxygen uptake in the European Coot (Fulica atra). The increase in oxygen uptake during cold exposure in this species is not matched by a similar increase in total ventilation, thus resulting in an increased oxygen extraction value (Brent et al., 1984).

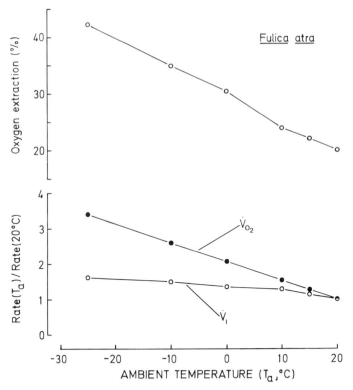

Fig. 1. Ventilation (V_I) and oxygen uptake (Vo_2) and the resultant oxygen extraction values (in %) for the European Coot plotted in relation to the ambient temperature (T_a). For comparative purposes, V_I and Vo_2 are expressed as a factor of the respective values at 20°C. Data from Brent et al. (1984).

That an increase in Eo_2 could function as a thermoregulatory mechanism at low ambient temperature in birds, was first mentioned by Bucher (1981) for the Linneated Parakeet (<u>Bolborynchus lineola</u>). She concluded that "these birds raised their mean Eo_2 thereby keeping their total ventilation as small as possible", and that: "elevated oxygen extraction efficiencies below T_{lc} help to reduce respiratory heat loss" (Bucher, 1981). Similar conclusions were drawn from the results of subsequent studies of the Kittiwake (<u>Rissa tridactyla</u>, Brent et al., 1983) and the Pekin duck (<u>Anas platyrhynchos</u>, Bech et al., 1984), as well as on the European Coot (Brent et al. 1984). However, some studies made of other species of birds have shown that no change occur in the ventilatory requirement at low ambient temperatures during cold exposure (Bernstein and Schmidt-Nielsen, 1974, Bouverot et al., 1976, Barnas and Rautenberg, 1984, Bech et al., 1985, Chappell and Bucher, 1987, Clemens, 1988, Stahel and Nicol, 1988a).

The techniques used to measure the ventilatory variables have not been standardized, however, in the different studies. In most studies, those in which an increase in oxygen extraction was demonstrated, the pneumotachographic method was used. It has consequently been claimed that the differences in cold response could have resulted because of the use of different methods (Barnas and Rautenberg, 1984). This was recently further expressed by Barnas and Rautenberg (1987), who stated that they were "unsure at this time whether these differences in responses depend on species or experimental methods." However, Bech et al. (1985), using the pneumotachographic method to study the ventilatory responses to cold in the pigeon, confirmed the results of previous studies based on other methods (Bouverot et al., 1976, Barnas and Rautenberg, 1984), viz. that in this species no change occurs in the oxygen extraction rate. Thus, at least in pigeons, the use of different methods to measure ventilation has no influence on the results. It is also of interest that, for the Pekin duck, no differences were found in the ventilatory parameters when measured by either the pneumotachographic method or the barometric method (Stahel and Nicol, 1988b).

The occurrence and importance of increased oxygen extraction at low ambient temperatures in birds is consequently still a matter of debate. To try to further clarify the changes occurring in the gas exchange system of ducks at low ambient temperatures, we therefore carried out a further series of experiments on Pekin ducks exposed to low ambient temperatures.

METHODS

Five domestic Pekin ducks were used in the study. Catheters were implanted, under a general anaesthesia (Equisithin, 2.5 ml kg^{-1}, given i.m.), in the brachial artery and vein. Additional catheters (PE90) were implanted into the abdominal and interclavicular air sacs.

During the experiments, the ducks were placed in a 17 litre capacity metabolic chamber, through which air was sucked. Oxygen uptake (V_{O_2}) was calculated from the air flow rate and the O_2-content of the effluent air, determined with a Servomex paramagnetic oxygen analyser (Type OA 272). Blood samples were drawn anaerobically into pre-heparinized 1 ml syringes and immediately analysed on Radiometer equipment (BMS3 Mk2). Air samples from the air sacs were slowly drawn into 100 ml glass syringes at a rate of 20 ml min^{-1} and the oxygen content determined by injecting the gas sample into a Servomex oxygen analyser. Resting values were first obtained at near thermoneutral conditions, i.e. around 20°C, whereafter the ambient temperature was lowered gradually (6-8 hours) to approximately -15°C. During such a cooling period, oxygen uptake was recorded at regular intervals and samples taken for measurement of blood gases and air sac oxygen content.

RESULTS AND DISCUSSION

A-V P_{O_2}-difference

One question that has been recently discussed (Barnas and Rautenberg, 1987), is whether any changes occur in the arterial-venous (A-V) oxygen difference. In pigeons, cold exposure is followed by a large increase in the A-V difference, mainly due in Pekin ducks, Bech et al. (1984) found that no change occurred in the A-V difference during cold exposure. In the present series of experiments on Pekin ducks this finding was confirmed. No significant relationship existed between ambient temperature and either arterial or venous P_{O_2}, mean arterial P_{O_2} being 77.2 torr (S.D.=9.7 torr, n=42) and mean venous P_{O_2} being 30.4 torr (S.D.=4.1 torr, n=33). Thus, the increase in oxygen uptake in response to cold exposure is mainly a result of an increase in cardiac output. It is not possible, at present, to decide whether the different response in the A-V-difference of pigeons and of ducks is related to the difference in their oxygen extraction responses to low ambient temperature.

Ventilatory oxygen extraction

The efficiency of the gas exchange system was assessed by calculating the oxygen extraction rate using the following formulae:

(I) $\qquad E_{O_2} = 100 \cdot (F_{in}O_2 - F_{ic}O_2)/F_{in}O_2$

(II) $\qquad E_{pb}O_2 = 100 \cdot (F_{ab}O_2 - F_{ic}O_2)/F_{ab}O_2$,

where $F_{ab}O_2$ and $F_{ic}O_2$ are the respective fractional oxygen contents of the abdominal and interclavicular air sacs, and $F_{in}O_2$ equals the fractional oxygen content of the inspired air. Equation I yields the total oxygen extraction rate (E_{O_2}). Assuming that the oxygen content of the interclavicular air sac equals that of the end-expired air (Powell et al., 1981), E_{O_2} is then the percentage of oxygen taken up from the inspired air during its way through the gas exchange apparatus. Equation II yields the parabronchial oxygen extraction ($E_{pb}O_2$), i.e. the percentage of oxygen taken up when the air flows from the abdominal air sac, through the parabronchi, into the interclavicular air sac.

There was a significant (p=0.01, linear regression) increase in E_{O_2} at low ambient temperatures, although a wide scatter in the values exists (fig. 2). This wide scatter is mainly due to differences in the oxygen uptake of the individual ducks at a given ambient temperature. This becomes evident from consideration of fig. 3, which indicates that a much stronger relationship (p<0.0001) exists between E_{O_2} and oxygen uptake. The increase in oxygen extraction found in response to cold exposure is thus more in the nature of a response to an increase in the oxygen uptake by the lung than an effect of the decrease in ambient temperature _per se_. The parabronchial oxygen extraction rate also increased during cold expose (fig. 4).

The actual oxygen contents of the abdominal and interclavicular air sacs are shown in fig. 5. While the O_2-content of the interclavicular air sac decreased, in consequence of an increase in E_{O_2}, the O_2-content of the abdominal air sac increased during cold exposure.

In the "classical" model of avian ventilation, air is believed to move from the abdominal air sac directly into the paleopulmonic parabronchi during exhalation, with only a very small proportion being directed out through the mesobronchus (Scheid, 1979, 1982). However, Hastings and Powell (1986)

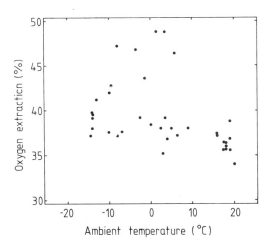

Fig. 2. Total oxygen extraction values for Pekin ducks plotted in relation to the ambient temperature (see text for the method used to calculate the oxygen extraction values).

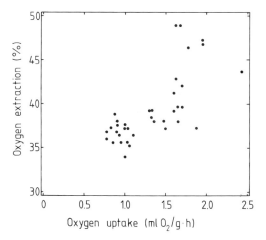

Fig. 3. Total oxygen extraction values for Pekin ducks plotted in relation to oxygen uptake (see text for details).

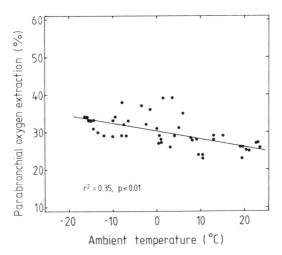

Fig. 4. Parabronchial oxygen extraction ($E_{pb}O_2$) values for Pekin ducks plotted in relation to the ambient temperature (see text for the method used to calculate the $E_{pb}O_2$ values). Line shows linear regression line.

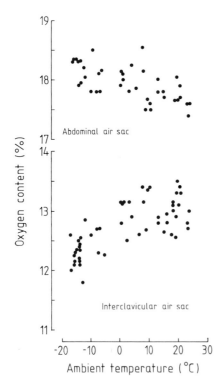

Fig. 5. Oxygen content (in %) of the abdominal and the interclavicular air sacs of Pekin ducks plotted in relation to the ambient temperature.

calculated the magnitude of this "mesobronchial shunt" in ducks and concluded that it may be greater than was previous believed. Regulation of the magnitude of the mesobronchial shunt could therefore form a mechanism for controlling effective parabronchial ventilation. If the magnitude of the mesobronchial shunt is decreased during cold exposure, this would mean that, during exhalation, a greater proportion of the air will be channelled through the paleopulmonic parabronchi during exhalation. The proportion of exhaled air passing through the mesobronchial shunt has been found to be greatest at the beginning of the expiration (Hastings and Powell, 1986). Consequently, the higher the tidal volume, the lower will the relative proportion of the mesobronchial shunt be. Because the increase in total ventilation during cold exposure in ducks is mainly due to an increase in tidal volume (Bech et al., 1984), the magnitude of the mesobronchial shunt may possibly be reduced during cold exposure. The effect of an increase in tidal volume would, in itself, be reflected as an increase in the oxygen content of the air in the abdominal air sac, which was in fact observed (fig. 5).

Oxygen diffusing capacity

The apparent oxygen diffusing capacity (Do_2) of the lung was calculated from $F_{ab}o_2$ and $F_{ic}o_2$ (expressed as partial pressures) and the arterial and venous Po_2 values, using equation 7 given in Burger et al. (1979). Fig. 6 shows that, as the oxygen uptake increases in response to low ambient temperature, there is a concomitant increase in the oxygen diffusing capacity. This increase in Do_2 does not necessarily reflect an increase in the true oxygen diffusing capacity of the lung, but may merely indicate that the apparent diffusing capacity increases in the direction of the true Do_2 value of the lung (Scheid, 1979).

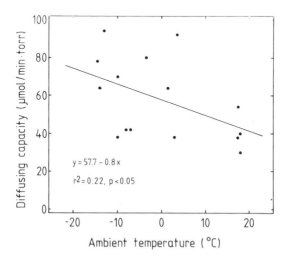

Fig. 6. Relationship between ambient temperature and the oxygen diffusing capacity of the lung of Pekin ducks. Line shows linear regression line.

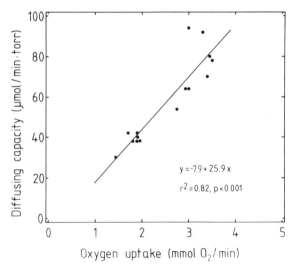

Fig. 7. Relationship between oxygen uptake and the oxygen diffusing capacity of the lungs of Pekin ducks. Increased oxygen uptake was induced by cold exposure. Line shows linear regression line.

Fig. 7 depicts the relationship between Do_2 and oxygen uptake. As in the case of oxygen extraction by the lung, the increase in the diffusing capacity of the lung is more strongly related to the increase in Vo_2 than to the decrease in ambient temperature.

The mechanism underlying an increase in the oxygen diffusing capacity of the lung is still unknown. One way of increasing the gas exchange efficiency could be to alter the relationship between ventilation and perfusion. Powell and Wagner (1982) found that, in a duck lung, there is a relatively large ventilatory shunt, of approx. 10%. If this is due to a number of unperfused, but nevertheless ventilated, parabronchi, then an increase in cardiac output, primarily "recruiting" new parabronchi to take part in gas exchange, could lead to an increase in the overall efficiency of gas exchange. Powell (1982) actually found, that in birds at rest some of the blood capillaries are in a collapsed state. It is of interest in this connection that in Pekin ducks, the increase found in O_2-uptake during cold exposure was mainly brought about by an increase in cardiac output (Bech et al., 1984).

An increase in oxygen uptake in ducks, whether induced by bodily activity (Kiley et al., 1985), by chemicals (Geiser et al., 1984, Hempleman and Powell, 1986) or by cold exposure (present study), is followed by an increase in the apparent diffusing capacity of the lung, possibly to prevent any limitation of gas exchange. In the present study, the slope of the regression line relating DO_2 to oxygen uptake (fig. 7), was 25.9 µmol/mmol/torr. This value is compared in table 1 with values derived from other studies on ducks. It is noteworthy, that, irrespective of the method used to induce an increase in the metabolic rate, the relationship between diffusing capacity and oxygen uptake is remarkably similar in all studies. This suggests that, for ducks, a constant relationship exists between the increase in oxygen diffusing capacity of the lung and the corresponding increase in oxygen uptake, and that this relationship is shown both during cold exposure as well as under other conditions that cause an increased oxygen demand.

Tab. 1. Diffusing capacity vs. oxygen uptake in ducks.

Condition	Slope (μmol/mmol/torr)	Reference
DNP-treated (A)	21.4	Hempleman and Powell (1986)
DNP-treated (A)	28.3	Geiser et al. (1984)
Exercise	18.4	Kiley et al (1985)
Cold exposure	25.9	This study

A: anaesthetized.

CONCLUSION

The thermoregulatory benefit gained from an alteration in the ventilatory requirement at low ambient temperatures may be limited. Thus, Chappell and Bucher (1987) in a study on the Chukar (*Alectoris chukar*), calculated that, during exposure to low ambient temperatures, only 2-3% of the total heat production would be saved even if the oxygen extraction rate was doubled. However, it now seems proven beyond doubt that, whether it has a thermoregulatory significance or not, the efficiency of the gas exchange system is increased at low ambient temperatures, at least in the Pekin duck. Such an increase in efficiency could be due to an alteration in the ventilatory pattern, or to a reduction in the mesobronchial shunt, or to an increase in the oxygen diffusing capacity of the lung.

REFERENCES

Barnas, G.M., and Rautenberg, W., 1984, Respiratory responses to shivering produced by external and central cooling in the pigeon. Pflügers Arch. ges. Physiol. 401:228.
Barnas, G.M., and Rautenberg, W., 1987, Temperature control, in: "Bird Respiration", T.J. Seller, ed., CRC Press, Boca Raton, Florida, Vol. I, p. 131.
Bech, C., Johansen, K., Brent, R., and Nicol, S., 1984, Ventilatory and circulatory changes during cold exposure in the Pekin duck. Respir. Physiol. 57:103.
Bech, C., Rautenberg, W., and May, B., 1985, Ventilatory oxygen extraction during cold exposure in the Pigeon (*Columba livia*). J. Exp. Biol. 116:499.
Bernstein, M.H., and Schmidt-Nielsen, K., 1974, Ventilation and

oxygen extraction in the crow. Respir. Physiol. 21:393.

Bouverot, P., Hildwein, G., and Oulhen, Ph., 1976, Ventilatory and circulatory O_2 convection at 4000m in pigeon at neutral or cold temperature. Respir. Physiol. 28:371.

Brent, R., Rasmussen, J.G., Bech, C., and Martini, S., 1983, Temperature dependence of ventilation and O_2-extraction in the Kittiwake Rissa tridactyla. Experientia 39:1092.

Brent, R., Pedersen, P.F., Bech, C., and Johansen, K., 1984, Lung ventilation and temperature regulation in the European Coot Fulica atra. Physiol. Zool. 57:19.

Bucher, T.L., 1981, Oxygen consumption, ventilation and respiratory heat loss in a parrot, Bolborynchus lineola in relation to ambient temperature. J. Comp. Physiol. 142:479.

Burger, R.E., Mayer, M., Graf, w., and Scheid, P., 1979, Gas exchange in the parabronchial lung of birds: experiments in unidirectionally ventilated ducks. Respir. Physiol. 36:19.

Chappell, M.A., and Bucher, T.L., 1987, Effects of temperature and altitude on ventilation and gas exchange in chukars (Alectoris chukar). J. Comp. Physiol. 157:129.

Clemens, D.T., 1988, Ventilation and oxygen consumption in rosy finches and house finches at sea level and high altitude. J. Comp. Physiol. 158:57.

Geiser, J., Gratz, R.K., Hiramoto, T., and Scheid, P., 1984, Effect of increasing metabolism by 2,4-dinitrophenol on respiration and pulmonary gas exchange in the duck. Respir. Physiol. 57:1.

Hastings, R.H., and Powell, F.L., 1986, Single breath CO_2 measurements of deadspace in ducks. Respir. Physiol. 63:139.

Hempleman, S.C., and Powell, F.L., 1986, Influence of pulmonary blood flow and O_2 flux on DO_2 in avian lungs. Respir. Physiol. 63:285.

Kiley, J.P., Faraci, F.M., and Fedde, M.R., 1985, Gas exchange during exercise in hypoxic ducks. Respir. Physiol. 59:105.

Powell, F.L., 1982, Diffusion in avian lungs. Fed. Proc. 41: 2131.

Powell, F.L., and Wagner, P.D., 1982, Ventilation-perfusion inequality in avian lungs. Respir. Physiol. 48:233.

Powell, F.L., Geiser, J., Gratz, R.K., and Scheid, P., 1981, Air flow in the avian respiratory tract: variations of the O_2 and CO_2 concentrations in the bronchi of the duck. Respir. Physiol. 44:195.

Scheid, P., 1979, Mechanisms of gas exchange in bird lungs. Rev. Physiol. Biochem. Pharmacol. 86:137.

Scheid, P., 1982, Respiration and control of breathing, in: "Avian Biology", D.S. Farner and J.R. King, eds., Vol. VI, Academic Press, New York, p. 406.

Schmidt-Nielsen, K., Hainsworth, F.R., and Murrish, D.E., 1970, Counter-current heat exchange in the respiratory passages: effect on water and heat balance. Respir. Physiol. 71:387.

Stahel, C.D., and Nicol, S.C., 1988a, Ventilation and oxygen extraction in the Little Penguin (Eudyptula minor), at different temperatures in air and water. Respir. Physiol. 71:387.

Stahel, C.D., and Nicol, S.C., 1988b, Comparison of barometric and pneumotachographic measurements of resting ventilation in the little penguin (Eudyptula minor). Comp. Biochem. Physiol. 89A:387.

RESPIRATORY RESPONSES OF THE MALLARD

TO EXTERNAL AND INTERNAL COOLING

Hege Johannesen and Claus Bech

Department of Zoology
University of Trondheim
N-7055 Dragvoll, Norway

INTRODUCTION

In some species of birds, simultaneous measurements of ventilation and oxygen consumption have shown that oxygen extraction increases with the increase in metabolic rate induced by cold exposure (Bucher, 1981; Brent et al., 1983, 1984; Bech et al., 1984). In addition, measurements of the oxygen content in the interclavicular air sac of Pekin ducks, indicate that oxygen extraction improves at low ambient temperatures (Bech et al., 1984). A more efficient gas exchange under cold conditions has an obvious thermoregulatory significance, because a decreased air convection requirement implies a reduction in the respiratory heat and water losses.

It is well documented that both total body cooling and spinal cooling induce appropriate thermoregulatory responses in birds (Simon et al., 1986). However, selective cooling of the spinal cord of Pekin ducks had no effect on oxygen extraction (Johannesen and Bech, 1988). In the present investigation we used an oesophageal thermode to study the effect of selective stimulation of peripheral, deep-body, temperature sensors. This method has previously been used by Mercer (1985) to determine the total body thermosensitivity of Eider ducks.

MATERIAL AND METHODS

Four male Mallards (Anas platyrhynchos) with body weights of between 1.2 and 1.5 kg were used in this study. The birds were housed indoors at room temperature and exposed to a 12 h light-12 h dark cycle. Prior to experimentation, the right abdominal air sac and the interclavicular air sac were cannulated, under a general anaesthesia (Equithesin, 2.5 ml kg^{-1} given i.m.). The catheters used were made from flexible Tygon tubing (i.d.=1.6 mm, o.d.=3.2 mm).

The experiments took place in a constant temperature cabinet, where the birds were exposed to either thermoneutral (25°C) or to cold (5°C) ambient conditions. The birds were placed in a 17 l plexiglass chamber, through which dry air was sucked at a rate of 7 l min^{-1}. The effluent air was dried and passed into an oxygen analyser (S-3A, Applied Electrochemistry). The output of the oxygen analyser was recorded continuously. By assuming an R_E of 0.75, oxygen consumption was calculated according to Withers (1977), and the values obtained converted to metabolic heat production (M) values. Body temperature (T_b) was measured in the colon, using a copper-constantan thermocouple inserted 9 cm into the colon through the cloaca. The fractional contents of oxygen in the two air sacs and in the inspired air were measured in 90 ml gas samples, drawn off by syringes at a rate of 30 ml min^{-1}. The gas samples were immediately injected into an oxygen analyser (Serwomex 1100 A). The overall oxygen extraction (E_{O_2}), i.e. the ratio of the amount of oxygen used to the amount of oxygen available, was calculated from simultaneous measurements of the fractional contents of oxygen in the inspired air and in the interclavicular air sac gas. The fractional contents of oxygen in the air sac samples were converted to partial pressures of oxygen (P_{O_2}), using the body temperature recorded during the gas sampling.

To achieve selective internal cooling, a hairpin-shaped thermode, made of polypropylene tubing (Portex, PP 190), was inserted 25 cm into the oesophageus, via the oral cavity. The thermode was perfused with thermostatically controlled water (5°C) at a rate of 14-22 ml min^{-1}. The temperatures of the inlet and outlet water were recorded, using chromel-alumel thermocouples placed within the thermode, just where it entered the bill.

RESULTS

External cooling

The Mallards maintained a stable body temperature (mean value 41.6°C) at ambient temperatures of 5°C and 25°C. Tab. 1 shows the partial pressures of oxygen in the right abdominal ($P_{ab}O_2$) and interclavicular ($P_{ic}O_2$) air sac, the overall oxygen extraction values and the metabolic heat production recorded at the thermoneutral temperature (25°C) and during cold stress (5°C). The $P_{ab}O_2$ value during cold exposure was significantly

Tab. 1. Respiratory and metabolic parameters in Mallards after a 2 h exposure to ambient temperatures of 5°C and 25°C. n = number of experiments.

Measurements	$T_a = 5°C$		$T_a = 25°C$	
	n	Mean±2SEM	n	Mean±2SEM
$P_{ab}O_2$ (torr)	9	140.3±0.7	6	106.9±3.1
$P_{ic}O_2$ (torr)	10	81.1±1.3	24	93.2±3.0
E_{O_2} (%)	10	42.8±1.1	24	32.5±2.4
M (W kg^{-1})	14	9.09±0.61	24	8.81±0.68

higher than at the thermoneutral temperature, but, more importantly, the mean $P_{i_C}o_2$ value decreased by 13% compared to the value recorded at 25°C, indicating a 32% increase in overall oxygen extraction during cold exposure.

Internal cooling

In these experiments, which were carried out at an ambient temperature of 25°C, the oesophageal thermode was perfused with water for periods of 30 min duration. During the internal cooling, heat was extracted at a rate of 8.90 W kg^{-1} (n=9, 2SEM= 0.95). Fig. 1 shows the results obtained in 9 experiments. At the start of thermode perfusion, a rapid rise in metabolic heat production occured, from 8.18 W kg^{-1} to 13.54 W kg^{-1}, followed by a slight fall in T_b. There were no significant changes in either M or T_b during further stimulation. The mean changes in $P_{ab}o_2$ and $P_{i_C}o_2$, measured 5 min after the onset of cooling, were an increase of 10% and a decrease of 7%, respectively. The concomitant change in Eo_2 was a rise from 32.8% to 37.3%, representing a 14% increase compared to the precooling value.

DISCUSSION

So far, results of direct measurements of oxygen extraction from the fractional contents of oxygen in the inspired air and in end-expired gas, have only been reported for experiments on Pekin ducks (Bech et al., 1984; Johannesen and Bech, 1988). The mean $P_{i_C}o_2$ and Eo_2 values at thermoneutral conditions for the Pekin ducks are in close agreement with the Mallard values shown in Tab. 1. In addition, the changes in $P_{i_C}o_2$ and Eo_2, which in the present study were induced by lowering T_a from 25°C to 5°C, are of the same magnitude as those found by Bech et al. (1984) for Pekin ducks when T_a was lowered from +20°C to -20°C. As illustrated in fig. 1, selective internal cooling in the Mallard induced similar, although less pronounced, changes in overall oxygen extraction as those induced by external cooling.

According to the present results, it is tempting to assume that the increase in gas exchange efficiency seen in ducks during cold exposure, is a result of the change in input from temperature sensors. During oesophageal cooling a decrease in T_b was always recorded, but the change was only of the order of -0.2°C, and there were no significant differences between the values recorded at 5°C and 25°C. Thus, changes in the deep body temperature are not very likely to be involved in eliciting the observed changes in Eo_2. Because the oesophageal thermode had been positioned in close vicinity to the spinal cord, changes in the spinal temperature may well have occured during internal cooling. In all the species of birds so far studied, thermal stimulation of the spinal cord is followed by appropriate thermoregulatory effector mechanisms (Simon et al., 1986), but spinal cooling had no effect on the $Eo2$ values of Pekin ducks (Johannesen and Bech, 1988). However, some of the increase in metabolic heat production during oesophageal cooling should probably be ascribed to thermal stimulation of the spinal cord.

Mercer (1985) concluded that the increase in metabolic heat production noted during oesophageal cooling in the Eider duck was due to stimulation of peripheral, deep-body, temperature sensors. The existence of thermosensory structures in the

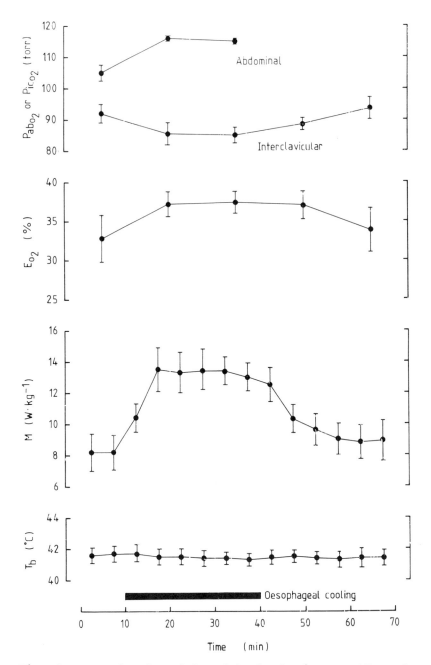

Fig. 1. Po_2 in the right abdominal air sac ($P_{ab}o_2$), Po_2 in the interclavicular air sac ($P_{ic}o_2$), parabronchial oxygen extraction (Eo_2), metabolic heat production (M) and body temperature (T_b) before, during and after a 30 min period of oesophageal cooling. Data shown are means±2SEM, based on the results of 9 experiments on 4 Mallards at an ambient temperature of 25°C.

deep-body tissues outside of the central nervous system has also been proven in the Goose (Helfmann et al., 1981) and in the Pekin duck (Inomoto and Simon, 1981), but their exact location has not yet been determined. In contrast, both warm- and cold-sensitive vagal afferents have been identified neurophysiologically in the mammalian digestive tract (Simon et al. 1986). According to El Ouazzani (1984), these vagal receptors do not respond to mechanical or chemical stimuli, but only to thermal stimuli, and can thus be considered to be true temperature receptors.

With respect to the importance of the ambient air temperature, Bech et al. (1988) found that the pigeon responds to selective cooling of the inspired air with a significant increase in metabolic heat production. It is therefore possible that peripheral temperature sensors, that react to changes in the temperature of the inspired air, are involved in mediating the observed changes in Eo_2. Sant'Ambrogio et al. (1985) have demonstrated the presence of cold receptors in the larynx of the dog. For all the receptors studied, the discharge rate was inversely related to the laryngeal temperature, independent of the air flow. If there are corresponding cold receptors present in the upper airways of birds, such receptors would be stimulated both by breathing in cold air and by experimental oesophageal cooling. The temperature of the inlet water during thermode perfusion in the Mallards was 5-7°C, which is close to the ambient air temperature experienced during external cooling. The greater effect of the external cooling, compared to internal, can be explained by assuming a strong influence of the general peripheral thermal state on oxygen extraction.

The higher partial pressure of oxygen in the abdominal air sac, recorded during both external and internal cooling, is probably due to an alteration in the ventilatory pattern. This implies that the increase in total ventilation is almost exclusively the result of an increased tidal volume, causing a smaller proportion of the total ventilation to be dead space ventilation. Such an effect of cold exposure on the ventilatory pattern in birds has been described previously in the kittiwake (Brent et al., 1983), the European coot (Brent et al., 1984) and the Pekin duck (Bech et al., 1984).

ACKNOWLEDGEMENT

This study was supported by grant D.65.46.138 from the Norwegian Research Council for Science and the Humanities.

REFERENCES

Bech, C., Johansen, K., Brent, R., and Nicol, S., 1984, Ventilatory and circulatory changes during cold exposure in the Pekin duck Anas platyrhynchos, Respir. Physiol., 57:103.

Bech, C., Rautenberg, W., and May-Rautenberg, B., 1988, Thermoregulatory responses of the pigeon (Columba livia) to selective changes in the inspired air temperature, J. Comp. Physiol., 157:747.

Brent, R., Pedersen, P. F., Bech, C., and Johansen, K., 1984, Lung ventilation and temperature regulation in the European coot Fulica atra, Physiol. Zool., 57:19.

Brent, R., Rasmussen, J. G., Bech, C., and Martini, S., 1983, Temperature dependence of ventilation and O_2-extraction in the kittiwake, Rissa tridactyla, Experientia, 39:1092.

Bucher, T. L., 1981, Oxygen consumption, ventilation and respiratory heat loss in a parrot, Bolborhynchus lineola, in relation to ambient temperature, J. Comp. Physiol., 142:479.

El Ouazzani, T., 1984, Thermoreceptors in the digestive tract and their role, J. Auton. Nerv. Syst., 10:246.

Helfmann, W., Jannes, P., and Jessen, C., 1981, Total body thermosensitivity and its spinal and supraspinal fractions in the conscious goose, Pflügers Arch., 391.60.

Inomoto, T., and Simon, E., 1981, Extracerebral deep-body cold sensitivity in the Pekin duck, Am. J. Physiol., 241:R136.

Johannesen, H., and Bech, C., 1988, Parabronchial oxygen extraction in ducks during selective cooling of the spinal cord, Acta Physiol. Scand., 132:563.

Mercer, J. B., 1985, Use of an oesophageal thermode for the determination of total body thermosensitivity in conscious Eider ducks (Somateria molissima), Acta Physiol. Scand., 124:S542.

Sant'Ambrogio, G., Mathew, O. P., Sant'Ambrogio, F. B., and Fisher, J. T., 1985, Laryngeal cold receptors, Respir. Physiol. 59:35.

Simon, E., Pierau, F.-K., and Taylor, D. C. M., 1986, Central and peripheral thermal control of effectors in homeothermic temperature regulation, Physiol. Rev., 66:235.

Withers, P. C., 1977, Measurements of Vo_2, Vco_2, and evaporative water loss with a flow-through mask, J. Appl. Physiol. 42:120.

THE RESPIRATORY PATTERN AND EXPIRATORY GAS CONCENTRATIONS

IN TORPID HUMMINGBIRDS COLIBRI CORUSCANS

Martin Berger and Kjell Johansen*

Westf. Museum für Naturkunde, Münster, Germany, and
Dept. Zoophysiol., Univ. of Aarhus, Aarhus, Denmark

INTRODUCTION

Most of the investigations on respiration in hummingbirds refer to oxygen consumption. Meanwhile there are many results about the metabolism during rest, flight and torpor in different thermal environment.

The need for oxygen in special situations requires adaptations of the respiratory and circulatory system to guarantee the necessary oxygen uptake and transport. Respiration frequency and ventilation have been determined in a few special cases: torpor (by plethysmography, Withers, 1977), hovering flight at different ambient temperatures (Berger and Hart, 1972) and at high altitude conditions (Berger, 1978). Much more informations can be obtained from direct measurements of the expiratory gas concentrations of single breaths. We were able to obtain such data by using a mass-spectrometer in unrestrained torpid birds. We obtained directly breath by breath expiratory carbon dioxide and oxygen content which allows the calculation of oxygen extraction, and with known oxygen consumption also pulmonary ventilation and tidal volume can be determined.

In this paper special interest is paid to torpor with its different stages and arousal with its high differences in oxygen consumption. The experiments require simultaneous and continuous measurement of the body temperature to be informed of the thermal stage of the bird at any time.

METHODS

We used birds that were imported from Ecuador and kept in indoor aviaries. The birds - Sparkling violeteat, Colibri coruscans - were fed with artificial nectar containing sugar, proteins and vitamins. Additionally they obtained regularly fruit flies. The weight of the 4 birds ranged between 6.5 and 8 g.

To collect continuously the expired air, a thin polyethylene tubing (inner diameter 0.3 mm) was bent over hot air and glued on the hummingbird's

---

* In common experiments at the university of Aarhus, Kjell Johansen, the "Viking and physiologist" gave invaluable stimuli to our investigations on hummingbirds. Kjell Johansen died in 1987.

beak. The end was about 0.5 - 1.0 mm inside the nostril opening that is protected by a skin fold and feathers (fig.1).

The flow rate through the tubing was about 4 ml/min which is in general small enough to ensure that during expiration no ambient air is collected. A resting bird (BW 7 g) with an oxygen uptake of 7 ml/(g.h) at an ambient temperature of 20°C and with an oxygen extraction coefficient of 0.3 would ventilate 7x7:(60x0.21x0.3) = 13 ml air per minute. This gives a total flow rate of about 26 ml/min (assuming equal inspiration and expiration times) and about 13 ml/min through each nostril. There are, however, cases in which the gas concentrations changed appreciably during expiration which might be caused by a reduced expiratory flow. This happened in long breaths in non-regulated torpor (see results).

Fig. 1. A Colibri coruscans with the polyethylene tubing at the nostril to collect the expired air.

The mass-spectrometer was adjusted to shorter response time to obtain also data at higher respiration frequencies. The oxygen consumption was measured in separate experiments by open system respirometry using a Servomex oxygen analyzer and Rota flowmeters. During torpor and arousal a funnel placed over the bird's head gave immediate response and accurate oxygen consumption data. The basis for comparison was the body temperature which was measured simultaneously in all experiments by thin thermocouples (copper-constantan) which were placed subcutaneously laterally over the pectoral muscle under the wing. The comparison is justified by a strong correlation between body temperature and oxygen consumption during arousal and by constant conductance values in regulated torpor (Berger and Johansen, in prep.).

The difference between inspired and expired oxygen content (DiffPO2) is measured, calibrated (dry gas mixtures, Woesthoff pumps) and expressed as mm Hg of oxygen. It is regarded that a lowered body temperature will lower the water pressure in the lungs and hence increase the inspired oxygen pressure PIO2 in the lungs. The PIO2 in a 16°C bird is 7.6 mm higher than at 38°C Consequently for the same expired PO2, the 16°C bird has a (5%) oxygen extraction value than the 38°C bird. The consequences of this effect have not been worked out here but they can be important for oxygen transport by blood in torpor.

In this paper the following abbreviations and dimensions are used:

oxygen consumption	VO2	ml(STPD)/(gxh)
respiration frequency	fR	1/min
oxygen pressure difference	DiffPO2	mm Hg
oxygen extraction	E = DiffPO2/PIO2	x100, %
ventilation	V = VO2(BTPS)xPB/DiffPO2	ml(BTPS)/(gxh)
tidal volume	VT = 16.7 V(BTPS)/fR	μl(BTPS)/g .

We wish to express our thanks to the DFG (Deutsche Forschungsgemeinschaft) for the financial support (Be 536) and appreciate the skilfull technical assistance of Einer Larsen.

RESULTS ANS DISCUSSION

According to metabolism VO2 and body temperature TB different stages of torpor can be distinguished. A semi-schematic diagram (fig.2) shows that during entry into torpor (stage 1) oxygen consumption is progressively reduced. It does not reach immediately its lowest value, but much earlier than does body temperature.

Non-regulated torpor (stage 2) occurs at medium or higher ambient temperatures TA, in C. coruscans at 16°C or higher. The body temperature is kept only slightly above ambient temperature (less than 2°C) and oxygen consumption shows low values of 1 ml/(gxh) or less.

If ambient temperature decreases below a certain level (about 16°C) the bird's body temperature does not follow, a behavior first described in hummingbirds by Hainsworth and Wolf (1970). Thus if in an experiment TA is lowered the point where regulation starts is indicated by increased oxygen uptake. This regulated torpor (stage 3) can occur during a whole torpor period at low TA but it can also be induced during non-regulated torpor by changing TA. The level of regulating appears to be characteristic for a given species (Berger and Johansen). The lower the ambient temperature the higher will be the oxygen consumption which can supercede resting values at higher TA.

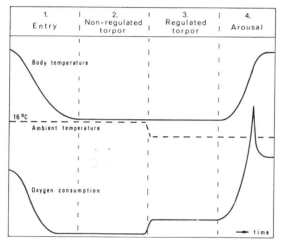

Fig. 2. The different stages of torpor with corresponding body temperatures and oxygen consumption. Note the change in ambient temperature between stage 2 and 3.

During arousal (stage 4) heat production is increased progressively and at the end oxygen consumption reaches a brief peak that can supercede 20 ml/(gxh). The body temperature increases rapidly and that much faster than it decreases during entry into torpor. Peak oxygen consumption occurs at body temperatures of about 35°C.

The description of the respiratory pattern refers to fig.3. When the bird starts <u>entering torpor</u> first the respiration frequency is gradually reduced along with the reduction of oxygen consumption. The oxygen extraction of respired air shows only slight decreases. At a body temperature of 30°C respiration frequency has dropped from over 100/min at TA = 24°C to less

181

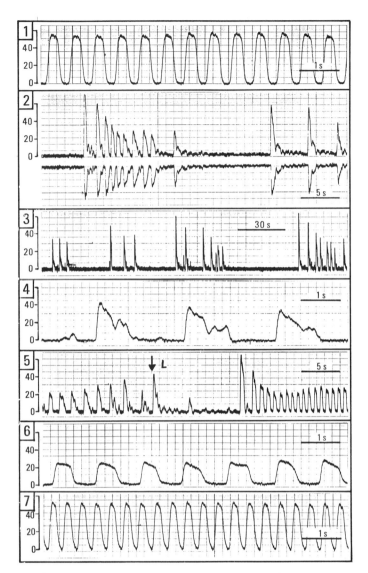

Fig. 3. Patterns of the respiratory oxygen content (in no.2 also carbon dioxide content) in single breaths, given as difference (mm Hg of oxygen) between inspired and expired air. Expiration = upward deflection.
1. Resting before entering torpor. TA = 18.0°C, TB = 34.9°C.
2. Entering torpor, heat production greatly reduced, phasic respiratory pattern. Note the high DiffPO2 (low PEO2) in the first breath after breathholding, the decrease in DiffPO2 in consecutive breaths and the relatively small change in DiffPCO2 (lower curve). TA = 12.8°C, TB = 25.0°C.
3. Torpor, not regulated. Apnea may last until 1.5 minutes. TA = 14.2°C, TB = 16.3°C.
4. Regulated torpor. TA = 11.6°C, TB = 15.9°C.
5. The respiratory pattern during regulated torpor is stopped when the light is switched on. After a short apnea the arousal is started by a regular breathing pattern. TA = 16.5°C, TB = 20.0 °C.
6. Arousal respiration at low TB (before the steep increase in TB). TA = 15.1°C, TB = 19.3°C.
7. Arousal respiration at high TB. TA = 16.6°C, TB = 33.0°C.

than half of the resting value. DiffPO2 is about 40 mm Hg compared to about 50 mm Hg at the start of entering torpor.

If heat production becomes more and more reduced finally the respiratory pattern can become irregular. This phasic pattern (fig.3.2) can start at body temperatures between 30 °C and 25 °C. The intervals between two breaths may last over 1 minute.

Non-regulated torpor (fig.3.3) is not very different from the previous pattern of the last part of entering torpor (note the different time scale between the figs. 3.2 and 3.3). A mean value for respiration frequency and DiffPO2 can hardly be given and therefore no estimation of ventilation and tidal volume was made. In this stage highest values of oxygen extraction are found (they reach 47 %), even if the arithmetic mean would give a lower value than during rest at moderate TA. The data demonstrate clearly that the peak values of oxygen extraction are found after periods of apnea (fig.4) and that in consecutive breaths the extraction is lowered. The shaded area of fig.4 gives the range of DiffPO2 if the interval between two expirations (inspiration included) is only 1 s or less. The lower values (dark shaded area) are from series from breaths following each other. The data of the light shaded area from second or third expirations after a breath with very high DiffPO2 after long apnea.

The high DiffPO2 values indicate a high depletion of oxygen reserves in the blood. Since during apnea the heart beats do not cease the whole blood's oxygen reserves are involved. If as in flight one heart stroke can transport 4 % of the whole blood, then with 30 beats/min in deep torpor the deoxygenated blood would circulate in the whole body in less than 1 min. We need confirmation for such calculations but can conclude a high tolerance to hypoxia and possibly regional differences in blood flow in the body (In separate experiments resting birds were able to extract from inspired oxygen of 50 mm Hg still 20 mm Hg, giving an expired oxygen pressure of 30 mm Hg).

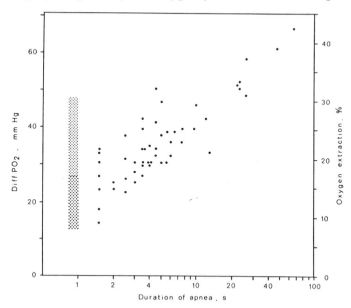

Fig. 4. DiffPO2 and oxygen extraction of single breaths in relation to the duration of apnea before the breath. Non-regulated torpor (cf. fig. 3.3). TA = 15.0 °C, TB = 16.4 °C. Shaded areas give the range for intervals of less than 1 s between two expirations (inspiration included).

In regulated torpor (fig.3.4) the respiration frequency is clearly lower than in resting birds. During one expiration expired oxygen content is lowest at the beginning and is increasing towards the end. The uneven coursa may be caused (additionally ?) by alterations in the expiratory flow rate. If this decreases below the flow rate in our PR-tubing (about 4 ml/min) we will suck in ambient air during expiration. The estimate for expiratory flow rate from oxygen consumption and peak DiffPO2 gives a value of about 4 ml/min, but we cannot exclude a high flow rate of the beginning of expiration and a lower one at the end. Further experiments must clearify also the possibility that (due to slow blood flow in the lungs ?) oxygen uptake from parabronchial air becomes reduced within the course of the expiration. Concerning the peak values of DiffPO2 and oxygen extraction at the beginning of the expiration the data are correct.

Arousal from torpor can occur spontaneously in the early morning hours but can also be induced in the experiment by light, noise or increased ambient temperature. The light induced aroesal starts immediately with a regular breathing pattern and with an increase in respiration frequency (fig.3.5). At 16.5 °C ambient temperature and 20.0 °C body temperature the torpid bird

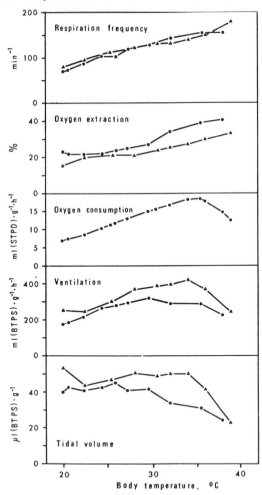

Fig. 5. Respiration frequency, oxygen extraction, oxygen consumption, ventilation and tidal volume at increasing body temperatures during arousal from torpor. 2 experiments; each point is the mean of 10 breaths.

has a respiration frequency of 40/min, and this can be switched immediately to 80/min. Frequency and DiffPO2 increase nearly linearly with body temperature in the course of arousal (fig.5). They continue to increase somewhat even if the oxygen consumption (and heat production) is reduced after a high peak to prevent overheating.

The convection requirement indicates how many liters (BTPS) of air have to be ventilated for the consumption of 1 mMol of oxygen. During arousal the values were lowest at high body temperatures. The convection requirement was falling during the course of arousal from about 0.80 to 0.35 µl -ir/mMol O2.

The respiration frequency during entering torpor is much lower than during arousal at the same body temperatures. At TB = 35 °C during entering torpor the respiration frequency is between 80 and 120/min, at TB = 30 °C between 40 and 70/min which is less than half of the arousal value (cf. fig. 5). Oxygen extraction values, however, are similar during entering torpor and arousal. At a TB of 35 °C the birds entering torpor have values between 30 % and 38 %, at 30 °C between 23 % and 29 %. (During entering torpor no precise data of oxygen consumption were obtained which prevents calculations of ventilation and tidal volume.)

When comparing our results with previous determinations of the ventilation we find some agreement even if completely different methods were used. Our calculations of the tidal volume of 40 - 50 µl/g (BTPS) during arousal correspond to 51 µl/g in rest at TA = 20 °C and to 80 µl/g in torpor at 20°C in Selasphorus sasis and Calypte anna, determined by plethysmography (Withers, 1977). Also the flight data of Colibri coruscans of 47 µl/g (sea level) and 56 µl/g (high altitude), calculated from expired water loss (Berger, 1978) are in this range, whereas a tidal volume of 110 µl/g in hovering Amazilia fimbriata (Berger and Hart, 1972) differs quite a lot. The low oxygen extraction values of 8.6 % for rest and 14 % for torpor in S. sasin ans C. anna (Withers, 1977) may reflect the difficulties in determining small volumes by plethysmography especially at small temperature differences.

With the analysis of the morphometrical data of the lung system in Colibri coruscans (Dubach, 1981) we find that a tidal volume of 50 µl/g makes 23 % of the total air sac volume of 1.59 ml. The lung air volume was determined to be 0.27 ml which is lower (by about 10 - 25 %) than the tidal volume. If in the unidirectional parabronchial air flow all respired air should pass the respiratory surface (either before entering the air sacs during inspiration or before leaving the lung during expiration) the tidal volume must be higher than lung air volume.

It has been discussed that in birds resting in the cold a high oxygen extraction and reduced ventilation will reduce the cooling of the body by the ventilated air (Bech et al., 1984, Brent et al., 1984). Apparently this effect does not play an important role in torpid hummingbirds since at moderate TA where the ventilation is reduced the possible cooling is minimal due to the very low temperature difference between body and environment. At low TA where a heat conserving mechanism would be important we find the birds in regulated torpor with high ventilation volumes and with oxygen extraction values lower than in rest. Apparently in this phase as well as during arousal the heat production by the muscular activity of respiratory movements is essential compared to the heat loss via respired air.

The oxygen consumption is regulated in Colibri coruscans by alterations of the ventilation and oxygen extraction. In non-regulated torpor with the necessity to avoid heat production and the apparent restrictions in muscular activity the oxygen extraction plays a predominant role. In regulated torpor and arousal, phases of enlarged heat production, increased ventilation mainly by respiration frequency is the important regulator.

SUMMARY

In torpid, unrestrained Colibri coruscans we measured by mass-spestrometry expired gas concentrations of single breaths.

During entering torpor and in non-regulated torpor a phasic respiration with intervals up to 1.5 min between two breaths was found. After such periods of apnea peak values of oxygen extraction up to 47 % were measured. Heart beats were regular (with lowest rates of about 30/min) by which the whole blood's oxygen reserves can be used.

During regulated torpor at low ambient tempareture the respiration frequency is reduced to about 30/min. When arousing from torpor the increased oxygen consumption at increased body temperature is the result of increased ventilation (by increased respiration frequency but unchanged and finally decreased tidal volume) and increased oxygen extraction.

KEY WORDS

Respiration - expired gas - oxygen extraction - ventilation - torpor - mass-spectrometry - hummingbirds - Colibri coruscans

REFERENCES

Bech, C., Johansen, K., Brent, R., Nicol, S., 1984, Ventilatory and circulatory changes during cold exposure in the Pekin duck Anas platyrhynchos, Resp. Physiol., 57: 103-112.

Berger, M., 1978, Ventilation in the hummingbirds Colibri coruscans during altitude hovering, in: "Respiratory function in birds, adult and embryonic," J. Piiper ed., Springer, Berlin.

Berger, M., and Hart, J. S., 1972, Die Atmung beim Kolibri Amazilia fimbriata während des Schwirrfluges bei verschiedenen Umgebungstemperaturen, J. comp. Physiol., 81:363-380.

Berger, M., and Johansen, K., (in prep.), The stages of torpor in hummingbirds.

Brent, R., Pedersen, P. F., Bech, C., and Johansen, K., 1984, Lung ventilation and temperature regulation in the European Coot Fulica atra, Physiol. Zool., 57:19-25.

Dubach, M., 1981, Quantitative analysis of the respiratory system of the house sparrow, budgerigar and violet-eared hummingbird, Resp. Physiol., 46: 43-60.

Hainsworth, F. R., and Wolf, L. L., 1970, Regulation of oxygen consumption and body temperature during torpor in a hummingbird, Eulampis jugularis, Science, 168:368-369.

Withers, P. C., 1977, Respiration, metabolism, and heat exchange of euthermic and torpid poorwills and hummingbirds, Physiol. Zool., 50: 43-52.

ENERGY METABOLISM AND PATTERNS OF VENTILATION IN EUTHERMIC AND TORPID HUMMINGBIRDS

Theresa L. Bucher[1] and Mark A. Chappell[2]

[1] Department of Biology, University of California
Los Angeles, CA 90024, USA
[2] Department of Biology, University of California
Riverside, CA 92521, USA

INTRODUCTION

Thermoregulation and energy metabolism in hummingbirds, the smallest of all avian species, have long been of interest to comparative physiologists. Early studies reported extremely high rates of oxidative metabolism during both rest and activity. Many species also undergo profound hypothermia, or torpor, often accompanied by long periods of apnea (Pearson, 1950; Bartholomew et al., 1957). Subsequent studies have established that regulated hypothermia (maintenance of stable body temperature via adjustments in metabolic heat production) commonly occurs during torpor (Hainsworth and Wolf, 1970; Wolf and Hainsworth, 1972). The respiratory system of hummingbirds must not only be able to support high rates of oxygen flow during activity, but must also accommodate very large changes in oxygen demand during the transition between torpor and euthermia. Somewhat surprisingly, hummingbird respiration has received relatively little attention from avian physiologists. Moreover, in the few studies of hummingbird ventilation to date, the birds have been restrained within the metabolic chamber (Lasiewski, 1964, 1967; Withers, 1977). These procedures almost certainly alter normal ventilation patterns (Bucher, 1981).

Since data on respiration in unrestrained individuals are now available for a number species of larger birds (Bucher 1985, Bucher and Bartholomew 1986), ventilation and gas exchange in unrestrained hummingbirds is of considerable comparative interest. Accordingly, we employed noninvasive, nonrestraining techniques to examine oxygen consumption, respiration frequency, tidal and minute volume, and oxygen extraction in two North American species, the rufous hummingbird *Selasphorus rufus* and the broad-tailed hummingbird *S. platycercus*. In their natural habitats, both species routinely experience a wide range of thermal conditions, including freezing nocturnal ambient temperatures, and both employ torpor. Therefore, we were able to obtain measurements during torpor and euthermia over a wide range of ambient temperatures. Since long-term captivity (Lasiewski, 1963) as well as measurement techniques may substantially affect metabolism and ventilation, we utilized only freshly caught individuals.

MATERIALS AND METHODS

Animals. Data were collected during July, 1987 at the Rocky Mountain Biological Laboratory in southwestern Colorado, USA (elevation 2900 m). Migrant rufous hummingbirds (five males and four females: mean mass 3.62±0.06 g) and territorial, resident broad-tailed hummingbirds (seven males: mean mass 3.48±0.09 g) were mist-netted in the early morning or shortly before dark. The mean mass of the two species did not differ significantly ($p=0.24$). Measurements began within one to three hours of capture and lasted six to twelve hr. Each individual was used only once, either during the day or at night. It was not always possible to collect data from the complete range of ambient temperatures (T_a; 0-30 °C) for every individual.

Birds were not necessarily post-absorptive during measurements because they were given the opportunity to feed before each experiment began and after measurements of body temperature (T_b). We followed this procedure because attempts to achieve post-absorptive conditions resulted either in increased restlessness or in a drop in T_b (as noted by Pearson 1950). Therefore, we refer to minimum rates of oxygen consumption ($\dot{V}O_2$) as resting, not standard, values.

Measurements. We measured metabolism using an open circuit respirometry system in which the animal chamber (volume ≈ 400 ml) also functioned as a whole-body plethysmograph. $\dot{V}O_2$, ventilation frequency (f) and tidal volume (V_T) were measured simultaneously (see Bucher, 1981 and Chappell and Bucher, 1987 for details of the system). Flow rates were 300 ml/min STPD for all euthermic individuals and 150 ml/min STPD for some torpid individuals with low $\dot{V}O_2$. Actual chamber pressure varied between 548 and 556 torr. T_a was maintained ± 0.5 °C between 0 and 30 °C and monitored by a thermocouple in the excurrent air line. The birds stood on a small perch in the center of the chamber. A thermocouple, attached to the perch and monitored continuously (T_{perch}), indicated if an individual left the perch or if it began to go torpid. After any period of stable $\dot{V}O_2$ accompanied by a stable T_{perch}, a deep esophageal T_b was obtained immediately (within 15-20 sec) upon removal of the bird from the chamber. We calculated V_T only for ventilation data collected during periods of stable oxygen consumption which were followed by T_b measurements.

Data Reduction and Statistics. Values of $\dot{V}O_2$ are corrected to STP. V_T and minute volume ($\dot{V}_I$; $=f \cdot V_T$) are calculated at BTPS and indicate actual volumes in the respiratory tract. STPD values for V_T and $\dot{V}_I$ are given in brackets [ ] and were used for computing percent oxygen extraction (EO_2) and for comparisons when appropriate.

Typically, several ventilation measurements were obtained at each combination of T_a and T_b. To avoid biasing results in favor of birds with many data points, single mean values of $\dot{V}O_2$, f, V_T, $\dot{V}_I$, and EO_2 were calculated for each individual at each T_a and T_b for which we had data. Results are given as mean±SEM. ANOVA's or t-tests were used to compare means. Regressions were fitted by the method of least squares and compared by analysis of covariance (ANCOVA). The significance level for all tests was $p<0.05$.

RESULTS AND DISCUSSION

Data were obtained from three individuals of *S. rufus* at night and six individuals during the day. One individual from each group entered torpor. Five *S.*

platycercus were used for night runs (three entered torpor), and two were studied during the day (one entered torpor). Torpor was operationally identified as the combination of depressed $\dot{V}O_2$ and $T_b < 35$ °C.

Body temperature. Mean euthermic T_b did not vary significantly with changing T_a (ANOVA: $p = 0.08$). However, at night the T_b of euthermic birds showed a weak, positive correlation with T_a ($T_b = 37.8 + 0.07[T_a]$; $r^2 = 0.20$, $t = 2.16$, $df = 19$). No correlation between T_b and T_a was present in the daytime data. The highest mean T_b was 41.0 ± 0.6 °C at 25 °C and the lowest mean T_b was 38.9 ± 0.4 °C at 10 °C. Mean euthermic T_b at night (39.0 ± 0.3 °C, n = 21) was significantly lower than mean daytime euthermic T_b (40.6 ± 0.2 °C, n = 24, $p < 0.001$). Individual values of T_b ranged from a high of 42.5 °C in two birds measured during the day (one at $T_a = 5$ °C and one at $T_a = 20$ °C) to a low of 35.6 °C at night (at $T_a = 10$ °C). The general relationships between T_b and T_a are similar to those reported in another small bird, the willow tit, *Parus montanus* (Reinertsen and Haftorn, 1983).

The T_b of torpid birds varied from a high of 34.9 °C at $T_a = 29$ °C to a low of 12.2 °C at $T_a = 2.2$ °C, both measured during the day. In torpid birds, $T_b - T_a$ varied from 3.9 to 18 °C. At T_a's > 10 °C there was no obvious relationship between T_a and the $T_b - T_a$ gradient. At T_a's < 10 °C there was a strong negative correlation ($[T_b - T_a] = 16.0 - 1.31[T_a]$; $r^2 = 0.90$. $t = -6.12$, $df = 4$) because T_b never fell below 12.2 °C. Individuals went torpid at T_a's < 10 °C on six occasions. The associated stable T_b's were 12.2, 13.1, 13.7, 15.4 and 18.0 °C, indicating that these hummingbirds "defend" a minimum T_b of about 12 °C during torpor, even at considerably lower T_a.

Oxygen consumption and thermal conductance. Although at T_a's ≥ 15 °C the mean resting $\dot{V}O_2$ averaged 5-16% higher during the day than at night, the difference was not significant at any single T_a. Also, the relationships between $\dot{V}O_2$ and T_a over the range of T_a's from 0 to 30 °C during the day and at night did not differ significantly in slope (ANCOVA: $F_{1,59} = 0.26$; $p = 0.50$) or in intercept ($F_{1,60} = 0.54$; $p = 0.47$). Because the two species did not differ significantly in mass or in daytime and nighttime resting $\dot{V}O_2$'s, all metabolic and ventilatory data were pooled for further analysis.

The mean $\dot{V}O_2$ we observed at 30 °C (0.553 ± 0.027 ml/min) may not be within the thermal neutral zone (TNZ), as it is considerably higher than predicted for 3.5 g hummingbirds. The expected basal $\dot{V}O_2$ calculated from a mass-independent measure of metabolism (MIM = $\dot{V}O_2/\text{mass}^{0.67}$) for the order Apodiformes (Bucher, 1986) is 0.221 ml/min, only 40% of our measured mean at 30 °C. Moreover, data of Lasiewski (1963) suggests that the TNZ may begin above 30 °C. He obtained a mean $\dot{V}O_2$ of 0.201 ml/min for *S. rufus* at $T_a = 34$ and 39 °C, but noted that T_b may have been low during his experiments.

The $\dot{V}O_2$ of euthermic birds was linearly related to T_a between 0 and 30 °C (Fig. 1; $\dot{V}O_2 = 1.328 - 0.024[T_a]$; $r^2 = 0.64$, $t = -10.2$, $df = 62$). At $T_a \leq 10$ °C, the $\dot{V}O_2$ of torpid individuals was also linearly related to T_a ($\dot{V}O_2 = 0.438 - 0.042[T_a]$; $r^2 = 0.88$, $t = -5.45$, $df = 4$). The slopes of these regressions are not significantly different (ANCOVA: $F_{1,66} = 1.12$; $p = 0.29$; common slope = 0.024). The intercepts of the lines (assuming a common slope) are not equal ($F_{1,67} = 203.8$; $p < 0.001$; intercepts = 1.33 and 0.34, respectively). These results indicate that torpid rufous and broad-tailed hummingbirds use regulated metabolic heat production to maintain stable T_b below T_a's of approximately 10 °C. This pattern is very similar to the pattern of regulation reported by Hainsworth and Wolf (1970) and Wolf and

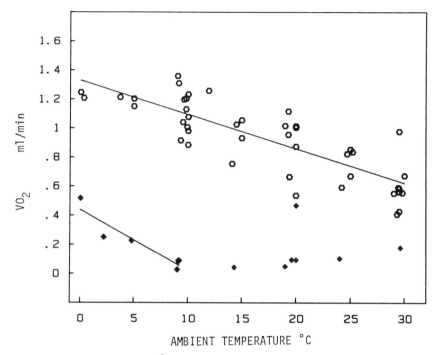

Fig. 1. Oxygen consumption ($\dot{V}O_2$) in relation to ambient temperature in rufous and broad-tailed hummingbirds. Open symbols represent euthermic individuals; closed symbols represent torpid individuals. See text for details on regression lines.

Hainsworth (1972) in three larger species of hummingbirds, and also resembles body temperature regulation in hibernating mammals.

We were impressed with the difference in metabolic intensity between euthermia and torpor. At T_a's of approximately 10 °C, the $\dot{V}O_2$ of one resting, euthermic individual was 37.6 times greater than the $\dot{V}O_2$ of the same individual when torpid. The importance of the $T_b - T_a$ gradient, and hence of T_a itself, on the efficacy of torpor as a mechanism for conserving energy is also apparent. At T_a's of 25 and 15 °C, the $T_b - T_a$ gradients were 4-6 °C and the associated changes in metabolism between torpor and endothermy were 9 to 27-fold. At lower T_a's (5 and 0 °C) the birds defended a relatively high T_b, the $T_b - T_a$ gradient was 8-18 °C, and the associated changes in metabolism between torpor and euthermia were only 2.3 to 5.5-fold. By comparison, Lasiewski and Lasiewski (1967) reported increases in metabolism as great as 59 fold at T_a's of 15-16 °C.

The $\dot{V}O_2$ is linearly related to T_b-T_a in both euthermic and torpid birds between 0 and 30 °C (Fig. 2: $\dot{V}O_2 = 0.307 + 0.028[T_b-T_a]$; $r^2 = 0.61$, $t = 10.17$, $df = 65$ and $\dot{V}O_2 = -0.062 + 0.033[T_b-T_a]$; $r^2 = 0.89$, $t = 9.32$, $df = 11$, respectively). The slopes of the two regressions do not vary significantly (ANCOVA: $F_{1,76} = 0.16$; $p = 0.50$; common slope = 0.029), but the intercepts are significantly different ($F_{1,77} = 25.5$; $p < 0.001$; intercepts = 0.30 and -0.03, respectively). The best

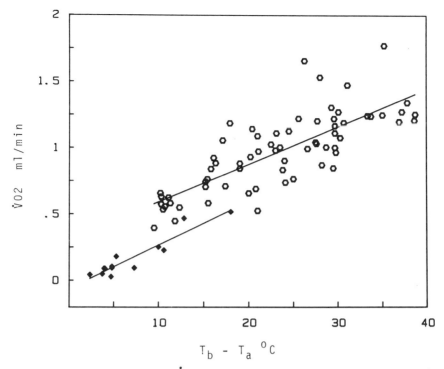

Fig. 2. Oxygen consumption ($\dot{V}O_2$) in relation to the gradient between body temperature and ambient temperature ($T_b - T_a$). Symbols as in Fig. 1; see text for data on regression lines.

explanation for the difference is a reduced thermal conductance (C) in torpid birds. We calculated C as $\dot{V}O_2/[T_b-T_a]$. In torpid individuals, mean C was 0.021 ± 0.002 ml $O_2/[°C\cdot min]$, less than half that of non-torpid birds (0.043 ± 0.001 ml $O_2/[°C\cdot min]$). In contrast, C was the same in non-torpid and torpid West Indian hummingbirds, *Eulampis jugularis* (Hainsworth and Wolf, 1970) and in the highland tropical species, *Eugenes fulgens* (Wolf and Hainsworth, 1972). The same authors report that in *Panterpe insignis*, another highland tropical species, mean C was greater in torpid than in non-torpid individuals.

Extrapolation of the regression of $\dot{V}O_2$ versus T_a in euthermic *S. rufus* and *S. platycercus* yielded an X-intercept of 56 °C (considerably above T_b), indicating that C is neither minimal nor constant at T_a's below the TNZ, even in euthermic individuals. In fact, C of euthermic birds ranged from 0.025 to 0.083 ml $O_2/[°C\cdot min]$ and was positively correlated with T_a between 0 and 30 °C (Fig. 3; $C = 0.030 + 0.0008[T_a]$; $r^2 = 0.40$, $t=5.31$, $df=43$). Schuchmann and Schmidt-Marloh (1979) indicate a constant C between 5 and 25 °C in two species of tropical hummingbirds. Also Lasiewski (1973) states that the slopes of the relationships between $\dot{V}O_2$ and T_a of six species are considered to be representative values of thermal conductance, thereby implying C is constant. He notes one exception: In *Stellula calliope* the zero metabolism extrapolation of the line was 45 °C, suggesting this species was lowering its T_b with decreasing T_a.

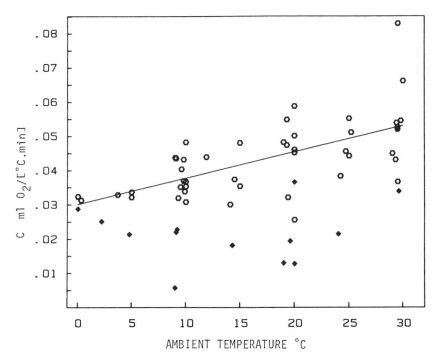

Fig. 3. Conductance (*C*) in relation to ambient temperature. Symbols as in Fig. 1. See text for data on regression line.

Ventilation. When $\dot{V}O_2$ was stable, mean V_T did not vary significantly between euthermic individuals (0.119±0.005 ml BTPS or 0.075±0.003 ml STPD) and torpid individuals (0.116±0.013 ml BTPS or 0.078±0.009 ml STPD). In euthermic birds, V_T was weakly, but significantly negatively correlated with T_a (V_T = 0.15 - 0.002[T_a]; r^2 = 0.18, t = -3.86, df = 65). The V_T at 30 °C was 0.054±0.002 ml (STPD), only 61% of the predicted standard value of 0.089 ml (equation 1 in Bucher and Bartholomew, 1986). The difference is partially due to relatively low ambient barometric pressure at our study site; the BTPS V_T is about 75% of predicted BTPS values. We know of few other published values for V_T in hummingbirds, and none for unrestrained individuals. Withers (1977) obtained a V_T of 0.18 ml at T_a = 20 °C in *S. sasis* and *Calypte anna* restrained in snugly-fitting plastic cylinders, but it is unclear if this is a BTPS or STPD value.

Not surprisingly, mean *f* was significantly higher (p < 0.001) in euthermic than in torpid individuals (209±6.9 br/min and 73.8±15.1 br/min, respectively). In resting, euthermic individuals mean *f* was minimal at 30 °C (173.4±15.3 br/min) and did not vary significantly between 20 and 30 °C (ANOVA: p = 0.61). There was a weak, but significant, negative correlation between *f* and T_a from 0 to 15 °C (*f* = 269.4 - 3.9[T_a]; r^2 = 0.11, t = -2.08, df = 36). The mean minimal resting values are 1.8 to 2.8 times predicted *f* (Calder, 1968 and Bucher, 1985, respectively). The *f* of resting individuals ranged from 98 to 360 br/min in euthermic birds, and from 6.5 to 151 br/min in torpid birds.

Torpid individuals with the lowest $T_b - T_a$ gradients (4-6 °C) occasionally showed periods of apparent apnea of up to five minutes duration, interspersed between bursts of ventilation consisting of one to ten quick breaths. Similar periods of apnea were reported by Bartholomew et al. (1957) and Lasiewski (1964). Paradoxically, the $\dot{V}O_2$ of these birds was usually stable despite the apparent periodicity of breathing, indicating that gas exchange was not interrupted during the apneas. If any breaths other than those occurring in periodic bursts occurred, the V_T must have been less than approximately 0.05ml BTPS (the resolution limit of our system at these $T_b - T_a$ gradients), or the time course of inhalation and exhalation must have been so long (i.e., several seconds) that there was no detectable pressure fluctuation in our open-circuit system.

Mean $\dot{V}_I$ was 26.0±1.7 ml/min [16.5±1.1 ml/min] in euthermic individuals and 6.9±1.1 ml/min [4.6±0.7 ml/min] in torpid individuals. These values were significantly different for both BTPS and STPD volumes ($p < 0.001$ in both cases). In euthermic birds $\dot{V}_I$ did not vary significantly at T_a's between 0 and 15 °C (ANOVA: $p = 0.27$) or between 20 and 30 °C (ANOVA: $p = 0.28$). However, for T_a from 0-30 °C there was a slight but significant negative correlation between T_a and $\dot{V}_I$ ($\dot{V}_I = 39.4 - 0.8[T_a]$; $r^2 = 0.25$, $t = -4.64$, $df = 65$). The mean $\dot{V}_I$ at 30 °C was 14.8±1.3 ml/min [9.4±0.8 ml/min]. This STPD value is 1.6 times the predicted standard value for a 3.5 g bird at 30 °C (equation 3 in Bucher and Bartholomew, 1986).

Oxygen extraction. We calculated oxygen extraction (EO_2; %) from STPD ventilation as $[\dot{V}O_2/(0.2095 \cdot \dot{V}_I)] \cdot 100$. Mean EO_2 was much higher (33.5±1.6%) in euthermic than in torpid birds (18.8±3.2; $p < 0.001$). In contrast, Withers (1977) reported higher EO_2 in torpid than in euthermic birds (13.8% and 8.59%, respectively), and both his values were lower than ours. In our data for both torpid and euthermic birds, EO_2 was highly variable, both within and among individuals. In euthermic birds mean EO_2 ranged between 30 and 38.5% and did not differ significantly at T_a's between 0 and 30 °C (ANOVA: $p = 0.58$). These values are fairly typical for birds, but the maximum observed EO_2's we observed in euthermic hummingbirds were 60 to 74%. These are among the highest yet reported for any avian species (see also Stahel and Nicol 1988) or other air-breathing vertebrate. Nevertheless, there was no indication that hummingbirds increase EO_2 at low T_a in order to minimize respiratory heat loss.

Ventilation parameters and oxygen extraction show considerable intra- and inter-individual variation in most birds (Bucher 1985, Chappell and Bucher 1987). Variability and lability are particularly evident in our data. This is hardly surprising in view of the differences in metabolic intensity and T_b in euthermic and torpid hummingbirds. However, even in euthermic individuals, EO_2 and ventilatory parameters varied by a factor of two- to four-fold at any T_a.

SUMMARY

The small size and limited insulation of hummingbirds provide little thermal inertia and results in a close "coupling" to the thermal environment, and their low mass and high rates of energy metabolism necessitate prudent use of available energy reserves. However, they can lower or raise T_b very quickly in response to changing environmental or physiological conditions (Bartholomew et al., 1957), which facilitates the use of hypothermia and hypometabolism as an energy conservation mechanism.

To accommodate changing oxygen demands at different T_a's, resting, euthermic hummingbirds adjust $\dot{V}_I$ instead of EO_2; the change in $\dot{V}_I$ is accomplished by approximately 1.5 fold increases in both f and V_T. The 3- to 40-fold change in $\dot{V}O_2$ between euthermy and torpor is accommodated by changes in both ventilation and oxygen extraction. V_T showed no consistent relationship with changing T_a or between torpor and euthermy. However, $\dot{V}_I$ of individuals varied by a factor of 2- to 29.4-fold between torpor and euthermy; this change was due primarily to similar changes in f. High $\dot{V}O_2$ was also supported by increased EO_2; in all individuals but one, EO_2 was substantially greater in euthermy than in torpor at similar T_a.

Acknowledgments. We are grateful for the friendly assistance of the staff at the Rocky Mountain Biological Laboratory, and especially to Dr. William Calder and Dana Bradley. Funding support was provided by National Science Foundation grants DEB 8202708 and DPP 8519107, and a UC Riverside Intramural Grant, to M.A.C.

BIBLIOGRAPHY

Bartholomew, George A., Thomas R. Howell, and Tom J. Cade. 1957. Torpidity in the white-throated swift, Anna hummingbird, and poor-will. *Condor* 59: 145-155.

Bucher, Theresa L. 1981. Oxygen consumption, ventilation and respiratory heat loss in a parrot, *Bolborhynchus lineola*, in relation to ambient temperature. *J Comp Physiol* 142: 479-488.

Bucher, Theresa L. 1985. Ventilation and oxygen consumption in *Amazona viridigenalis* - a reappraisal of 'resting' respiratory parameters in birds. *J Comp Physiol* 155B: 269-276.

Bucher, Theresa L. 1986. Ratios of hatchling and adult mass-independent metabolism: a physiological index to the altricial-precocial continuum. *Respir physiol* 65: 69-83.

Bucher, Theresa L. and George A. Bartholomew. 1986. The early ontogeny of ventilation and homeothermy in an altricial bird, *Agapornis roseicollis* (Psittaciformes). *Respir Physiol* 65: 197-212.

Calder, William A. 1968. Respiratory and heart rates at rest. *Condor* 70: 358-365.

Chappell, Mark A. and Theresa L. Bucher. 1987. Effects of temperature and altitude on ventilation and gas exchange in chukars (*Alectoris chukar*). *J Comp Physiol* 157B: 129-136.

Hainsworth, F. Reed and Larry L. Wolf. 1970. Regulation of oxygen consumption and body temperature during torpor in a hummingbird, *Eulampis jugularis*. *Science* 168: 368-369.

Lasiewski, Robert C. 1963. Oxygen consumption of torpid, resting, active, and flying hummingbirds. *Physiol Zool* 36: 122-140.

Lasiewski, Robert C. 1964. Body temperature, heart and breathing rate, and evaporative water loss in hummingbirds. *Physiol Zool* 37: 212-223.

Lasiewski, Robert C. and Richard J. Lasiewski. 1967. Physiological responses of the blue-throated and Rivoli's hummingbirds. *Auk* 84: 34-48.

Lasiewski, Robert C., Wesley W. Weathers and Marvin H. Bernstein. 1967. Physiological responses of the giant hummingbird, *Patagona gigas*. *Comp Biochem Physiol* 23: 797-813.

Pearson, Oliver P. 1950. The metabolism of hummingbirds. *Condor* 52: 145-152.

Reinertsen, Randi E. and Svein Haftorn. 1983. Nocturnal hypothermia and metabolism in the willow tit *Parus montanus* at 63º N. *J Comp Physiol* 151B: 109-118.

Schuchmann, Karl L. and Dagmar Schmidt-Marioh. 1979. Temperature regulation in non-torpid hummingbirds. *Ibis* 121: 354-356.

Stahel, C. D. and Stuart C. Nicol. 1988. Ventilation and oxygen extraction in the little penguin (*Eudyptula minor*), at different temperatures in air and water. *Resp Physiol* 71:387-398.

Wither, Philip C. 1977. Respiration, metabolism, and heat exchange of euthermic and torpid poorwills and hummingbirds. *Physiol Zool* 50: 43-52.

Wolf, Larry L. and F. Reed Hainsworth. 1972. Environmental influence on regulated body temperature in torpid hummingbirds. *Comp Biochem Physiol* 41A: 167-173.

RESPIRATION BY BIRDS AT HIGH ALTITUDE AND IN FLIGHT

Marvin H. Bernstein

New Mexico State University
Las Cruces, New Mexico, U. S. A.

INTRODUCTION

When they fly, birds use more energy per unit of time than any other exercising vertebrate of similar size. Oxygen transfer to their tissues must therefore be especially efficient. Some birds visit and others reside at high altitudes, where environmental temperatures and available O_2 are below the tolerance limits for many mammals. At 7,000 meters above sea level (ASL) where several birds are found, for example, barometric pressure (P_B) and O_2 partial pressure (PO_2) fall to 308 Torr and 65 Torr, respectively. Mean air temperature (T_a) reaches an average of -30°C at the same height.

What are the adaptations that confer high-altitude tolerance to resting and flying birds? This question has attracted considerable recent attention, but direct information about avian physiology during flight at high altitude is still almost completely lacking. Much of what can be inferred, therefore, is based on studies of hypoxic birds at rest and of normoxic birds in flight, none of which included extreme cold exposure.

PHYSIOLOGY OF FLYING BIRDS

Information is available for several bird species flying freely or in wind tunnels. An example is the 0.4-kg white-necked raven (<u>Corvus cryptoleucus</u>), for which representative experiments are summarized in Figure 1. Respiratory gas exchange and evaporation all increased within seconds of take-off, O_2 uptake and CO_2 release rising from less than 20 to more than 100 ml/min STPD and respiratory evaporation increasing from about 50 to about 200 mg/min. During the next 5 min these values typically decreased to a steady state. Body temperature (T_b) stabilizes in about the same time at a plateau about 2°C higher than at rest. Preflight breathing rates of about 40/min gave way to rates of about 180/min nearly at the instant of take-off, whereas tidal volume increased only slightly. Increased breathing rate in this species thus accounts almost entirely for the elevations in gas exchange and respiratory evaporation. Under the conditions of this experiment, ravens achieved pulmonary O_2 extractions of 35-40%.

Steady-state data such as those for the raven have been obtained for

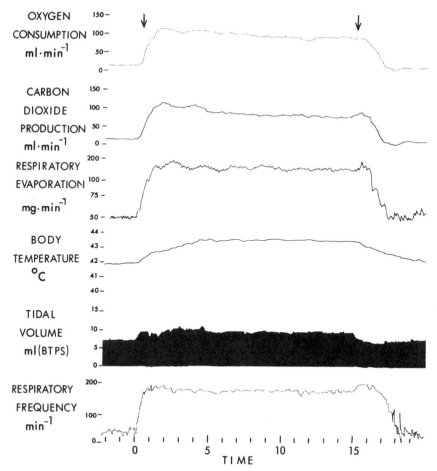

Fig. 1. Representative recordings of respiratory gas exchange and body temperature in a 0.4-kg white-necked raven (<u>Corvus cryptoleucus</u>) before, during, and after a horizontal flight at a speed of 10 m/s in a wind tunnel at an air temperature of 20°C. Flight onset and termination are indicated by arrows. Time is in minutes; O_2 and CO_2 data are expressed under STPD conditions. heads. Body temperature was recorded by a thermocouple implanted in the cloaca. Respiratory gas exchange was determined from the gas contents of air continuously sampled from a ventilated, open-ended mask worn by the bird and analyzed by in-line gas analyzers. Additional drag conferred by thermocouple leads and the gas-sampling tube trailed by the bird were compensated by tilting the wind tunnel with its upstream end below horizontal. Tidal volume and frequency were recorded in the same bird, in a separate but identical flight, as the breath-by-breath integration of the signal from a linear, temperature-compensated, hot-film anemometer enclosed in a mouthpiece. Anemometer leads were trailed during flight and their drag compensated by tilting the wind tunnel. Data from Hudson and Bernstein (in preparation). See also Hudson and Bernstein (1981; 1983).

a variety of species flying at moderate temperatures and low altitude, and have been shown to vary exponentially with body size, as in birds at rest. For example, in flying birds O_2 uptake (in ml/min STPD), respiratory evaporation (in mg/min), and tidal volume (in cm^3 BTPS) increased with body mass (M_b, in kg) according to $172\ M_b^{0.73}$; $259\ M_b^{0.80}$; and $27.8\ M_b^{0.89}$, respectively (Bernstein, 1987). The corresponding allometric relationships for resting birds are $11.3\ M_b^{0.72}$ (Lasiewski and Dawson, 1967), $24.2\ M_b^{0.61}$ (Crawford and Lasiewski, 1968), and $16.9\ M_b^{1.05}$ (Boch et al., 1979). Thus O_2 uptake scales with body size similarly in both resting and flying birds, and most flying birds would thus increase resting O_2 uptake by about 15-fold. Respiratory evaporation does not scale similarly in resting and flying birds, however; in a hypothetical flying bird weighing 0.4 kg evaporation would increase by 9-fold over the resting level. A 0.4-kg bird taking to flight would increase tidal volume by only about 2-fold, so, as observed in ravens, the increases in pulmonary ventilation supporting the large increases in O_2 uptake and evaporative water loss during flight must be due principally to increased breathing rate.

EFFECTS OF HYPOXIA

Respiration. The only data available on the physiology of birds flying under hypoxia are those of Berger (1974; 1978) for hummingbirds (*Colibri coruscans* and *Amazilia fimbriata fluviatilis*) and of Torre-Bueno (1985) for starlings (*Sturnus vulgaris*). In flying hummingbirds O_2 uptake increased 8%, from sea level to 4 km ASL, and respiratory evaporation increased by 38%, from which it was calculated that ventilation increased by 36%. In flying Starlings metabolic rate increased by about 9% from sea level to 3.5 km ASL.

To approach more closely the question of high altitude and low-temperature tolerance in birds, it is useful to ask, using allometric relationships, how respiration might change during high-altitude flight. For example, the tidal volumes of 6.5 and 12.3 cm^3 BTPS calculated for a hypothetical, 0.4-kg bird at rest and in flight at sea level comprise 10 and 18% of the respiratory system volume calculated for a bird of this size, respectively. If such a bird were to inspire the same BTPS volume at 7 km ASL, then the molar quantity of O_2 it would take in with each breath would be just one-third that at sea level. If, to compensate completely for the reduced O_2, the bird were to increase only its tidal volume, it would have to inspire 26% of its respiratory-system volume with each breath at rest, and 51% in flight. These are unreasonably high tidal volumes; it is therefore more likely that pulmonary O_2 flow is maintained by combined increases in tidal volume and frequency in birds at high altitude. Indeed the 36% increase in ventilation observed in flying hummingbirds when exposed to hypoxia (Berger, 1978) was due nearly equally to increases in breathing rate and tidal volume.

Such increases in pulmonary ventilation probably increase the rate of respiratory CO_2 loss, engendering an hypocapnic alkalosis similar to that observed during hypobaric hypoxia in several avian species at rest. Increased ventilation would also be expected to elevate evaporation and thus the rate of heat loss. At the low temperatures found at high altitude, however, it is probably advantageous to conserve heat, even during flight. Therefore, as pointed out by Chappell and Bucher (1987), hyperventilation to compensate for reduced inspired PO_2 is probably accompanied by a reduction in exhaled-air temperature. The mechanism described by Schmidt-Nielsen et al. (1970) accounts for such a reduction.

Figure 2 shows the calculated effects of exhaled air temperature (T_e) on heat loss in a resting and a flying, 0.4-kg bird, with and without increased ventilation. A bird at rest having a T_e equal to T_b would dissipate 5% of its metabolic heat production by respiratory evaporation.

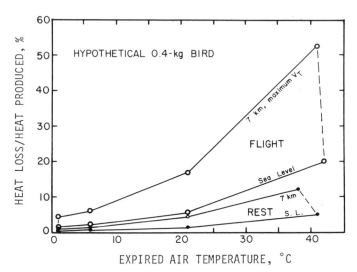

Fig. 2. Percentage ratio of heat lost by respiratory evaporation to heat produced metabolically, in relation to temperature of expired air, in a hypothetical, 0.4-kg bird at rest or in flight, at sea level (S. L.) or at an altitude of 7 km above sea level (ASL). Evaporative water loss at sea level was calculated from respiratory ventilation, calculated in turn using previously reported allometric equations for resting and flying birds (Bernstein, 1987) and assuming that the expirate is saturated with water vapor at its temperature. Evaporative water loss at 7 km ASL was calculated similarly, but with the additional assumption that respiratory ventilation increased above its sea-level value sufficiently to maintain pulmonary O_2 flow at the sea-level value. Heat loss was calculated from evaporation rate by using latent heat values given by List (1951) for the expired air temperatures used, and by assuming that 1 cal is equivalent to 4.184 J. Heat production was calculated from allometric expressions for O_2 uptake in resting and flying birds (Bernstein, 1987). It was assumed that at 7 km ASL O_2 uptake during flight increases by 10% over corresponding values at sea level, as approximated by flying hummingbirds (Berger, 1974) and starlings (Torre-Bueno, 1985). Further assumptions were that steady-state flight metabolism is entirely aerobic and that a liter of O_2 consumed corresponds to 20.1 kJ. Dashed lines connect sea-level and 7 km values at highest expired air temperatures for resting birds and for flying birds. The highest expired air temperature for each condition were taken as equal to the corresponding body temperature measured previously in resting or flying pigeons (0.4 kg) at sea level.

At 7 km ASL the same bird with T_e equal to T_b would evaporate away 12% of its metabolic heat. If, instead, T_e were cooled to 21°C, respiratory evaporation would dissipate only 1% of metabolic heat production at low altitude and 5% at 7 km ASL. Further expirate cooling would reduce respiratory heat loss nearly to zero. Thus birds resting at high altitude probably reduce T_e moderately to prevent an increase in the fraction of their metabolic heat lost by evaporation.

As also illustrated in Figure 2, birds flying at low altitude and having T_e equal to T_b would increase cooling by 4-fold over resting birds having the same T_e, and would thereby lose 20% of their metabolic heat by respiratory evaporation. In contrast, a 0.4-kg bird flying at 7 km ASL with T_e equal to T_b would increase its respiratory heat loss to more than 50% of its metabolic heat production. Reduction in T_e to 21°C by a bird flying at sea level would bring metabolic heat loss nearly within the range observed for a resting bird with the same T_e, a potentially undesirable circumstance. In a bird flying at 7 km ASL, however, reducing T_e to 21°C would set evaporative heat loss back only to 20%. Even at a T_e of 6°C respiratory evaporation at 7 km ASL would still dissipate 7% of metabolic heat.

Thus for a bird to fly at high altitude without sustaining a significant evaporative loss of metabolic heat, the increase in ventilation undertaken to sustain respiratory O_2 flow probably requires a simultaneous cooling of the expirate nearly to freezing. Excess metabolic heat, beyond that stored during the initial rise in T_b, will then be lost by convection. In the unlikely event that convection is insufficient, despite the subfreezing T_a, the bird would simply not reduce its T_e to the same extent. Measurement of the contributions by convection and evaporation to heat loss for birds flying in extreme cold remains for future study.

Oxygen extraction. In coping with high altitude, birds might be expected to increase their pulmonary and tissue O_2 extractions. The few available data suggest that they do. In both bar-headed geese (Anser indicus) (Black and Tenney, 1980; Faraci et al., 1984) and pigeons (Columbia livia) (Weinstein et al., 1985) the difference between inspired and arterial PO_2 at rest decreased with decreasing P_B, implying improved pulmonary O_2 extraction. In hummingbirds hovering at 4 km ASL, pulmonary O_2 extraction was 32%, similar to the sea-level value for ravens. Hummingbirds at sea level extracted 24% of inspired O_2 (Berger, 1978).

The O_2 content of anaerobic, mixed-venous blood samples in resting pigeons was immeasurably low during hypobaric hypoxia simulating altitudes of 7 or 9 km ASL (Weinstein et al., 1985). This suggests a tissue O_2 extraction approaching 100%. In flying pigeons, tissue O_2 extraction was 61% (Butler et al., 1977). Pulmonary and tissue O_2 extraction values in birds flying at high altitude have not as yet been measured.

Body temperature and oxygen transport. Along with their reduction of T_e, birds resting at high altitude would be expected to cool their bodies. Pigeons reduced their T_b by several degrees when subjected to hypobaric hypoxia; at 7 km ASL and -6°C T_a the decrease was from 41 to 38°C (Bernstein, this volume). This suggests that hypoxic birds may reduce their metabolic rate in order to conserve energy, to minimize O_2 demand, and to reduce heat loss to the environment. The effects of temperature and acid-base status on blood O_2 affinity were quantified for pigeon blood by Pinshow et al. (1985). Using their results and data on temperature and blood gases in pigeons at high altitude (Weinstein et al., 1985), they calculated that at altitudes of 7 and 9 km ASL arterial O_2 saturation would be 54 and 42% respectively. In the absence of the observed reduction in T_b, however, blood O_2 affinity would have increased much less; the acid-base changes alone, which probably reflect both metabolic-acid production and hyperventilation, would have resulted in O_2 saturations of only 35 and 19% at 7 and 9 km ASL, respectively. The reduced T_b in birds resting at high altitude therefore seems not to represent a partial failure of thermoregulation, but a means to increase arterial O_2 saturation, and is thus apparently part of a suite of adaptive responses optimizing O_2 transport.

<u>Brain temperature and brain oxygen delivery</u>. In nearly all bird species studied, brain temperature remains below the temperatures measured in the colon or heart, even at very low T_a (Bernstein, 1989). For example, brain temperature in Pigeons remained near 39.5°C, compared with a T_b of about 41°C, over a range of T_a extending from -6 to 40°C (Bernstein, this volume). The explanation lies in the countercurrent cooling of arterial blood within the ophthalmic retia of the head. These bilateral vascular networks are comprised of ramified branches of the external ophthalmic artery, intertwined with branches of the veins that drain blood from the evaporative cooling surfaces of the oropharyngeal and nasal cavities and the orbit. The retial veins thus serve as a heat sink for blood flowing to the brain (Kilgore, et al., 1976; Bernstein, et al., 1976).

The moist mucosa from which the venous blood in the ophthalmic retia originates have extensive surface areas, and are richly vascularized (Midtgård, 1984). It therefore seemed appropriate to ask whether O_2 and CO_2, in addition to water vapor, could move between the blood and the atmosphere at these sites. If so, the venous blood arriving in the ophthalmic retia would have elevated PO_2 and reduced PCO_2, lessening the corresponding arteriovenous gradients. At the least this could minimize or prevent the loss of O_2 and the uptake of CO_2 by arterial blood as it traversed the retia. The loss of CO_2 and heat from mucosal blood to the air would increase blood O_2 affinity and thus O_2 saturation, so in the extreme the mucosal blood might become even more arterialized than arterial blood. On arrival at the retia such blood could act as a source of O_2 and as a CO_2 sink for arterial blood, enhancing O_2 transport to the brain.

To test the idea of a gas pathway between atmosphere and mucosal blood, H_2 was used as a marker because it is physiologically inert and is detectable by a simple electrode. It has a molecular diameter approximating that of O_2, and is therefore likely to diffuse like O_2 (Kawashiro, et al., 1975; Bernstein, 1989). It was predicted that H_2 introduced into the inspirate would be taken up in the lung and detected by indwelling electrodes in the carotid artery and the brain, whereas H_2 passing only over the nasal mucosa but prevented from entering the respiratory tract would appear only in the brain but not in the carotid.

The results in normoxic, lightly anesthetized pigeons, as shown in Figure 3, were as predicted above (Pierce, et al., in preparation). The data were interpreted to mean that H_2, when introduced into the nasal passages only, was taken up and carried by venous blood to the ophthalmic retia, where it diffused into counter-currently flowing arterial blood that carried it to the brain. Brain H_2 was subsequently washed out and carried to the pulmonary circuit where most of it was removed in one pass, preventing its appearance in the carotid.

To test whether the ophthalmic retia can amplify arterial PO_2 (PaO_2) in cephalic blood, it would be desirable to measure PaO_2 in samples obtained from pre- and post-retial segments of the external ophthalmic artery. The difficulty in obtaining anaerobic, post-retial blood samples has so far precluded this experiment. It was reasoned, however, that O_2 in post-retial blood may become equilibrated with O_2 in cerebrospinal fluid (CSF). PO_2 in CSF sampled anaerobically from the dorsal ventricle of the brain should therefore be representative of post-retial PaO_2. Such CSF samples were therefore obtained via indwelling glass cannulas from awake pigeons; carotid arterial blood was also sampled to represent pre-retial blood. The samples were obtained from unoperated control birds and from experimental birds in which respiratory ventilation proceeded via a tracheostomy and in which the eyes, nares, and mandibles were closed and sealed, preventing contact between air and the eyes, and between air and the oronasal surfaces (Bernstein, et al., 1984).

In the experimental group of tracheostomized birds gas tensions and pH in both arterial blood and CSF were indistinguishable, PO_2 averaging 83

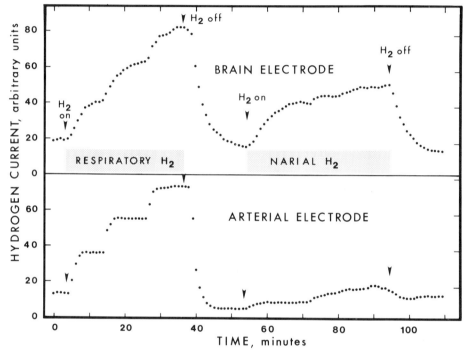

Fig. 3. Current recorded from indwelling, H_2-sensing electrodes in the hypothalamus and in a carotid artery of a pigeon, upon introduction of H_2 into respiratory tract alone (time interval indicated by arrowheads at 3 and 37 min) or into external nares alone (time interval indicated by arrowheads at 53 and 94 min). The left half of the figure shows that H_2 administered in the lungs appeared in both brain and carotid. When pulmonary H_2 was increased stepwise, rapid corresponding increases in the H_2 were recorded at both recording sites. These results indicate that the absorbed gas was transported to the brain via the pulmonary and systemic circulations. The right half of the figure shows that H_2 administered into the nasal cavity, without inhalation into the respiratory tract, appeared in brain only. This suggests that H_2 was absorbed across the nasal epithelium and was transported to the brain via an exclusively cephalic pathway. Data from Pierce, et al. (in preparation).

Torr. In the control group PaO$_2$ was 82 Torr, but mean PO$_2$ in the CSF was 114 Torr. If, in the controls, this increase in CSF PO$_2$ were due to acid production by choroid plexus cells during CSF secretion and to the consequent fall in O$_2$ affinity of the blood perfusing the plexus, this should have occurred in the experimental birds also, elevating their CSF PO$_2$. That it did not suggests that a different mechanism enhances PO$_2$ in CSF, one that is inhibited when contact between the moist cephalic membranes and the air is prevented as in the experimental group. Such a mechanism is not likely to be related to choroidal acid production. The results, rather, are consistent with what would be predicted if, in control birds, venous PO$_2$ in the ophthalmic retia exceeded PaO$_2$ owing to O$_2$ uptake across cephalic membranes, and if arterial blood thereby acquired a higher PO$_2$ before flowing to the brain.

A rise in PaO$_2$ would have the additional advantage of increasing O$_2$ saturation, since at 82 Torr pigeon blood is not normally O$_2$-saturated. A reduction in arterial temperature to, say, 36°C would enhance the latter effect further by increasing blood O$_2$ affinity. The system would work most effectively if all arterial blood flowing to the brain were to pass first through the ophthalmic retia, and if countercurrent gas exchange there, like heat transfer (Midtgård, 1983; Clair, 1985), were complete.

To analyze the potential role of this system in birds at high altitude, it must be taken into account that blood temperature at the mucosal surfaces may be as low as 6°C, as suggested above. Even at the most extreme subfreezing T_a, blood flow in the mucosa is likely to continue to prevent tissue freezing. Blood returning to the retia is therefore likely to be many degrees colder than arterial blood. Based on data of Pinshow et al. (1985), the increase in O$_2$ affinity at low blood temperatures in pigeons combined with O$_2$ uptake directly from air would cause mucosal blood, and therefore venous blood flowing to the ophthalmic retia, to become O$_2$-saturated even at low ambient PO$_2$.

On absorbing heat from arterial blood in the retia, the venous blood would then undergo a rise in PO$_2$, due to its decreasing O$_2$ affinity. Brain cooling, however, was observed to cease in pigeons resting at high altitude (Bernstein, this volume), suggesting that blood flow through exchange arteries in the ophthalmic retia effectively halted. This would prevent arteriovenous heat and gas transfer. It therefore seems unlikely that O$_2$-saturated venous blood from cephalic evaporative surfaces serves an adaptive function in resting birds at high altitude.

In birds flying at high altitude, however, it is possible that retial flow could again increase, as it apparently did in low-flying American kestrels (<u>Falco sparverius</u>) (Bernstein, et al., 1979). Arterial blood might then take up O$_2$ from saturated venous blood in the retia, supplementing the brain's O$_2$ supply. Cerebral blood flow (CBF) increases dramatically during hypoxia in resting birds (Grubb et al., 1978; Faraci et al., 1984); during hypoxic flight an increase in the O$_2$ content of the cerebral arteries might reduce this. Since CBF during flight is probably a minor fraction of cardiac output, it remains unclear whether a reduction in CBF, permitted by O$_2$-saturated cerebral arterial blood, would significantly reduce a flying bird's cardiac-output requirement. In any case, an increase in cerebral PaO$_2$ would significantly improve brain O$_2$ diffusion.

CONCLUSIONS

As in other forms of exercise, flight is probably accompanied by high muscle temperatures that maximize both the efficiency of energy conversion and the delivery of O_2; cooler lung temperatures improve blood O_2 loading. At high altitude birds at rest have lower body temperatures than those at low altitudes; apparently this causes an increase in blood O_2 saturation. The advantage in keeping muscle temperatures as high as possible during flight probably applies even during hypoxia. It is therefore likely that in the extreme cold of high altitudes birds with spread wings and high convective heat loss attempt to minimize heat loss. One way to do this is to decrease expired air temperature and thus respiratory evaporation. This is especially important because hypoxic birds probably always hyperventilate during flight in order to maintain pulmonary O_2 flow, to shift blood acidity favoring O_2 loading, and to compensate for metabolic acid production.

If, as seems true in normoxic pigeons at rest, hypoxic birds in flight take up O_2 across cephalic mucosal surfaces and transfer it to cerebral arterial blood in the ophthalmic retia, it may also be advantageous to keep blood flow through retial exchange vessels high. This entails the risk, however, that the cold venous blood returning from mucosal surfaces could cause excessive brain cooling. A bird flying at high altitude may therefore optimize blood flow through retial exchange vessels to maintain an appropriate brain temperature on the one hand and to supplement the brain's O_2 supply on the other.

REFERENCES

Bech, C., Johansen, K., and Maloiy, G. M. O., 1979, Ventilation and expired gas composition in the Flamingo, Phoenicopterus ruber, during normal respiration and panting, Physiological Zoology, 52:313.
Berger, M., 1974, Energiewechsel von Kolibris beim Schwirrflug unter Hohenbedingungen, Journal für Ornithologie, 115:273.
Berger, M., 1978, Ventilation in the hummingbird Colibri coruscans during altitude hovering, in: "Respiratory Function in Birds, Adult and Embryonic," J. Piiper, ed., Springer, New York.
Bernstein, M. H., 1987, Respiration in flying birds, in: "Bird Respiration," T. J. Sellers, ed., CRC Press, Boca Raton.
Bernstein, M. H., 1989, Temperature and oxygen supply in the avian brain, in: "Comparative Pulmonary Physiology," S. C. Wood, ed., Dekker, New York.
Bernstein, M. H., Curtis, M. B., and Hudson, D. M., 1979, Independence of brain and body temperatures in flying American Kestrels Falco sparverius, American Journal of Physiology, 237:R58.
Bernstein, M. H., Duran, H. L., and Pinshow, B., 1984, Extrapulmonary gas exchange enhances brain oxygen in pigeons, Science, 226:564.
Bernstein, M. H., Sandoval, I., Curtis, M. B., and Hudson, D. M., 1979, Brain temperature in pigeons: Effects of anterior respiratory bypass, Journal of Comparative Physiology, 129:115.
Black, C. P., and Tenney, S. M., 1980, Oxygen transport during progressive hypoxia in high-altitude and sea-level waterfowl, Respiration Physiology, 39:217.

Butler, P. J., West, N. H., and Jones, D. R., 1977, Respiratory and cardiovascular responses of the pigeon to sustained, level flight in a wind-tunnel, Journal of Experimental Biology, 71:7.

Chappell, M. A., and Bucher, T. L., 1987, Effects of temperature and altitude on ventilation and gas exchange in chukars Alectoris chukar, Journal of Comparative Physiology, 157:129.

Clair, P. M., 1985, The rete mirabile ophthalmicum of the double-crested cormorant (Phalacrocorax auritus): its form and function, M.S. Thesis, New Mexico State University.

Crawford, E. C., Jr., and Lasiewski, R. C., 1968, Oxygen consumption and respiratory evaporation of the Emu and Rhea, Condor, 70:333.

Faraci, F. M., Kilgore, D. L., Jr., and Fedde, M. R., 1984, Oxygen delivery to the heart and brain during hypoxia: Pekin Duck vs. Bar-headed Goose, American Journal of Physiology, 247:R69.

Grubb, B., Colacino, J. M., and Schmidt-Nielsen, K., 1978, Cerebral blood flow in birds: effect of hypoxia, American Journal of Physiology, 234:H230.

Hudson, D. M., and Bernstein, M. H., 1981, Temperature regulation and heat balance in flying White-necked Ravens, Corvus cryptoleucus, Journal of Experimental Biology, 90:267.

Hudson, D. M., and Bernstein, M. H., 1983, Gas exchange and energy cost of flight in the white-necked raven, Corvus cryptoleucus, Journal of Experimental Biology, 103:121.

Kawashiro, T., Campos Carles, A., Perry, S. F., and Piiper, J., 1975, Diffusivity of various inert gases in rat skeletal muscle, Pflügers Archiv, 219.

Kilgore, D. L., Jr., Boggs, D. F., and Birchard, G. F., 1979, Role of the rete mirabile ophthalmicum in maintaining the body to brain temperature difference in pigeons, Journal of Comparative Physiology, 129:119.

Lasiewski, R. C., and Dawson, W. R., 1967, A re-examination of the relation between standard metabolic rate and body weight in birds, Condor, 69:13.

List, R. J., 1951, Smithsonian Meteorological Tables, Smithsonian, Washington, DC.

Midtgård, U., 1983, Scaling of the brain and the eye cooling system in birds: A morphometric analysis of the rete ophthalmicum, Journal of Experimental Zoology, 225:197.

Midtgård, U., 1984, Blood vessels and the occurrence of arteriovenous anastomoses in cephalic heat loss areas of Mallards, Anas platyrhynchos (Aves), Zoomorphology, 104:323.

Pinshow, B., Bernstein, M. H., and Arad, Z., 1985, Effects of temperature and PCO_2 on O_2 affinity of pigeon blood: implications for brain O_2 supply, American Journal of Physiology, 249:R758.

Schmidt-Nielsen, K., Hainsworth, F. R., and Murrish, D. E., 1970, Counter current heat exchange in the respiratory passages: effect on water and heat balance, Respiration Physiology, 9:263.

Torre-Bueno, J. R., 1985, The energetics of avian flight at altitude, in: "Biona Report 3: Bird Flight - Vogelflug," W. Nachtigall, ed., Gustav Fischer, Stuttgart.

Weinstein, Y., Bernstein, M. H., Bickler, P. E., Gonzales, D. V., Samaniego, F. C., and Escobedo, M. A., 1985, Blood respiratory properties in pigeons at high altitudes: effects of acclimation, American Journal of Physiology, 249:R765.

BODY AND BRAIN TEMPERATURES IN PIGEONS AT SIMULATED HIGH ALTITUDES*

Marvin H. Bernstein

New Mexico State University
Las Cruces, New Mexico, U. S. A.

INTRODUCTION

Most birds maintain the brain temperature (T_b) below body-core temperature (T_c), and regulate T_b and T_c independently. During exercise or exposure to high ambient temperature (T_a) the temperature difference between the body core and brain (dT = T_c - T_b) is greater than at rest or thermal neutrality, due mainly to elevated T_c. By keeping stored heat in the body core and away from the brain, birds reduce their water requirements while protecting the brain from thermal damage.

Brain cooling depends on the ophthalmic retia (OR), vascular heat exchangers in the temporal regions in most birds. Venous blood enters the anterior of these structures after perfusing the head's cooling surfaces (Bernstein, et al., 1979; Midtgård, 1984a). Warmer arterial blood enters the posterior, gives up heat to the veins, then flows to the brain and eye (Kilgore, et al., 1979; Midtgård, 1984a). Each OR has a sympathetically innervated bypass artery (Midtgård, 1985) that apparently regulates flow through the heat-exchange vessels and thereby helps regulate T_b. Regulation of T_b also depends on venous cooling, as modulated by oropharyngeal evaporation rate (Bernstein, et al., 1979).

At the high altitudes where many birds occur, the need to maintain T_c and T_b and to sustain cerebral oxygenation suggests that hypoxic birds may employ thermoregulatory strategies and cerebral blood-flow routes different from those used during normoxia, and that both brain and core temperatures may reflect this. Since no information on brain temperature is available for hypoxic birds, it seemed of interest to test this idea at simulated high altitudes. For comparison with normoxic thermal stress, T_c and T_b were measured during exposure to high as well as to low T_a.

MATERIALS AND METHODS

Domestic Pigeons (<u>Columba livia</u>, mean mass 0.45 kg) were selected for study because they were easy and economical to obtain and maintain, they are representative of strongly flying birds, and detailed information is available about their thermoregulation and responses to hypoxia.

---
*Supported by N. S. F. grant PCM-8402659.

Commercially acquired birds were maintained on pigeon feed and water within individual cages in a windowless room on a 12L:12D photoperiod at 23°C, 25% RH, and at the barometric pressure (660 Torr) prevailing at the local altitude of 1.2 km above sea level (ASL).

A previously calibrated, electrically insulated, copper-constantan thermocouple (0.2 mm OD) was implanted, under local anesthesia, into the anterior hypothalamus of each bird. Thermocouple position was confirmed by post-mortem examination. Leads were attached to a copper-constantan connector fixed to the scalp. After at least 24 h of recovery, a bird was kept from access to food overnight, then copper and constantan wires were attached between the connector and a digital thermometer with an internal ice-point reference. The output, accurate to $\pm 0.1°C$, was continuously recorded. A similar thermocouple (1 mm OD) was inserted into the cloaca, fastened to the rectrices, and connected to a recorder.

The bird was then placed in a temperature-controlled ($\pm 0.5°C$) hypobaric chamber (± 2 Torr) having interior walls painted flat black, and exposed to a T_a of -6, 9, 24, 35, or 40°C, as recorded with a third thermocouple suspended in the air near the bird's head. The pressure was kept at 660 Torr until a thermal steady state was achieved (about 45 min), after which continuous temperature data were recorded and averaged over 10-15 min. The bird was then subjected to about 30 min of gradual pressure reduction to 308 Torr (equivalent to an altitude of 7 km ASL) or to 231 Torr (9 km ASL). After re-establishment of a steady state, data were again recorded. Finally, to determine whether hypoxia had had any lasting effect, the bird was returned to 1.2 km ASL and a final set of steady-state data recorded.

Results at each T_a-pressure combination were evaluated by analysis of variance; differences were considered significant when the probability of the null hypothesis was 5% or less.

RESULTS

Based on data in Table 1, T_c at 1.2 km ASL averaged 41.0°C, T_b averaged

Table 1. Core temperature (T_c), brain temperature (T_b), and core-brain temperature difference (dT) of Pigeons in relation to air temperature (T_a) and altitude. Data are means $\pm$ SE. <u>n</u> is number of data for each condition.

T_a, °C	-6	9	24	35	40
1.2 km ABOVE SEA LEVEL:					
T_c, °C	41.2±0.3	40.4±0.2	41.2±0.3	41.0±0.1	41.2±0.2
n	12	15	10	6	5
T_b, °C	39.5±0.2	39.4±0.2	39.8±0.2	39.8±0.2	39.9±0.5
7 km ABOVE SEA LEVEL:					
T_c, °C	38.2±0.8	38.9±0.2	39.6±0.3	40.2±0.1	40.7±0.2
n	7	8	8	6	5
T_b, °C	37.4±1.2	38.3±0.3	39.3±0.3	39.5±0.2	40.1±0.3
9 km ABOVE SEA LEVEL:					
T_c, °C		36.5±0.5	38.8±0.5	39.5±0.2	
n		5	5	6	
T_b, °C		36.3±0.5	39.0±0.6	39.2±0.3	

39.7°C, the difference between them was significant, and neither was affected by T_a.

At 7 km ASL, T_c decreased until it was indistinguishable from T_b, and at 40, 35, and 24°C neither was statistically different from T_b at 1.2 km. At T_a of 9 and -6°C, T_c and T_b were significantly lower than the T_b observed at 1.2 km.

At 9 km ASL, pigeons could not tolerate T_a of -6 or 40°C. At intermediate T_a, T_c was indistinguishable from T_b; both averaged about 36°C at T_a of 9°C, compared with about 39°C at T_a of 24 or 35°C. At the latter two T_as, T_c and T_b were not distinguishable from the corresponding values at 7 km, nor from the T_b values at 1.2 km. On repressurization of the chamber, T_c and T_b always returned to within 0.2°C of their initial, normoxic levels.

DISCUSSION

When exposed to thermoneutral and cold air, normoxic Pigeons maintained the T_c and the body-brain temperature difference observed at high T_a. Thus, not only was T_c kept constant at T_a below freezing, but brain cooling was also maintained. Brain cooling may thus signify more than just an adaptation allowing heat storage and water conservation during heat stress or exercise. It may also signify that the temperature optimum for brain is significantly below that for other tissues, necessitating independent thermal regulation even during cold exposure. It is therefore likely that, even at subfreezing T_a, perfusion of oronasal cooling surfaces and of the OR continue unabated, and that venous blood arrives at the OR at a lower temperature than arterial blood, just as during panting and/or gular flutter at high T_a.

This point of view is consistent with the observation that, in Pigeons exposed to simulated high altitudes and to moderate or high air temperature, core temperature decreased but brain temperature did not change, compared with normoxic values. It is as if an advantage accrued to maintaining T_b at about 39°C, irrespective of temperature in other tissues. The decrease in T_c to T_b implies that metabolism and O_2 consumption decreased in most body tissues, but not in brain.

Abolition of the body-brain temperature difference at high altitude suggests one or more of the following changes: (1) blood flow through the heat-exchanging vessels of the OR decreased; (2) perfusion of the evaporative surfaces of the oral and pharyngeal cavities decreased; (3) cooling of the oropharyngeal surfaces, and therefore of the venous return to the OR, decreased. It is likely that the oral and nasal tissues, and therefore the blood returning from them, would be substantially colder below 24°C than at or above this T_a. Condition (3) is therefore unlikely, and vascular adjustments, such as (1) and (2), are more likely to effect the disappearance of dT.

The further fall in T_c and T_b at T_a below 24°C, and at 7 or 9 km ASL, suggests that hypoxic Pigeons may further reduce metabolism, saving more energy and limiting O_2 demand. This also decreases the body-environment temperature gradient, reducing heat loss, but may require abandonment of optimal brain temperature.

It has been suggested that in the cold, the OR transfers heat from arteries to veins to minimize heat loss from the head (Frost, et al., 1975). This seems unlikely, since the arterial outflow from the OR apparently perfuses only a small fraction of the cephalic integument (Midtgård, 1984b), most flowing to brain and orbit.

It is instructive to consider the effect of reduced T_b and T_c during hypoxia on O_2 transport. First, if these effects are ignored, then arterial O_2 saturations at T_a of 9°C and altitudes of 1.2, 7, and 9 km ASL are 89, 35, and 19%, respectively, as calculated by the method of Pinshow, et al. (1985). This assumes that arterial PO_2 for resting Pigeons at these altitudes are the same as reported by Weinstein, et al. (1985). When T_c is taken into account, however, using the same arterial PO_2 values, arterial O_2 saturation at 7 and 9 km ASL become 54 and 42%, respectively.

Thus, by reducing T_c, Pigeons may improve arterial saturation from 35 to 54% at 7 km ASL and from 19 to 42% at 9 km ASL, thereby improving tissue O_2 delivery. The additional effects of hypocapnic alkalosis, associated with hyperventilation, or of increased lactic-acid production on blood O_2 affinity and saturation remains to be determined.

In conclusion, normoxic Pigeons maintained brain temperature below body temperature even at subfreezing T_a. When hypoxic at thermal neutrality or above, they reduced the temperature of the body core until it was equal to that of the brain. This implies that they also abolished cooling of their cephalic arterial blood. By doing this they probably conserved energy, lowered O_2 demand, and maintained the brain's thermal optimum. When exposed to both cold stress and environmental hypoxia, Pigeons further reduced body and brain temperature, probably at the expense of optimal brain function, but by so doing they may have improved O_2 delivery to their tissues.

REFERENCES

Bernstein, M. H., Sandoval, I., Curtis, M. B., and Hudson, D. M., 1979, Brain temperature in pigeons: Effects of anterior respiratory bypass, Journal of Comparative Physiology, 129:115-118.

Frost, P. G. H., Siegfried, W. R., and Greenwood, P. J., 1975, Arteriovenous heat exchange systems in the Jackass Penguin Spheniscus demersus, Journal of Zoology, 175:231-241.

Kilgore, D. L., Jr., Boggs, D. F., and Birchard, G. F., 1979, Role of the rete mirabile ophthalmicum in maintaining the body to brain temperature difference in pigeons, Journal of Comparative Physiology, 129:119-122.

Midtgård, U., 1984a, The blood vascular system in the head of the Herring Gull (Larus argentatus), Journal of Morphology, 179:135-152.

Midtgård, U., 1984b, Blood vessels and the occurrence of arteriovenous anastomoses in cephalic heat loss areas of Mallards, Anas platyrhynchos (Aves), Zoomorphology, 104:323-335.

Midtgård, U., 1985, Innervation of the avian ophthalmic rete, Fortschritte der Zoologie, 30:401-404.

Pinshow, B., Bernstein, M. H., and Arad, Z., 1985, Effects of temperature and PCO_2 and O_2 affinity of pigeon blood: implications for brain O_2 supply, American Journal of Physiology, 249:R758-R764.

Weinstein, Y., Bernstein, M. H., Bickler, P. E., Gonzales, D. V., Samaniego, F. C., and Escobedo, M. A., 1985, Blood respiratory properties in pigeons at high altitudes: effects of acclimation, American Journal of Physiology, 249:R765-R775.

CIRCULATORY ADAPTATIONS TO COLD IN BIRDS

Uffe Midtgård

Institute of Cell Biology and Anatomy
Universitetsparken 15
DK-2100 Copenhagen, Denmark

INTRODUCTION

The maintenance of high body temperatures in homeothermic animals is to a large extent dependent on an effective insulation like feathers, fur, or subcutaneous fat. However, in most animals the insulative covering is more or less incomplete, and in birds the beak, eyes, feet, and underside of wings are generally naked or poorly feathered. Together with the upper part of the respiratory tract, these surfaces constitute potential heat loss areas and have therfore been characterized as thermal windows (Schmidt-Nielsen, 1983). There is an obvious advantage in reducing heat loss from the naked extremities in a cool environment, but this is complicated by the circulating blood which will enevitably carry heat to these areas from the body core by means of internal convection. In contrast, a surplus of heat flow to unprotected areas may be of vital importance in order to prevent the tissue from freezing during exposure to subzero ambient temperature.

Birds have evolved several structures and mechanisms which enable them to cope with the conflicting demands of reducing heat loss to the environment and avoiding frostbite in exposed tissues. The circulatory adaptations to cold that will be discussed in this paper are: 1) arteriovenous (AV) heat exchange systems in which warm arterial blood from the body core is cooled by countercurrent heat exchange with cool venous blood returning from the periphery; 2) arteriovenous anastomoses (AVAs), which are large shunts that allow high rates of peripheral blood flow; and 3) circulatory responses to cold, especially cold induced vasodilatation (CIVD).

ARTERIOVENOUS HEAT EXCHANGE SYSTEMS

The arteries and veins usually run side by side throughout the body, and in the peripheral parts of the animal the two blood streams will often have different temperatures. Therefore, countercurrent heat exchange can theoretically occur within any set of peripheral artery and vein. This was realized already in 1876 by Claude Bernard and later demonstrated by recordings of intravascular temperatures in the arm of humans (Bazett et al., 1948). Scholander (1955) was the first to point out the importance of this phenomenon in reducing heat loss from the naked extremities of homeothermic animals, and it was realized that veins grouped around a central artery (venae comitantes system) or vascular networks (retia mirabilia), which had previously been considered only as anatomical peculiarities, were in fact complex AV heat exchange systems. Consequently, environmental physiologists became interested in these structures and information on the vas-

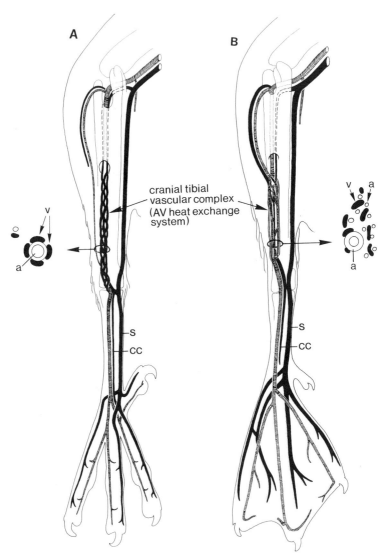

Fig. 1. Drawings showing the structure and location of the AV heat exchange system in birds with a venae comitantes system (A) and a tibiotarsal rete (B). Abbreviations: a, artery; cc, deep countercurrent veins; s, superficial collateral vein. From Midtgård (1986).

cular arrangement in the extremities of mammals and birds was retrieved from the old anatomical literature, which unfortunately led to some eroneous statements. However, there is now sufficient comparative studies available to give an accurate and detailed account of the structure and distribution of some of the AV heat exchange systems in birds.

The AV heat exchange system in the hind limb

The heat exchange system in the leg is located in the tibiotarsal region near the feather margin, and it serves to reduce heat loss from the feet. The arterial blood from the body gives up its heat to the incoming venous blood, and by this mechanism the temperature of the foot can be maintained near that of the environment. In support of the heat conserving function of the vascular arrangement is the observation that the zone of AV contact is most extensive in the region where the steepest temperature gradients have been recorded (e.g. Kahl, 1963; Ederstrom and Brumleve, 1964). However, more direct experimental evidence for the heat conserving function comes from calorimetric experiments showing that heat loss from the feet of ducks increased after surgical blockade of the heat exchanger in the leg (Midtgård 1980a).

In a comparative study of more than 60 species of birds from 21 orders, Midtgård (1981) found that the heat exchangers generally could be classified either as a rete mirabile type or venae comitantes system (Fig. 1). However, to be more correct, the variation in the heat exchanger constitutes a continuum ranging from retia with more than 100 blood vessels, over single artery-single vein arrangements, to well-developed venae comitantes systems. Table 1 summarizes the distribution of the two types of heat exchangers. In contrast to what it says in many textbooks, it is obvious that a rete mirabile is not unique to long-legged wading birds. It is also clear that the presence of retia is not particularly related to the habitat or the geographical distribution of the birds. Since the heat exchanger in the leg appears to be equally well-developed in arctic and tropically distributed species, this vascular structure probably represents and adaptive advantage in all climates. However, there is a tendency for some arctic birds to have more extensive contact between the arteries and veins in the legs than their close relatives in warmer areas (Midtgård, 1981).

The measurements of heat loss from birds feet that have been conducted so far suggest that a venae comitantes system is as efficient in heat conservation as a rete mirabile (compare Steen and Steen, 1965; Kilgore and Schmidt-Nielsen, 1975; Midtgård, 1980a). It has been suggested that a rete mirabile might be disadvantageous in cold environments since it renders the feet susceptible to frost bite. However, the tibiotarsal rete of birds is equipped with both an arterial and venous shunt that allows the blood to bypass the heat exchanger when there is need for distribution of warm blood to the feet.

The AV heat exchange system in the wing

Information about the association of arteries and veins in the wing is limited to very few species. A rete mirabile with a presumed heat conserving function has so far only been recorded in the upper part of the flippers in penguins. The number of

Table 1. Structure of the AV heat exchanger in the hind limb of short-legged and long-legged birds. From Midtgård (1986) with additional data from Hyrtl (1864).

Heat exchanger type	short-legged birds	long-legged birds
well-developed rete	ducks geese swan cormorant grebes anhinga	flamingo wood storks
	galliforms kiwi rhea emu ostrich	
simple rete	penguin gannet owl	cranes bustard ibis storks
	rails	
venae comitantes system	loon guillemots shearwater diving petrel gulls pigeon	heron
	shorebirds falcons passerines	
	humming birds	

arteries involved in the formation of the rete ranges from 3-5 in the smaller species to about 15 in the Emperor penguin (Aptenodytes forsteri) (Watson, 1883; Trawa, 1970; Frost et al., 1975). The presence of a rete mirabile in the flippers of penguins could be associated with the under-water "flight" of these animals; an idea that could be supported if retia were found in other species using the wings for propulsion during diving (e.g. diving petrels and auks). The finding of a very well-developed venae comitantes system in the wing of the Turkey vulture (Arad, Midtgård and Bernstein, in prep.) suggests that the variability in the structure of the AV heat exchanger in the wing will be found to equal that seen in the hind limb.

The AV heat exchange system in the head

The most conspicuous AV heat exchangers in the head are the ophthalmic retia which are located bilaterally in the temporal fossa. There is strong physiological and anatomical evidence that the arterial blood leaving the rete has been cooled by countercurrent heat exchange with venous blood returning from various heat loss areas (eye, nasal cavity, beak, and palate) and is destined for the brain (Bernstein et al., 1979; Kilgore et al., 1979; Midtgård, 1986). By this mechanism the brain remains cooler than the rest of the body, which is of great importance in preventing heat stroke during an internal or external heat load. However, rather than being a structure which has its primary function during heat exposure, Frost et al. (1975) suggested that the ophthalmic rete is important in reducing heat loss from the naked and poorly feathered areas of the head. The vascular anatomy suggests that this function would chiefly concern the eyes and periorbital skin. Surely, animals like birds would seem in particular need of structures which could reduce the heat loss from the eyes during exposure to the extreme forced convection associated with flight or diving. Thus, the ophthalmic rete may have a dual function in temperature regulation.

ARTERIOVENOUS ANASTOMOSES

Arteriovenous anastomoses are muscular shunts with a luminal diameter which in the dilated state is 6-10 times that of capillaries. Direct communication between arteries and veins were identified more than 250 years ago (for review cf. Clara, 1956), but it was not until late in the 19th century with the development of histological techniques that information on AVAs started to accumulate. In spite of extensive anatomical evidence the existence of AVAs were denied by many, or they were at best considered only as pathological structures. It is now acknowledged that the AVAs are naturally occuring structures, and in birds they are present in high numbers in the oral and nasal mucosa and in the naked and poorly feathered skin (feet, brood patch, beak, wattles, eyelids) (Midtgård 1986).

Ever since their discovery, the cutaneous AVAs have been ascribed a thermoregulatory role and this has been confirmed during the last two decades by using the radioactive microsphere technique in both mammals (Hales, 1985) and birds (Hillman et al., 1982, Wolfenson, 1983). Hillman et al. (1982) found that AVA blood flow in the feet of chickens constitutes about 8% of total blood flow to the foot at $5^{o}C$ and increases to 63% at an abient temperature of $36^{o}C$. Thus, it appears that the AVAs form the structural basis for high rates of peripheral blood flow and resultant transport of heat from the body core to the periphery. As discussed below, there is also physiological evidence that the AVAs dilate when the tissue is exposed to subzero ambient temperatures.

CIRCULATORY ADJUSTMENTS TO COLD

Peripheral adjustments

Most of the information on the vascular response to cold and peripheral vasomotor control in birds comes from experiments with ice immersed feet. The characteristic temperature pattern seen upon immersion of the feet of ducks and chickens in ice water was first described by Grant and Bland (1931). Immediately after immersion, the temperature of the foot fell to near that of the water, but a few minutes later periodic increases in the temperature was observed. This temperature response, which closely resembled the "hunting reactions" described in the human finger by Lewis (1930), was best displayed in areas rich in AVAs, and it was therefore suggested that these structures were in part responsible for the cold induced vasodilatation (CIVD).

An important contribution to the understanding of the vascular response to cold and its adaptive value was provided by Johansen and Millard (1973) and later Murrish and Guard (1977). Both employed electromagnetic flowmeters and recorded intravascular temperatures in the feet of the Antarctic Giant fulmar (Macronectes giganteus) during different thermal stimulations. They found that ice immersion of the feet, as experienced by the bird in its natural environment, produced an immediate increase in blood flow followed by vasoconstriction and later on by periodic vasodilatation (hunting reactions). The initial increase in blood flow, which was termed the "cold flush" (Johansen and Millard, 1973), was accompanied by increases in the temperature of the arterial and venous blood in the web (Fig. 2). The observation of pulsatile venous pressure in the web and the absence of and AV oxygen difference during peak blood flow suggest that AVAs participate in the cold flush. The CIVD of the feet in the Giant fulmar differs from that previously observed in the extremities of other animals in that the increase in flow in the fulmar is not preceeded by vasoconstriciton and the hunting reactions appear more quickly.

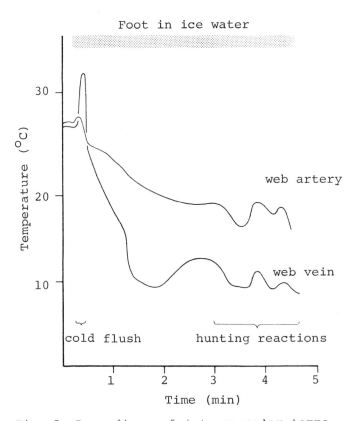

Fig. 2. Recordings of intravascular temperatures in the ice immersed foot of the Giant fulmar. Redrawn from Johansen and Millard (1973).

CIVD is not restricted to the feet of birds. Midtgård et al. (1985) measured blood flow in the brood patch of bantam hens by the 133-xenon washout technique and found that blood flow increased in response to cooling (Fig. 3). By using the radioactive microsphere technique, it has recently been shown that dilatation of AVAs appears to be responsible for the increase in blood flow observed during cooling of the brood patch (Hales, Midtgård and Fawcett, in prep.). While quite low temperatures are required to elicit CIVD in the feet, the brood patch responds with vasodilatation after moderate cooling. The importance of CIVD in the brood patch is obvious: it secures a high rate of blood flow and transport of heat to the patch during incubation in a cold environment, and it will also speed up warming of a cold clutch when the hen returns to the nest from feeding excursions.

Control mechanisms

The peripheral blood vessels in birds are well supplied with adrenergic nerves (e.g. Midtgård, 1980b; Molyneux and Harmon, 1982), and variations in blood flow can in part be ascribed to variation in adrenergic vasoconstrictor tone. However, active, neurogenic vasodilatation may also occur. For example, Johansen and Millard (1974) reported that CIVD in the feet of the Giant fulmar consists of an initial short phase (the cold flush), which

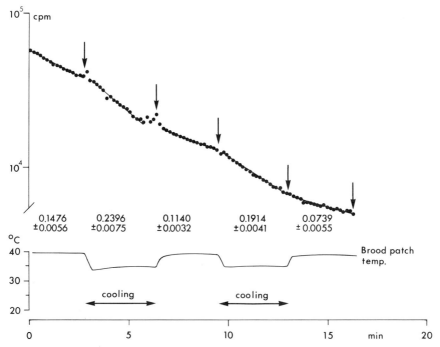

Fig. 3. Effect of cooling on brood patch blood flow in a restrained bantam hen. Cooling the brood patch increases the washout rate of 133-xenon (top curve). The numbers below the washout curve are the rate constants which are proportional to blood flow. From Midtgård et al. (1985).

is cholinergic, and a long lasting phase which is atropine resistant. On the other hand, Murrish and Guard (1977) concluded that the cold flush was due to beta-adrenergic stimulation of AVAs. Pharmacoligcal studies in domestic ducks and chickens have confirmed the presence of neurogenic vasodilatation in the feet, but it is neither adrenergic nor cholinergic in nature (McGregor 1979). Hillman et al. (1982) used radioactive microspheres to distinguish between capillary and AVA blood flow in the feet of chickens. They found that blood flow through capillaries, but not through AVAs, increased after beta-adrenergic stimulation. AVA blood flow increased with nerve stimulation during adrenergic blockade, whereas in a warm environment, with high AVA blood flow, nerve blockade decreased blood flow through AVAs. Furthermore, they found that adenosine triphosphate selectively increased AVA blood flow, and consequently they suggested that the active vasodilatation of the AVAs was mediated by purinergic nerves. This is apparently supported by the observation that some nerve terminals around AVAs contain a high proportion of large dense-cored vesicles (Molyneux and Harmon, 1982), which is postulated to be characteristic of purinergic nerves (Burnstock, 1972). However, vasoactive intestinal polypeptide (VIP) is also stored in large dense-cored vesicles (Larsson et al., 1977), and recently this neuropeptide has been found in nerves surrounding the AVAs of the brood patch (Midtgård 1988), feet and wattles (unpublished observations). Since VIP plays an important role in neurogenic vasodilatation in mammals (e.g. Lundberg et al., 1982), it is reasonable to suggest that VIP also may be involved in CIVD in the feet and brood patch of birds.

In addition to the afore-mentioned neurogenic vascular response to cold, blood flow in peripheral areas also changes due to direct temperature effects in the exposed tissues. First of all, the viscosity of the blood increases during cooling and this in itself reduces blood flow considerably (e.g. Millard and Reite, 1975; Guard and Murrish, 1975). Secondly, the vasoconstrictor response to noradrenaline and perivascular nerve stimulation in the feet of ducks has been found to decrease during cooling (Millard and Reite, 1975; Reite et al., 1977). This indicates a direct temperature effect on the vascular smooth muscles and on nerve transmission and transmitter release. Thirdly, acute cooling has been reported to have an excitatory effect on smooth muscle cells in some blood vessels (Smith 1952), whereas in others it apparently decreases the myogenic (inherent) tone (Winquist and Bevan, 1980). Finnaly, with low blood flow levels there is a tendency for accumulation of vasodilatory substances in the tissue and this may eventually lead to relaxation. For example, accumulation of locally produced prostaglandins has recently been suggested to be responsible for CIVD, since administration of drugs which interfere with the synthesis of prostaglandins almost eliminates the hunting reactions in the cat paw (Franz, 1985). The hunting reactions seen in ice immersed feet can in part be explained as cold blockade of nerves and reduced sensitivity of vascular smooth muscles to noradrenaline at low tissue temperatures. That is, cooling inititates vasoconstriction and as the temperature decreases nerve blockade follows and the smooth muscle cells become unresponsive to noradrenaline and consequently relax. The tissue warming which is associated with vasodilatation relieves the cold blockade and renders the blood vessels responsive to the transmitters again (see Folkow et al., 1963 for further discussion). The suggestion that hunting reactions may be due to local events is apparently supported by the fact that CIVD can be elicited in the acutely denervated cat paw (Folkow et al., 1963) and in fingers of sympathectomized humans (Lewis, 1930; Greenfield et al., 1951).

Summarizing this part, it seems that the peripheral blood vessels is supplied with both vasoconstrictor and vasodilator nerves, and the response to cold depends on the interaction of several mechanisms. CIVD involves AVAs and may be ascribed to release of adrenergic tone (due to cold blockade of nerves and/or reduced contractability of smooth muscles) and to active vasodilatation, possibly mediated by acetylcholine and VIP.

Regulation of AV heat exchange

The vascular arrangement in the head, wings and feet suggests that the venous return may bypass the heat exchangers when the situation calls for heat dissipation rather than heat conservation. Furthermore, species with a rete mirabile type of heat exchanger also have a shunt for arterial bypass of the heat exchange units. Based on observations of the muscularity and innervation of the blood vessels in the hind limb of ducks, Midtgård (1980) pointed to these shunts as possible sites for regulation of AV heat exchange. Later, it was found that the arterial shunt, in line with its muscularity and dense adrenergic innervation, was capable of complete closure (Midtgård and Bech, 1981). A similar sphincteric action of the venous shunt could, however, not be demonstrated.

Based on anatomical and physiological observations, Murrish and Guard (1977) presented a conceptual model for blood flow in the feet of the Giant fulmar. This model holds that the AVAs are closed at low ambient temperature and blood from capillaries returns to the AV heat exchange system via the countercurrent veins, but when heat dissipation is neeeded, the AVAs are open and blood returns to the body via the superficial collateral vein. However, Hillman et al. (1982) measured blood flow in the deep and superficial metatarsal veins of chickens and found no evidence for selective return of blood from capillaries and AVAs. Although the large superficial vein appears to conduct most of the blood from the AVAs in a warm environment, there is still a considerable fraction of AVA flow in the countercurrent veins. Furthermore, they found that at low ambient temperature, blood flow in the superficial vein was about double that in the countercurrent veins. It is still possible, however, that constriction of the tibial shunt vein may divert the blood towards the heat exchanger, but the regulatory mechanism remains to be determined.

Central adjustments and acclimation to cold

Among the central responses to cold is a 2-3 fold increase in heart rate seen during shivering evoked either by low environmental temperature (e.g. Aulie) or by cold eggs in incubating birds (Gabrielsen and Steen, 1979; Tøien et al., 1986). Although there is no direct observations on blood flow in shivering muscles or in thermogenic tissues in general in birds, the increase in heart rate which is associated with shivering suggests a redistribution of circulating blood during cold exposure. The increased heart rate may first of all be coupled with an increased nutritive demand of the shivering muscles, but it may also be important in distributing the heat generated (Aulie, 1976).

The circulatory adjustments that have been discussed so far are those which occur in response to acute cold. However, the cardiovascular system also responds to long-term cold exposure. Rautenberg (1969) found that cold acclimated pigeons had higher foot temperatures than controls in a similar environment. This is suggestive of an increased peripheral circulation, which may be ascribed to an elevation in blood pressure, as reported for cold acclimated chickens (Sturkie, 1967), or to an increased vascularity. Heroux and St. Pierre (1957) found that the number of blood vessels in the ears of rats increased 12-fold after exposure to a $6^{\circ}C$ environment for 4 weeks. In birds, there is no direct evidence for an increased capillarity of the skin after cold exposure, but it appears that the density of AVAs increases after cold acclimation (Midtgård, unpublished). An increase in vascularity obviously reduces the risk of cold injury in exposed tissue, as exemplified by the observation that gulls kept indoors during winter immediately get frostbite in the feet when released to the outside at $-20^{\circ}C$ (Scholander et al., 1950).

ACKNOWLEDGMENT

Supported by the Carlsberg Foundation.

REFERENCES

Aulie, A., 1976. The shivering pattern in an arctic (willow ptarmigan) and a tropical bird (bantam hen). Comp. Biochem. Physiol., 53A:347.

Bazett, H.C., Love, L., Newton, M., Eisenberg, L., Day, R., and Foster, R., 1948. Temperature changes in blood flowing in arteries and veins in man. J. Appl. Physiol., 1:3.

Bernstein, M.H., Sandoval, I., Curtis, M.B., and Hudson, D.M., 1979. Brain temperature in pigeons: Effects of anterior respiratory bypass. J. Comp. Physiol., 129:115.

Burnstock, G., 1972. Purinergic nerves. Pharmacol. Rev., 24:509.

Clara, M., 1956. Die arterio-venösen Anastomosen. Anatomie, Biologie, Pathologie. Springer, Wien.

Ederstrom, H.E., Brumleve, S.J., 1964. Temperature gradients in the legs of cold-acclimatized pheasants. Am. J. Physiol., 207:457.

Folkow, B., Fox, R.H., Krog, J., Odelram, H., and Thoren, O., 1963. Studies on the reactions of cutaneous vessels to cold exposure. Acta Physiol. Scand., 58:342.

Franz, D.R., 1985. The effect of indomethacin on cold-induced vasodilatation in the cat. T. Therm. Biol., 10:245.

Frost, P.G.H., Siegfried, W.R., and Greenwood, P.J., 1975. Arterio-venous heat exchange systems in the Jackass penguin, Spheniscus demersus. J. Zool., Lond., 175:231.

Gabrielsen, G., and Steen, J.B., 1979. Tachycardia during egg-hypothermia in incubating ptarmigan (Lagopus lagopus). Acta Physiol. Scand., 107:273.

Grant, R.T., and Bland, E.F., 1931. Observations on arteriovenous anastomoses in human skin and in the bird's foot with special reference to the reaction to cold. Heart, 15:385.

Greenfield, A.D.M., Shepherd, J.T., and Whelan, R.F., 1951. The part played by the nervous system in the response to cold of the circulation through the finger tip. Clin. Sci., 10:347.

Guard, C.L., and Murrish, D.E., 1975. Effects of temperature on

viscous behavior of blood from antarctic birds and mammals. Comp. Biochem. Physiol., 52:287.
Hales, J.R.S., 1985. Skin arteriovenous anstomoses, their control and role in thermoregulation. In: "Cardiovascular Shunts", K. Johansen and W. Burggren, eds., Munksgaard, Copenhagen.
Hales, J.R.S., Midtgård, U., and Fawcett, A.A. Arteriovenous anastomoses are the target of cutaneous cold-induced vasodilatation. In prep.
Heroux, O., and St. Pierre, J., 1957. Effect of cold acclimation on vascularization of ears, heart, liver and muscles of white rats. Am. J. Physiol., 188:163.
Hillman, P.E., Scott, N.R., and van Tienhoven, A., 1982. Vasomotion in chicken foot: dual innervation of arteriovenous anastomoses. Am. J. Physiol., 242:R582.
Hyrtl, J., 1864. Neue Wundernetze und Geflechte bei Vögeln und Säugethieren. Denksch. d. Kaiserl. Akad. d. Wissensch. Wien, 22:113.
Johansen, K., and Millard, R.W., 1973. Vascular responses to temperature in the foot of the Giant fulmar, Macronectes giganteus. J. Comp. Physiol., 85:47.
Johansen, K., and Millard, R.W., 1974. Cold-induced neurogenic vasodilatation in the skin of the Giant fulmar, Macronectes giganteus. Am. J. Physiol., 227:1232.
Kahl, M.R., 1963. Thermoregulation in the Wood stork, with special reference to the role of the legs. Physiol. Zool., 36:141.
Kilgore, D.L., Jr., and Schmidt-Nielsen, K., 1975. Heat loss from ducks' feet immersed in cold water. Condor, 77:475.
Kilgore, D.L., Jr., Boggs, D.F., and Birchard, C.F., 1979. Role of the rete mirabile ophthalmicum in maintaining the body-to-brain temperature difference in pigeons. J. Comp. Physiol., 129:119.
Larsson, L.-I., 1977. Ultrastructural localization of a new neuronal peptide (VIP). Histochemistry, 54:173.
Lewis, T., 1930. Observation upon the reactions of the vessels of the human skin to cold. Heart, 15:177.
Lundberg, J.M., Anggård, A., and Fahrenkrug, J., 1982. VIP as a mediator of hexamethonium-sensitive, atropine-resistant vasodilation in the cat tongue. Acta Physiol. Scand., 116:387.
McGregor, D.D., 1979. Noncholinergic vasodilator innervation in the feet of ducks and chickens. Am. J. Physiol., 237:H112.
Midtgård, U., 1980a. Heat loss from the feet of mallards Anas platyrhynchos and arterio-venous heat exchange in the rete tibiotarsale. Ibis, 122:354.
Midtgård, U., 1980b. Blood vessels in the hind limb of the mallard (Anas platyrhynchos): Anatomical evidence for a sphincteric action of shunt vessels in connection with the arterio-venous heat exchange system. Acta Zool., Stockh., 61:39.
Midtgård, U., 1981. The rete tibiotarsale and arterio-venous association in the hind limb of birds: A comparative morphological study on counter-current heat exchange systems. Acta Zool., Stockh., 62:67.
Midtgård, U., 1986. The peripheral circulatory system in birds. Thesis, University of Copenhagen, Copenhagen.
Midtgård, U., 1988. Innervation of arteriovenous anastomoses in the brood patch of the domestic fowl. Cell Tissue Res., 252:207.
Midtgård, U., and Bech, C., 1981. Responses to catecholamines and nerve stimulation of the perfused rete tibiotarsale and

associated blood vessels in the hind limb of the Mallard (Anas platyrhynchos). Acta Physiol. Scand., 112:77.

Midtgård, U., Sejrsen, P., and Johansen, K., 1985. Blood flow in the brood patch of Bantam hens: evidence of cold vasodilatation. J. Comp. Physiol. B., 155:703.

Millard, R.W., and Reite, O.B., 1975. Peripheral vascular response to norepinephrine at temperatures from 2 to 40°C. J. Appl. Physiol., 38:26.

Molyneux, G.S., and Harmon, B., 1982. Innervation of arteriovenous anastomoses in the web of the foot of the domestic duck, Anas platyrhynchos. Structural evidence for the presence of non-adrenergic non-cholinergic nerves. J. Anat. 135:119.

Murrish, D.E., and Guard, C.L., 1977. Cardiovascular adaptations of the Giant petrel, Macronectes giganteus, to the antarctic environment. In: "Adaptations within Antarctic Ecosystems", G.A. Llano, ed., Smithsonian Institute, Washington D.C.

Rautenberg, W., 1969. Untersuchungen zur Temperaturregulation wärme- und kälteakklimatisierter Tauben. Z. Vergl. Physiol., 62:221.

Reite, O.B., Millard, R.W., and Johansen, K., 1977. Effects of low temperature on peripheral vascular control mechanisms. Acta Physiol. Scand., 101:247.

Schmidt-Nielsen, K., 1983. Animal Physiology. University Press, Cambridge.

Scholander, P.F., 1955. Evolution of climatic adaptations in homeotherms. Evolution, 9:15.

Scholander, P.F., Walters, V., Hock, R., and Irving, L., 1950. Body insulation of some arctic and tropical mammals and birds. Biol. Bull., 99:225.

Smith, D.J., 1952. Constriction of isolated arteries and their vasa vasorum produced by low temperatures. Am. J. Physiol., 171:528.

Steen, I., and Steen, J.B., 1965. The importance of the legs in the thermoregulation of birds. Acta Physiol. Scand., 63:285.

Sturkie, P.D., 1967. Cardiovascular effects of acclimatization to heat and cold in chickens. J. Appl. Physiol., 22:13.

Trawa, G., 1970. Note preliminaire sur la vascularisation des membres des spheniscides de Terre Adelie. L'Oiseau Rev. Franc. Ornithol., 40:142.

Tøien, Ø., Aulie, A., and Steen, J.B., 1986. Thermoregulatory responses to egg cooling in incubating bantam hens. J. Comp. Physiol. B., 156:303.

Watson, M., 1883. Report on the anatomy of the Spheniscidae. Report on the scientific results of the voyage of H.M.S. Challenger during the years 1873-76, London, Zoology Vol., 7:1.

Winquist, R.J., and Bevan, J.A., 1980. Temperature sensitivity of tone in the rabbit facial vein: Myogenic mechanism for cranial thermoregulation. Science, 207:1001.

Wolfenson, D. 1983. Blood flow through arteriovenous anastomoses and iths thermal function in the laying hen. J. Physiol., Lond., 334:395.

CONTROL OF CARDIORESPIRATION DURING SHIVERING THERMOGENESIS
IN THE PIGEON

Werner Rautenberg

Ruhr-Universität Bochum
Faculty of Biology, AG. Temperaturregulation
D-4630 Bochum, Postbox 102148, F.R.G.

INTRODUCTION

Shivering thermogenesis is the most important source of extra heat production in resting birds to maintain deep body temperature at a constant level under cold conditions. The main heat by shivering is produced by the large breast muscle whose mass amounts to 15-25% of avian's body weight. The frequency of electrical activity recorded in the pectoral muscle is about 200 Hz and seems to be independent of the intensity of tremor (Hohtola 1982). The cold tremor in pigeons is obviously due to contractions of white, fast-twitch glycolytic muscle fiber types (George, 1984). Hohtola (1982) had previously demonstrated a close correlation between the integrated EMG during shivering and the oxygen consumption in pigeons. No increase of blood lactate acid was found during violent shivering which resulted in up to five times of basic metabolism in birds and mammals (cf. Bligh, 1983). In contrast to severe physical work no oxygen debt has been observed during shivering. The increase of oxygen consumption with increasing intensity of cold tremor refers to a precise cooperation between temperature regulation and the cardiorespiratory system. How this interaction of the autonomic feedback control systems may function, shall be described in this paper.

METHOD

Animals. We used conscious pigeons (*Columba livia domestica*), which were equipped with chronic vertebral thermodes, arterial (*A. brachialis*) and venous (*V. cutaneous ulnaris*) blood catheters. The tip of the last one was positioned in the right heart ventricle to obtain mixed venous blood. The birds were fitted with an air-tight plastic helmet, thermocouples and placed in a body plethysmograph, whose temperature could by controlled (for detailed method s. Barnas et al., 1984); Barnas and Rautenberg, 1984).

---
* The study was supported by the DFG (Ra 167/9-1).

Measurements: Continuous recordings were made of ambient (T_a) and body temperatures, further, of respiratory oxygen consumption (V_{O_2}), of arterial blood pressure (BP_a), heart rate (HR), tidal volume (V_T), respiratory frequency (f_{resp}), original and integrated EMG of pectoralis muscle. Partial pressures of oxygen and carbon dioxide of arterial (P_{aO_2} and P_{aCO_2}) and of mixed venous blood ($P_{\bar{v}O_2}$ and $P_{\bar{v}CO_2}$), acid-base status (pH), O_2 content of blood (C_{O_2}) and hematocrit (Ht) of each blood samples were analysed immediately (Barnas et al., 1984). Cardiac output (CO) and minute ventilation ($\dot{V}_E$) were calculated.

Protocol: Various intensities of shivering were produced by external coolings at 5°C T_a, by selective spinal cord coolings at neutral T_a 28°C and by a combination of external and central coolings. Spinal cord temperature (T_{vc}) was lowered to about 36°C. The various coolings were randomly made to avoid any adaptations. Blood samples were taken during rest and during shivering when birds has reached a steady state level of V_{O_2} and T_b at constant cooling.

RESULTS

The relationship between shivering and cardiorespiration may be demonstrated by spinal cord cooling of neutral T_a 28°C. A lowering of T_{vc} from 41.2 to 36.8°C evokes shivering within few seconds. Start and increase of cold tremor is shown by EMG and its integration (Fig. 1). V_T, f_{resp} and HR synchronously increase with increasing EMG.

The original recordings in figure 1 impressively show the simultaneous responses of respiratory and circulatory effector actions (V_T, f_{resp}, HR) with onset and increase of muscle tremor. The blood pressure, however, is unchanged during this dynamic phase of shivering thermogenesis. The blood pressure is an important controlled variable in the circulatory feedback system whose signals control the heart action via peripherally located baro-receptors. The results clearly show a drive of effector or control action of circulation without any stimulations of its receptors or feedback elements, respectively. Similar responses also occured during external cooling. In this case, however, shivering does only slowly increase and a dynamic phase is less demonstrable. The control actions of respiration, V_T and f_{resp}, also responded at onset of shivering. Analyses of blood gases and pH in this transient state of cold tremor never showed deviations of these controlled variables of the respiratory feedback system.

Analyses of all parameters (s. method) were done in the steady state when V_{O_2} and T_b had reached a nearly constant level during the various cold stimulations. Figure 2 firstly shows the cardiovasculatory responses during shivering. The various increases of heat production ($\dot{M}$) by different cold stimulations well correspond to increases of cardiac output (CO), the important measurement of control action in the circulatory feedback system. The increase of CO was closely related to $\dot{M}$ (Fig. 3) and mainly caused by increase of HR. However, the important controlled variable MBP_a which normally adjusts the action of CO via signals of baro-receptors seems not be affected by shivering. We suggest, therefore, a direct

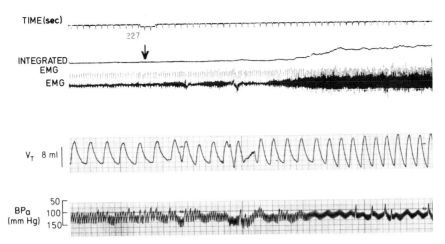

Fig. 1. Original recordings of EMG and cardiorespiratory responses in the transient state of shivering evoked by spinal cord cooling (ΔT_{VC} 4.4°C) at neutral T_a 28°C. Traces from top to bottom: time marker (s), arrow indicates start of spinal cooling; integrated EMG; EMG; tidal volume (V_T) and breathing frequency, arterial blood pressure (BP_a) and heart rate (after Barnas and Rautenberg, 1984 with permission).

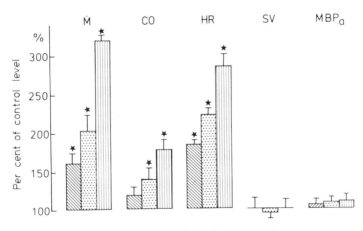

Fig. 2. Changes of heat production by shivering and cardiovascular responses in per cent of control level at neutral T_a 28°C. Mean values of 12 awake pigeons. $\dot{M}$: heat production, CO: cardiac output, HR: heart rate, SV: stroke volume, MBP_a: mean arterial blood pressure. Cold stresses: ambient cooling (T_a 5°C) ◨, spinal cord cooling (ΔT_{VC} 4.0-4.5°C) ▦, combination of spinal cord and ambient coolings (ΔT_{VC} 4.0-4.5°C, T_a 5°C) ▥. Vertical bars indicate standard error (SE), asterisks significancy to controls (2 $P < 0.02$). (After Barnas et al., 1984 with permission).

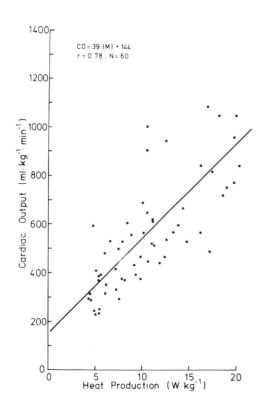

Fig. 3. Correlation between heat production and the cardiac output during rest and shivering of 12 awake pigeons. (After Barnas et al., 1984 with permission.)

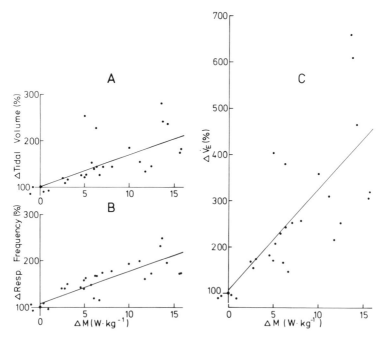

Fig. 4 A-C. Correlation between heat production by shivering and (A) tidal volume (V_T), (B) breathing frequency (f_{resp}) and (C) minute ventilation ($\dot{V}_E$) in 10 pigeons. (After Barnas and Rautenberg, 1984 with permission.)

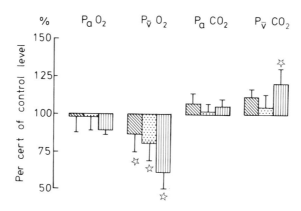

Fig. 5. Blood gases at various cold stresses (for columns see Fig. 2) of 10 pigeons in per cent of control levels. Control values of pH_a were 7.468 and $pH_{\bar{v}}$ 7.441. Changes during shivering were less than ± 1%. Vertical bars indicate standard error (SE), asterisks significancy to controls (T_a 28°C, $2P < 0.02$). (After Barnas and Rautenberg, 1984 with permission.)

neuronal connection from the shivering muscles to the circulatory controller who drives the effector action.

A similar interference may exist between shivering and respiration. V_T and f_{resp} increase with increasing heat production by shivering (Fig. 4). Both control actions and their product V_E of the respiratory feedback system are regulated by blood gases and acid-base status. A strong ventilatory drive are especially deviations of arterial P_{CO_2} and pH. However both variables are not changed during shivering (Fig.5). That means a drive of effector action occured without any disturbances of the respiratory feedback system. A similar neuronal control mechanism seems to exist as described above for the cardiovasculatory system.

DISCUSSION AND CONCLUSION

How can the interaction of these three autonomic feedback systems be interpreted? A possible explanation is given by a simplified block diagram (Fig. 6) involving feedback loops of temperature regulation (left hand) and cardiorespiration (right hand). The diagram may demonstrate processes under cold conditions. A cold load decreases the heat content of an animal and affects by this a decrease of mean body temperature, the controlled variable. The fall of T_b activates thermoreceptors or feedback elements which drive via central control elements some cold defense actions. That may eventually result in shivering thermogenesis by which extra heat is produced in a proportional extent to the heat loss as long as deep body temperature can be stabilized. The effector actions of cardiorespiration, cardiac output (CO) and minute ventilation (V_E), do immediately increase with shivering and are closely related to intensity of cold tremor without sig-

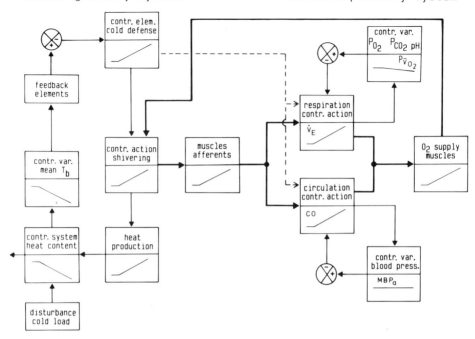

Fig. 6. A simplified block diagram of the interaction between thermoregulatory and cardiorespiratory feedback systems. For explanation see text.

nificant deviations of their important controlled variables, blood pressure, pH and arterial P_{CO_2}.

We suggest, therefore, a direct neuronal regulation of the cardiorespiratory control actions by afferents from shivering muscles. This nervous connection is drawn by thick solid lines in the diagram (Fig. 6). This neural control seems to be due on muscle receptors of group III and IV showing various sensitivities. One of these is contraction sensitive and may be responsible for adjustments of circulation and respiration during physical muscle work (Mense and Meyer, 1988). Sivering is also caused by muscle contraction, and one should assume the same receptor activation as during locomotion. Apart from these peripheral afferents a hypothalamic locomotor region has been recently described whose electric stimulation also affected cardiorespiratory responses in cats (Waldrop et al., 1986). These hypothalamic units may also be of importance for the adjustment of cardiorespiratory effector actions during shivering. This efferent connection is marked by dotted lines (Fig. 6).

Though direct evidences of such neural adjustments don't exist at time with respect of shivering thermogenesis in birds and mammals, the presented results on pigeons support the assumption that the cardiorespiratory actions are triggered by muscle afferents and/or by hypothalamic locomotoric efferents during shivering as described for locomotion. This mode of interaction of the autonomic feedback systems provides the active muscles with oxygen to such an extent as those organs require oxygen for an aerobic metabolism. This may explain why shivering never shows an oxygen debt nor an increase of blood lactate. This hypothesis is supported by studies on vagotomized pigeons showing the same cardiorespiratory responses as intact animals (Gleeson et al., 1986a) and by the fact that shivering is reduced or canceled during hypoxia (Gautier et al., 1987; Gleeson et al., 1986b).

REFERENCES

Barnas, G. and Rautenberg, W., 1984, Respiratory responses to shivering produced by external and central cooling in the pigeon, Pflügers Arch., 401:228.

Barnas, GM., Nomoto, S., Rautenberg, W., 1984, Cardiovascular and blood-gas response to shivering produced by external and central cooling in the pigeon, Pflügers Arch., 401:227.

Bligh, J., 1973, Temperature regulation in mammals and other vertebrates, North-Holland Publishing Company - Amsterdam.

Gautier, H., Bonora, M., Schultz, S. A., and Remmers, J. E., 1987, Hypoxia-induced changes in shivering and body temperature, J. Appl. Physiol., 62:2477.

George, J. C., 1984, Thermogenesis in birds, in: Thermal Physiology, J. R. S. Hales, ed., Raven Press, New York.

Gleeson, M., Barnas, G. M., and Rautenberg, W., 1986a, Cardiorespiratory responses to shivering in vagotomized pigeons during normoxia and hypoxia, Pflügers Arch., 407:664.

Gleeson, M., Barnas, G. M., and Rautenberg, W., 1986b, The effects of hypoxia on the metabolic and cardiorespiratory responses to shivering produced by external and central cooling in the pigeon, Pflügers Arch., 407:312.

Hohtola, ESA., 1982, Thermal and electromyographic correlates of shivering thermogenesis in the pigeon, Comp. Biochem. Physiol., 73A:159.

Mense, S., and Meyer, H., 1988, Bradykinin-induced modulation of the response behaviour of different types of feline group III and IV muscle receptors, J. Physiol., 398:49.

Waldrop, T. G., Mullins, D. C., and Henderson, M. C., 1986, Effects of hypothalamic lesions on the cardiorespiratory responses to muscular contraction, Respir. Physiol., 66:215.

SLEEP, HYPOMETABOLISM, AND TORPOR IN BIRDS

H. Craig Heller
Department of Biological Sciences
Stanford University
Stanford, CA 94305   USA

Introduction

Endothermy is energetically expensive, yet it has enabled birds and mammals to exploit a wide variety of extreme habitats where energy demands are high and food is frequently scarce. Increased energy demand coupled with decreased opportunity to feed may occur on a daily or on a seasonal basis. It is not surprising, therefore, that endotherms have evolved adaptations to reduce energy expenditure on daily and seasonal time scales. The oldest and most ubiquitous of these adaptations may be sleep. The argument has been advanced that specifically slow wave sleep, (SWS), also known as non-rapid eye movement (NREM) sleep evolved in parallel with endothermy as a means of reducing energy expenditure during the portion of the day that the animals were inactive (Walker & Berger, 1980). More extreme forms of adaptive hypometabolism, and shallow, daily torpor, and deep hibernation, may have evolved from the general mammalian and avian phenomenon of sleep (Heller et al, 1978).

Adaptive hypometabolism has been studied mostly in mammalian species. In the seven international symposia held on this subject since 1959, only a small number of papers have dealt with studies of avian species. Common opinion has been that there is only one true avian hibernator, the poorwill (Jaeger, 1949; Marshall, 1955; Howell & Bartholomew, 1959; Ligon, 1970). However, other caprimulgid species readily enter shallow, daily torpor upon food deprivation (Lasiewski & Dawson, 1964; Dawson & Fisher, 1969). Shallow, nocturnal torpor in hummingbirds has long been recognized as a general adaptation in this family due to their very small size and specialized ecological niche. In 1970 Dawson and Hudson reviewed the literature and could not conclude that the occasional observations of moderate to severe hypothermia observed in species other than caprimulgids and hummingbirds had the same adaptive physiological basis. Such observations were usually associated with severe inanition and could have been pathological. Indeed, it was not clear in 1970 whether or not inanition was the only factor inducing adaptive torpor in birds other than the poorwill. The perceived rarity of torpor as an energy conserving adaptation in birds except as a pathological response to starvation seemed specious given the high body temperature, high metabolism, and small body size of so many avian species, and the fact that birds have successfully exploited most extreme environments on earth.

A recent comprehensive review of facultative torpor in avian species presented an impressive list covering six avian orders (Reinertsen, 1983). Torpor appears to be a common adaptation for energy conservation in birds as well as in mammals. True hibernation is still much rarer in birds, however. Perhaps this is related to their greater ability to migrate seasonally, but another possible reason will be advanced at the end of this paper. The major focus of this paper will be on what is known about physiological mechanisms underlying avian torpor and how they compare with our understanding of the mechanisms of torpor in mammals.

Mechanisms of Torpor in Mammals

A brief summary of the physiology of torpor and hibernation in mammals will provide useful background for a discussion of these phenomena in birds. Mammalian torpor and hibernation are regulated declines in body temperature. The major site of thermoregulatory integration in mammals is the pre-optic and anterior hypothalamic nuclei (POAH), and the dominant feedback signal to this regulatory system is the POAH temperature (T_{hy}). The thermoregulatory system of a mammal under a given set of conditions can be characterized by heating and cooling the POAH by means of indwelling thermodes and measuring a thermoregulatory response such as metabolic heat production (MHP). The threshold T_{hy} for inducing an increase in MHP over basal levels can be considered a thermoregulatory set-point. This set-point decreases during entrance into torpor, is low during torpor, and rises during arousal. Anytime during torpor if the POAH is cooled below the threshold T_{hy}, an increase in MHP will be elicited. The conclusion of these studies is that body temperature is continuously regulated during torpor and the turning down of the POAH set-point for the MHP response is a primary regulatory mechanism in the suite of adaptations known as mammalian torpor. (Heller et al, 1978; Heller, 1979, 1988).

How did this prodigious capacity for readjustment of a homeostatic regulatory system evolve? To answer this question, it seemed reasonable to focus on general mammalian adaptations which involve alterations in the regulated body temperature (T_b), namely sleep and circadian rhythms. The T_b of mammals cycles on a daily basis, and this cycling free runs in an environment without time cues. The results of a number of studies permit the conclusion that the daily cycle of T_b is regulated and it is due to both sleep related and circadian rhythm related inputs to the thermoregulatory integrative system (Heller & Glotzbach, 1985). The sleep related adjustments have been studied at the hypothalamic level in both hibernators and non-hibernators. In both, SWS is accompanied by a decrease in hypothalamic thermosensitivity and REM sleep is accompanied by a complete absence of hypothalamic thermosensitivity (Glotzbach & Heller, 1976). Since hibernation is a downward adjustment and not an absence of thermoregulation, it seemed that SWS was a reasonable candidate for a pre-adaptation that selection could have shaped into the adaptation of torpor. This speculation has been supported by electrophysiological studies showing that shallow torpor and hibernation consist mostly of SWS with REM sleep disappearing as T_b falls (Walker et al, 1977, 1979, 1980).

The question of the possible involvement of circadian adjustments in thermoregulatory mechanisms of hibernation was unresolved. Species undergoing shallow torpor generally do so with a 24-hour periodicity (French, 1977). Generally, deep hibernation does not show a definite daily rhythmicity. The only evidence for the maintenance of free-running circadian rhythms during deep hibernation comes from bats (Menaker, 1961). Recent autoradiographic studies, however, strongly implicated the circadian control system in hibernation. The suprachiasmatic nuclei (SCN), the putative circadian time keeper, show elevated uptake of 2-deoxyglucose (2DG) relative to other brain structures during entrance into and throughout bouts of hibernation (Heller & Kilduff, 1985). Since it has been shown in other species that the SCN has a circadian rhythm of 2DG uptake, these results suggested that the hibernation bout could be an extended circadian cycle. The transition from the tight 24 hour rhythm of daily, shallow torpor to the extended bouts of deep torpor could have been achieved evolutionarily by relaxing the temperature compensation of the circadian control system. In support of this hypothesis is the quantitative relationship between T_b and length of hibernation bouts (Twente & Twente, 1965; French, 1982). The lower T_b falls during a bout, the longer the bout lasts, and this relationship projects to the usual length of the daily sleep period at a T_b of 37° C.

In summary, mammalian torpor, both shallow and deep, appear to be extensions of the downward regulation of T_b associated with NREM sleep. In the case of daily, shallow torpor, the occurrence of the deepened episodes of NREM sleep is controlled by a temperature compensated circadian control system. The occurrence of bouts of deep torpor or hibernation may also be controlled by a circadian control system, but one which is incompletely temperature compensated, so the lower T_b falls, the longer the bout lasts.

Thermoregulation, Sleep and Rhythms in Birds

To compare mechanisms of torpor in birds with what we know about these mechanisms in mammals, it is necessary to recognize differences which exist in the underlying physiological systems of birds and mammals. The basal metabolic rates of birds are higher than those of similar sized mammals, and birds regulate their body temperatures 2 to 4° higher than the normal mammalian range of 36 to 38°. Birds do not have well developed mechanisms of non-shivering thermogenesis as do mammals, but depend on shivering for thermoregulatory thermogenesis. In birds the POAH remains the major integrative structure in the thermoregulatory system, but the temperature of the spinal cord and not the temperature of the POAH is the dominant feedback signal (Dawson & Hudson, 1970; Rautenberg et al, 1972).

The cortical electrographic characteristics of arousal states appear to be homologous in birds and mammals (Berger & Walker, 1972; Walker & Berger, 1972). In most avian species sleep occurs mostly during the hours of darkness and consists of a much lower percentage (<10%) of REM sleep than is common in mammals. Moreover, REM sleep epochs in birds are very short, almost always less that 30 seconds and usually less than ten seconds. Birds may sleep with their eyes open except when exposed to very cold ambient temperatures.

Birds, like mammals, have strong circadian rhythms of activity, sleep, body temperature, and other variables. A very significant difference between the circadian rhythm control systems of birds and mammals is that the avian system is under strong photoperiodic control at the level of the pineal (Cassone & Menaker, 1984). In mammals, the photoperiodic input to the system is via the eyes to the SCN which in turn controls the pineal. This difference may be extremely important in influencing the ease of evolution of deep torpor or hibernation from shallow, daily torpor as will be discussed below.

The circadian rhythms of body temperature in birds have a larger amplitudes than seen in most mammals. It is not unusual for a bird to exhibit a $2^{\circ}C$ fluctuation in body temperature between day and night. These circadian cycles of body temperature are regulated. At a warm ambient temperature which will not elicit panting during the day, a bird may pant all night long to keep its T_b at the lower nighttime level (Graf, 1980a). In experiments in which the spinal canal of pigeons could be heated and cooled by means of an indwelling water perfused thermode, the spinal canal was periodically cooled to the same level across the circadian cycle. The increases in metabolic heat production stimulated by these uniform coolings of the spinal cord were much lower during the night than during the day (Graf, 1980b). Of course, since sleep in birds has a strong circadian organization, neither of the experiments just described enable us to discriminate between lowerings of the regulated T_b which are due to sleep and which are due to circadian rhythm controlling mechanisms.

The only study which has quantitatively distinguished the direct influence of circadian control versus the influence of sleep on the daily rhythms of T_b was done on pigeons (Heller et al, 1983). Pigeons chronically prepared with EEG and EMG electrodes as well as spinal canal thermodes were held on a 12:12 light/dark cycle. It was therefore possible to quantify the metabolic heat production and the panting responses to manipulations of vertebral canal temperature (T_{vc}) in sleeping and waking birds both during the dark and during the light portions of the daily cycle. The results clearly showed that the awake bird had a daily rhythm of T_{vc} thresholds for thermoregulatory responses and that these thresholds were further suppressed by sleep either during L or D portions of the daily cycle (figure 1). The conclusion from these experiments on pigeons was that the 2°C daily fluctuations in the T_b of the pigeon is about equally due to inputs to the thermoregulatory system from sleep control and from circadian rhythm control mechanisms.

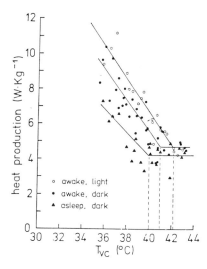

Figure 1. Relation of metabolic responses to spinal cord temperature (T_{vc}) of a pigeon during light or dark portion of a 12:12h light/dark cycle. For the dark period, data are represented separately for when the bird was awake or asleep. Data points represent average metabolic rates measured over a minimum time period of 20 min while T_{vc} was clamped at level shown on abscissa. (Reprinted with premission from Heller et al, 1983.)

Mechanisms of Torpor in Birds

Adaptive torpor has been long known and well studied in hummingbirds and a few other species of small passerines (Dawson & Hudson, 1970; Hainsworth & Wolf, 1970; Wolf & Hainsworth, 1972; Carpenter, 1974; Withers, 1976; Hainsworth et al, 1977; Prinzinger et al, 1981). Several general conclusions can be drawn from such studies as long as we consider the poorwill as a special case. First, torpor is under strong circadian control. Second, avian torpor is facultative and requires some degree of energetic stress to be expressed. Third, body temperature is regulated during torpor so that it remains above a lower limit. If it falls below that limit, spontaneous rewarming cannot occur.

The poorwill continues to be the only avian species shown to naturally undergo multiday bouts of deep torpor and arouse spontaneously (Jaeger, 1949; Marshall, 1955; Howell & Bartholomew, 1959; Ligon, 1970; French, pers. com). Moreover, the poorwill enters torpor without prior food deprivation and weight loss (Ligon, 1970). We can consider the poorwill as a true hibernator, and in nature it returns year after year to the same spot to hibernate (Jaeger, 1949; French, pers. com.). Other caprimulgids are prone to exhibit daily torpor upon food deprivation, but not spontaneous multiday bouts of torpor (Lasieweski & Dawson, 1964; Dawson & Fisher, 1969; Peiponen, 1970).

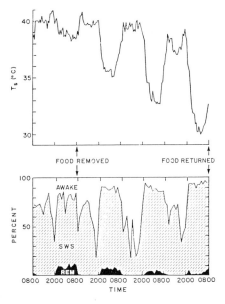

Figure 2. Subcutaneous temperature (T_s) and hourly wakefulness, slow wave sleep (SWS) and rapid eye movement sleep (REM) percentages for a dove prior to and throughout a 3-day fast. Note the 9° drop in T_s and virtual disappearance of REM sleep during the last 24 hours of the fast. (Reprinted with permission from Walker et at, 1983.)

The fact that the natural occurrence of nocturnal torpor in birds requires some degree of energetic stress led to the investigation of the effects of food deprivation on the daily cycle of body temperature in species not generally known to enter torpor in nature. It must be noted, however, that recent efforts were preceded by similar investigations more than 100 years ago. The French physiologist Chossat (1843), described daily cycles of body temperature in pigeons and noted that the magnitude of these cycles could be enhanced by food deprivation. It is now known in a variety of species of birds that food deprivation enhances daily cycles of body temperature by inducing lower and lower nocturnal levels (Biebach, 1977; Walker et al, 1983; Graf et al, 1988; Ivacic & Labisky, 1973; MacMillen & Trost, 1967; Ketterson & King, 1977; Peiponen, 1970). We assume that nocturnal hypothermia induced experimentally by food deprivation in such species as blackbirds, pigeons and doves is homologous with natural daily torpor in smaller species such as nightjars and hummingbirds. These larger species can be easily maintained in the laboratory and more readily subjected to various physiological methods of investigation. They, therefore, are valuable model systems for studies of mechanisms of avian torpor.

The dove was used to answer the question of whether avian torpor is continuous with euthermic NREM sleep (Walker et al, 1983). Doves chronically prepared with cortical EEG, EOG, and neck muscle EMG electrodes were subjected to food deprivation. On consecutive nights they dropped to lower and lower T_b's (figure 2). The decline in T_b began each evening soon after the onset of sleep, about 30 minutes before lights off. The end point of the experiment for each bird was when it fell to 20 to 22% of its initial body weight. The average nocturnal T_b minimum for these doves on the last night of fasting was 35.0°C, almost 4°C lower than the average T_b minimum under conditions of ad libitum food availability. Each day they returned to a typical avian euthermic T_b. Immediately following the return of food ad libitum, they displayed daily T_b cycles indistinguishable from baseline values.

The question being asked in this study on doves was whether the nocturnal hypothermia induced by food deprivation was continuous with euthermic NREM sleep. The answer was ambiguous due to the fact that the nocturnal phase of the daily cycle of doves is mostly NREM sleep even under baseline conditions. Although there were no statistically significant changes in total sleep time (% 24 hour period) throughout the fast, the doves entered torpor during sleep and the periods of torpor were almost continuous NREM sleep (figure 2). The birds showing the lowest nocturnal temperatures were the ones with the highest total sleep times, and in all of the doves torpor was characterized by a progressive reduction in REM sleep as T_b fell to lower levels as reported in mammals. It appears in birds as in mammals that torpor is homologous with NREM sleep.

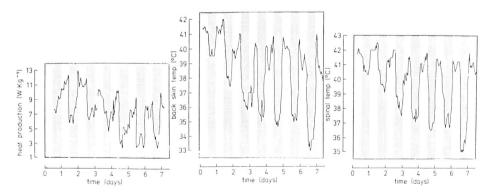

Figure 3. Continuous records of spinal temperature, backskin temperature, and metabolic rate for one pigeon during 6 days of food deprivation. The daily light cycle was 12L:12D and the D phase of the cycle is indicated by stippled bars (from Graf et al, 1989).

As discussed above, birds regulate their body temperatures lower during NREM sleep than during wakefulness, and this is evidenced by lowed T_{vc} thresholds for thermoregulatory responses (Heller et al, 1983). Does nocturnal torpor involve an extension of these sleep related

237

adjustments in CNS thermosensitivity? To answer this question pigeons were chronically prepared with cortical EEG and neck muscle EMG electrodes as well as vertebral canal themodes. After recovery from surgery, the pigeons were food deprived to induce nocturnal torpor. On consecutive nights they displayed progressively lower body temperatures, skin temperatures, and metabolic rates (figure 3). When the central nervous thermoregulatory systems of these birds were characterized by measuring metabolic responses to manipulations of T_{vc} it was clear that the central nervous thermoregulatory system was active during torpor, but the T_{vc} threshold for the shivering response was much lower than the typical nocturnal levels seen in fed birds. Moreover, when the measurements of spinal thermosensitivity were done on consecutive days and nights of food deprivation, the progressive lowering of the nocturnal T_b was associated with a progressively deeper suppression of the T_{vc} threshold for the metabolic heat production response. Each day this T_{vc} threshold returned to the normal daytime level (figure 4).

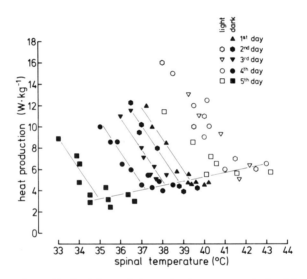

Figure 4. Relationships between metabolic rate and spinal cord temperature in a pigeon during successive days of food deprivation. The spinal temperature threshold for the shivering response falls to a lower level each successive night, but returns to approximately the same daytime level (from Graf et al, 1989).

The results shown in figures 3 and 4 are important because they demonstrate that the food deprivation protocol has not compromised the birds' abilities to generate thermogenic responses. The slopes of the spinal thermosensitivity curves during day and night and between the first and last day of food deprivation are not significantly different. The deprivation has resulted in a downward readjustment of the central nervous thermosensitivity and not an impairment of the ability to regulate. Obviously, prolonged inanition will eventually cause depletion of all energy reserves and result in pathological hypothermia. But, a significant, and presumably widespread avian adaptation which staves off the event is regulated, nocturnal hypothermia.

We presume that this regulated, nocturnal hypothermia is homologous to and continuous with the lesser sleep and circadian rhythm related nocturnal suppressions of T_b on the one hand, and the more pronounced bouts of daily torpor seen in hummingbirds and caprimulgids.

Induction of Nocturnal Hypothermia

The concept that avian species require food deprivation and loss of body mass before entering nocturnal hypothermia as been called into question by recent studies which have shown that some small species enter nocturnal hypothermia on a regular basis even when abundant food is available. These include at least 18 species of hummingbirds (Prinzinger & Schuchmann, 1986; Carpenter, 1974) and the willow tit (Reinertsen & Haftorn, 1983, 1984).

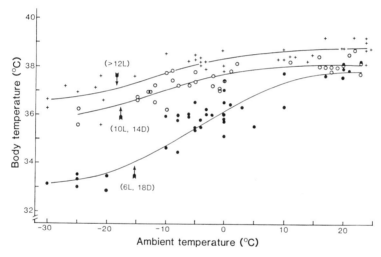

Figure 5. The relationship between natural body temperature and ambient temperature, for willow tits during different seasons of the year and consequently on different photoperiods. Each point represents the minimum value recorded on a single night. The birds were tested in midwinter (6L, 18D; filled circles), in late autumn or early spring (10L, 14D; open circles) and summer (>12L; crosses). The curves are fitted by hand. (Reprinted with permission from Reinertsen and Haftorn, 1983).

The willow tits are remarkable in that they use multiple cues to precisely regulate their use of nocturnal hypothermia. Willow tits are quite small (10-15g), but they cache food, so in nature they always have a source of food in the morning. In laboratory experiments they had food <u>ad libitum</u>. Although the willow tits enter nocturnal hypothermia regularly, they do not use this adaptation to achieve maximal metabolic savings; rather, they regulate hypothermia to survive each night at the highest possible T_b. On any one night, the depth of hypothermia is a function of season (hence night length) and ambient temperature (figure 5). Even more remarkably, the metabolic rate during any one night is a function of the body mass at the time of roosting (figure 6). This relationship is not simply due to running out of fuel sooner if you start the night with lower body mass, because entrance to nocturnal hypothermia in this species is characterized by a rapid drop to a specific T_b and metabolic rate which are then maintained for the rest of the night. Clearly, the nervous system of the willow tit integrates a variety of information in the regulation of nocturnal hypothermia.

The question of why birds do not use nocturnal hypothermia or torpor to achieve maximal energy conservation can only be answered with speculation. Birds roost in very exposed locations, and the more hypothermic a bird becomes, the less capable it is of responding rapidly and vigorously to the threat of predation. The absolute metabolic savings derived from a given lowering of T_b decreases as T_b falls. The greatest savings are obtained from the elimination of thermoregulatory heat production. Once the regulated T_b has fallen so far that the T_a is within the thermoneutral zone, further reductions in T_b result in decreases in the MR according to an exponential relationship, and therefore the savings in MR per degree fall in T_b decreases as T_b falls. With progressively deeper hypothermia the cost due to susceptibility to predation increases as the benefit due to energy conservation decreases.

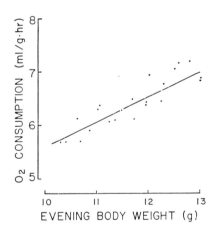

Figure 6. The relationships between average metabolic rate during bouts of nocturnal hypothermia in willow tits and body weight at roosting time. Ambient temperature was held at $0^\circ C$. (Redrawn with permission from Reinertsen and Haftorn, 1984.)

From Daily to Multiday Bouts of Torpor

The physiological mechanisms of adaptive torpor in birds appear to be homologous with those seen in mammals, and their regulation is just as sophisticated. Why then, has the evolution from daily to multiday bouts of torpor occurred, to the best of our knowledge, in only the poorwill? We can only speculate. As mentioned earlier in this paper, circadian rhythms of mammals are controlled by the suprachiasmatic nuclei which are entrained by photic cues transduced through the retina. In the absence of photic cues, the daily rhythms of the mammal free run with a periodicity of about 24 hrs (Turek, 1985).

It is generally assumed that circadian oscillators are temperature compensated. Therefore, a mammal entering shallow torpor should still retain an approximately 24 hr periodicity of its sleep/activity rhythm. One way to produce multiple day bouts of torpor using the same physiological oscillator is to reduce its temperature compensation so that the lower the T_b, the longer the bout. This line of reasoning must be tested experimentally, but it seems like a reasonable hypothesis. Can we apply a similar line of reasoning to birds?

In birds, as in mammals, the SCN are integral parts of the circadian rhythm controlling system. However, the pineal plays a much greater role in avian than in mammalian circadian control. (Cassone & Menaker, 1984). Pinealectomy in mammals has no effects on free running circadian rhythms, but in many avian species including estrillid, fringillid and ploceid passerines pinealectomy totally abolishes free-running locomotor rhythms. In sturnid passerines pinealectomy severly disrupts free-running rhythms, but in galliformes it has little if any effect. Another important difference between the role of the pineal in birds and mammals is that in birds the pineal serves a phototransducing function. In many species of birds, the pineal may be directly responsible for photic entrainment of the circadian system. The pineal can therefore be both an important circadian oscillator and a phototransducer involved in the control of avian rhythms. Coupling these facts with the observation that the roosting sites of birds rarely remove them from photoperiodic stimuli leads to the realization that it maybe simply more difficult mechanistically to evolve a temperature sensitive, freerunning circadian control system in species of birds in which the pineal is the dominant oscillator and phototransducer. It would be very interesting to know if the caprimulgids are more like galliformes or passerines in their response to pinealectomy.

Conclusion

Adaptive hypothermia or torpor is widely distributed among avian species and is now known to occur in 6 orders. Avian torpor has a circadian organization, and since most birds are day active, the phenomenon is frequently referred to as nocturnal hypothermia. It is normally expressed under conditions of food deprivation or energetic stress due to low ambient temperatures. However, its regulation may involve integration of various cues including photoperiod and body mass. Even species not observed to display adaptive torpor in nature can be induced to do so in the laboratory with food deprivation. The resulting hypothermia is regulated and retains a strong circadian organization. We know of only one species, the poorwill, which enters multiday bouts of torpor, and therefore can be considered a true hibernator.

There is strong evidence for physiological homology between all forms of avian and mammalian torpor or hibernation. Torpor is a regulated decrease in body temperature which normally occurs during the phase of the daily cycle when the animal would be inactive and spending much of its time asleep. Electrophysiological studies reveal that torpor consists mostly of NREM sleep. There is very little REM sleep or wakefulness during a bout of torpor. Both circadian rhythms and sleep control systems influence the regulation of body temperature in birds and mammals not undergoing torpor. Therefore, it is impossible to say in any case whether the extreme suppression of regulated T_b occurring during torpor is due to adaptive amplification of the circadian or the sleep related influence on the regulation of T_b, or both. It would be of interest to see whether birds made arrhythmic by pinealectomy or SCN lesions can be induced to enter torpor.

We can only speculate as to why multiday torpor is not a more common phenomenon in birds. The evolutionary transition from daily (or nocturnal) torpor to multiday bouts of torpor may have involved a relaxation of the temperature compensation of the circadian rhythms control system. Such a transition would only be possible if photoperiod cues were eliminated or their influence on the circadian oscillator was weakened. Either of these possibilities may be difficult to achieve in avian species which do not burrow and have strong involvement of the pineal in both phototransduction and generation of a circadian rhythm.

Acknowledgements

I greatly appreciate the help of Ms. Amy Donohue with preparation of this manuscript. Research contributions from my laboratory to the topics discussed in this paper were supported by grant NS-16317 from U.S.N.I.C.D.S., grant BN582-16981 from USNSF, a grant from the Upjohn Company and DFG grant Gr 724/1-1 to Rudolf Graf.

References

Berger, R. J. and Walker, J. M., 1972, Sleep in the burrowing owl (Speotyto cunicularia hypugaea). Behavioral Biology. 7:183.
Biebach, H., 1977, Das winterfett der Amsel (Turdus merula). J. Orn. 18:117.
Carpenter, F. L., 1974, Torpor in an Andean hummingbird: its ecological significance. Science. 183:545.
Cassone, V. M. and Menaker M., 1984, Is the avian circadian system a neuroendocrine loop? J. Exp. Zool. 232:539.

Chossat, C., 1943, Recherches experimentales sur l'inanition. Des effets de l'inanition sur la chaleur animale. Ann. Sci. naturelles 2 Serie. 20:293.

Dawson, W. R. and Fisher, C. D., 1969, Responses to temperature by the spotted nightjar (Eurostopodus guttatus). Condor. 71:49.

Dawson, W. R. and Hudson, J. W., 1970, Birds. In: "Invertebrates and Nonmammalian Vertebrates: Comparative Physiology of Thermoregulation," Vol. 1. Whittow, G. C., ed. Academic Press, New York.

French, A. R., 1977, Periodicity of recurrent hypothermia during hibernation in the Pocket mouse, (Perognathus longimembris). J. Comp. Physiol. 115:87.

French, A. R., 1982, Effects of temperature on the duration of arousal episodes during hibernation. J. Appl. Physiol.:Respirat. Environ. Exercise Physiol. 52(1):216.

Glotzbach, S. F. and Heller, C. H., 1976, Central nervous regulation of body temperature during sleep. Science. 194:537.

Graf, R., 1980a, Diurnal changes of thermoregulatory functions in pigeons. Pflugers Arch. 386:173.

Graf, R., 1980b, Diurnal changes of thermoregulatory functions in pigeons. Pflugers Arch. 386:181.

Graf, R. Krishna, S. and Heller H. C., 1988, Regulated nocturnal hypothermia induced in pigeons by food deprivation. In press.

Hainsworth, F. R., Collins, B. G. and Wolf, L. L., 1977, The function of torpor in hummingbirds. Physiol. Zool. 50:215.

Hainsworth, F. R. and Wolf, L. L., 1970, Regulation of oxygen consumption and body temperature during torpor in a hummingbird (Eulampis jugularis). Science. 168:368.

Heller, C. H., 1979, Hibernation: neural aspects. Ann. Rev. Physiol. 41:305.

Heller, H. C., Walker, J., Florant, G., Glotzbach, S. F. and Berger, R. J., 1978, Sleep and hibernation: Electrophysiological and thermoregulatory homologies. In: "Strategies in Cold: Natural Torpidity and Thermogenesis," Wang, L. C. H. and Hudson, J. W. eds., Academic Press, New York.

Heller, H. C., Graf, R. and Rautenberg, W., 1983, Circadian and arousal state influences on thermoregulation in pigeons. Am. J. Physiol. 245:R321.

Heller, H. C. and Glotzbach, S. F., 1985, Thermoregulation and sleep. In: "Heat Transfer in Medicine and Biology Analysis and Applications," A. Shitzer and R. C. Eberhard, eds., Plenum Press, New York.

Heller, H. C. and Kilduff, T. S., 1985, Neural control of mammalian hibernation. In: "Circulation, Respiration and Metabolism," Gilles, R., ed., Springer-Verlag, Berlin Heidelberg.

Heller, H. C., 1987, Sleep and hypometabolism. Can J. Zool. 66:61.

Howell, T. R. and Bartholomew G. A., 1959, Further experiments on torpidity in the poorwill. Condor. 61:180.

Ivacic, D. C. and Labisky, R. F., 1973, Metabolic responses of mourning doves to short-term food and temperature stresses in winter. The Wilson Bull. 85:182.

Jaeger, E. C., 1949, Further observations on the hibernation of the poorwill. Condor. 51:105.

Ketterson, E. D. and King, J. R., 1988, Metabolic and behavioral responses to fasting in the White-crowned sparrow (Zonotrichia leucophrys gambelii). Physiol. Zool. 50(2):115.

Lasiewski, R. C. and Dawson W. R., 1964, Physiological responses to temperature in the common nighthawk. Condor. 66:477.

Ligon, J. D., 1970, Still more responses of the poorwill to low temperatures. Condor. 72:496.

Marshall, J. T., 1955, Hibernation in captive goatsuckers. Condor. 57:129.

MacMillan, R. E. and Trost, C. H., 1967, Nocturnal hypothermia in the Inca dove. Scardefella inca. Comp. Biochem. Physiol. 23:243.

Menaker, M., 1961, The free-running period of the bat clock: seasonal variations at low body temperatures. J. Cell. comp. Physiol. 57:81.

Peiponen, V. A., 1970, Body temperature fluctuations in the nightjar (Caprimulgus e. europaeus L.) in light conditions of southern Finland. Ann. Zool. Fennici. 7:239.

Prinzinger, R., Goppel, R., Lorenz, A., and Kulzer, E., 1981, Body temperature and metabolism in the red-backed mousebird (Colius castantus) during fasting and torpor. Comp. Biochem. Physiol. 69A:689.

Prinzinger, R. and Schuchmann, K. L., 1986, Torpor in hummingbirds (Trochilidae), XIX Congressus Internationalis Orinithologius, Ottawa, Abst. 908.

Rautenberg, W. R., Necker, R. and May, B., 1972, Thermoregulatory responses of the pigeon to changes of the brain and spinal cord temperatures. Pflugers Arch. 338:31.

Reinertsen, R. E., 1983, Nocturnal hypothermia and its energetic significance for small birds living in the arctic and subarctic regions. Polar Res. n.s. 1:264.

Twente, J. W. and Twente, J. A., 1965, Regulation of hibernation periods by temperature. Proc. N. A. S. 54:1058.

Turek, F. W., 1985, Circadian neural rhythms in mammals. Ann. Rev. Physiol. 47:49.

Walker, J. M. and Berger R. J., 1972, Sleep in the domestic pigeon. Behav. Biol. 7:195.

Walker, J. M. and Berger, R. J., 1980, Sleep as an adaptation for energy conservation functionally related to hibernation and shallow torpor. Prog. in Brain Research. 53:255.

Walker, J. M., Garber, A., Berger, R. J. and Heller, H. C. 1979, Sleep and estivation (shallow torpor): continuous processes of energy conservation. Science. 204:1098.

Walker, J. M., Haskell, E., Berger, R. J. and Heller, H. C., 1977, Sleep and hibernation in ground squirrels (Citellus spp.): electrophysiological observations. Am. J. Physiol. 233:R213.

Walker, J. M., Haskell, E., Berger, R. J. and Heller, H. C., 1981, Hibernation at moderate temperatures: A continuation of slow wave sleep. Experientia. 37:726.

Walker, L. E., Walker, J. M., Palca, J. W. and Berger, R. J., 1983, A continum of sleep and shallow torpor in fasting doves. Science: 221:195.

Withers, P. C., 1976, Respiration, metabolism and heat exchange of euthermic and torpid poorwills and hummingbirds. Phys. Zool: 50:43.

Wolf, L. L. and Hainsworth, F. R., 1972, Environmental influence on regulated body temperature in torpid hummingbirds. Comp. Biochem. Physiol. 41:167.

ADAPTIVE CAPACITY OF THE PIGEON'S DAILY BODY TEMPERATURE RHYTHM

Rudolf Graf, H.Craig Heller*, Sindu Krishna*,
Werner Rautenberg**, and Beate Misse**

MPI für Neurologische Forschung, D-5000 Köln 91, FRG; *Dept. Biol. Sci., Stanford University, Stanford, Ca 94305-2493, USA
** Inst.für Tierphysiologie, Ruhr-Universität, D-4630 Bochum FRG

INTRODUCTION

Daily cycles of deep body temperature (Tb) have been discovered in both mammals and birds already in the nineteenth century (Gierse, 1842 and Chossat, 1843 in Aschoff, 1970). Chossat mentioned in his study in pigeons that the magnitude of the temperature cycle was enhanced under the influence of food deprivation. This early report gives possibly a first hint at the adaptive function of body temperature temporal organization as accomplished by circadian temperature control.

Studies of temperature regulation in pigeons show that the generation of daily temperature cycles is related to daily alterations of thermoregulatory effector activity and spinal thermosensitivity as documented by threshold changes for metabolic responses to spinal cooling (Graf, 1980a and 1980b). Furthermore, manipulations of spinal temperature as a function of arousal state and time of day demonstrated that there is sleep related suppression of spinal thermosensitivity superimposed upon a daily cycle of spinal thermosensitivity seen in awake birds (Heller et al., 1983). Thus, separate influences on CNS regulation of daily body temperature cycles have to be considered emanating from mechanisms subserving arousal state and circadian rhythm control.

The questions we would like to follow in the present paper are 1) whether the body temperature rhythm deriving from sleep and/or circadian influences is altered as a function of various factors, namely ambient temperature, light intensity, food deprivation and fever in an adaptive manner and 2) whether these adaptations are achieved by active thermoregulatory adjustments. We will mainly discuss own experiments in pigeons and refer briefly to reports given for other species.

AMBIENT TEMPERATURE

Studies in pigeons concerning ambient temperature (Ta) influences (Graf and Necker, 1979; Graf, 1980a) were performed in a light-dark regime (LD 12:12 h). The metabolism chamber we used could be clamped at different Ta's for the entire daily period or cooled for short episodes in both L and

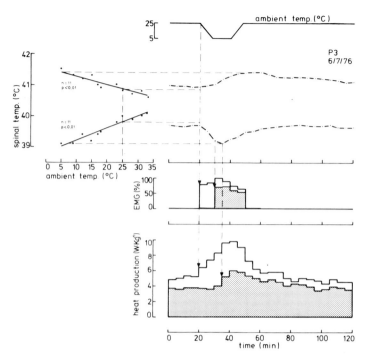

Fig. 1. Effect of ambient temperature alterations on spinal temperature, shivering and metabolic heat production in the course of an LD cycle. Left side: Spinal temperature means obtained in 5 birds during L and D at different ambient temperatures which were kept constant over the entire cycle. Right side: Response of spinal temperature, shivering (integrated EMG) and heat production to short pulses of lowered ambient temperature (5 C) during L and D (spinal temperature in D: lower curve; EMG and heat production in D: stippled). Note the delayed metabolic response to cold in D following the drop of spinal temperature to a level characteristic under steady state conditions for this portion of the cycle.

D. Under steady state conditions, skin temperature rhythms were shifted downwards the lower Ta's were, and the magnitude of the cycles increased slightly. In contrast, core temperatures (spinal temperature, see Fig. 1, left side) were decreased only during D while during L means increased slightly in the cold. A similar divergency of core temperatures in L and D in response to ambient cooling has been found in other bird species (Trost, 1972; van Kampen, 1974), and it has been shown in chicken (van Kampen, 1974) that with ambient warming, core temperatures in D exceed those in L. Thus, even though ambient temperature influences on core temperatures are small they may uncover the adaptability of day-night body temperature control. This hypothesis is supported by results deriving from short cold pulse experiments (Fig. 1, right side). Lowering of ambient temperature during L resulted in an almost immediate shivering, and metabolic response followed by a slight increase of spinal temperature. In contrast, night-time cold pulses led to a temporary drop of spinal temperature until shivering and metabolism started to increase. The core temperature levels reached in L and D in response to the cold pulses resemble those obtained under steady state conditions in the lower Ta's, and it seems reasonable to interpret this result as an adaptive thermoregulatory setting of the temperature cycle to cold stress.

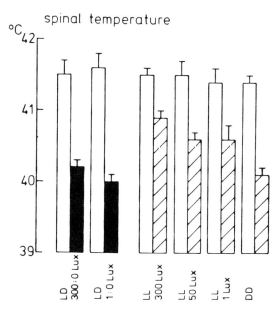

Fig. 2. Circadian spinal temperature alterations in various light-dark or continuous light regimes. Means (SD) of 6 pigeons were calculated for the middle 6 hours of L and D or activity and rest time, respectively.

LIGHT INTENSITY

To our knowledge, studies concerning direct influences of light on circadian temperature control in birds have not been performed. Light influences on Tb cycle waveform have been described, however. Veghte (1964) showed that alterations of the photoperiod result in an elongated temperature maximum if the light portion of the LD cycle is prolonged (summer). It is remarkable that in such condition, the rise and fall of Tb in the morning and evening appears to be much faster than with short L (winter). Effects of constant illumination (LL) as compared to constant darkness (DD) have been described in sparrows (Binkley et al., 1971). In this species, LL almost eliminated Tb rhythms. Body temperatures leveled off at day-time values of normal LD cycles and remained there continuously over several periods. In pigeons, we performed experiments on light intensity influences under both LD and constant light conditions. The birds were kept in a temperature controlled box in small single cages so that thermocouple measurements of spinal temperature could be obtained continuously. After 1 day of adaptation to the experimental situation (LD 12:12 h, 1:0 Lux), the light regime was either kept for another 2 periods or changed to other LD or continuous light regimes (Fig. 2). The magnitude of the daily cycle was systematically influenced in continuous light in that bright light (300 Lux) almost extinguished the day-night difference whereas constant darkness left the rhythm intact. In LD, temperature cycles were not significantly affected by light intensity alterations in L. An according report on long-term influences of bright light on Tb rhythms in pigeons is given by Berger and Philips in this volume. The authors found a complete cessation of core temperature cycles after several days. This cessation in LL may be interpreted as an adaptation of circadian body temperature control to extreme summer conditions where no need of a downward regulation of body temperature during rest time exists.

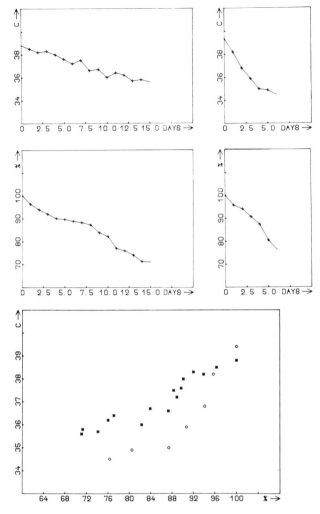

Fig. 3. Food deprivation induced decline of nocturnal body temperature and body weight in birds with an initially high (left side, 480 ± 53g, n=3) and low (right side, 321 ± 62g, n=5) body weight and rate of body temperature decline per body weight in the heavy (closed symbols) and light (open symbols) birds.

FOOD DEPRIVATION

Evidence for the hypothesis of circadian temperature control adaptability is provided mainly by studies on Tb-rhythm in food deprived birds. As mentioned above, a first example of body temperature cycle alteration due to starvation was given by the french physiologist Chossat as early as 1843. Similar enhancements of the day-night body temperature difference have been found in other species (MacMillan and Trost, 1967; Peiponen, 1970; Biebach, 1977) and resemble the naturally occuring daily torpor characteristic for various nectar and insect feeding birds which can also be induced or facilitated by food deprivation (for review: Dawson and Hudson, 1970; Reinertsen, 1983). These alterations are achieved by a night to night downward adjustment of Tb as body weight drops while day-time temperatures stay nearly at fed state levels. This downward adjustment seems to depend on the initial body weight. In pigeons studied under single cages in a climatic chamber

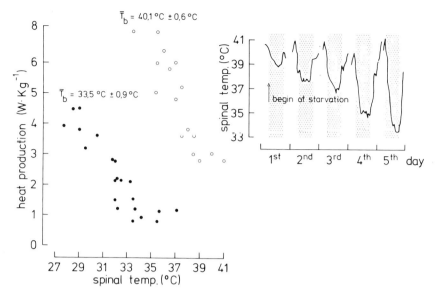

Fig. 4. Deep body temperature cycles during consecutive days of food deprivation and metabolic responses to spinal cooling during the fifth day of starvation. Spinal coolings were performed in three separate experiments on one bird. Means of body temperature (rectal temperature) during these cycles are indicated.

set at thermoneutral temperatures an initial body weight of about 500g was correlated with a slow decline of body weight and temperature in consecutive nights (Fig. 3, left side) as compared to an initial weight of about 300g which was correlated with much faster drops (Fig. 3, right side). Additionally, the rate of temperature decline per body weight loss was higher in the lighter birds (Fig. 3, bottom). Since the lighter birds were wild captured whereas the heavy ones came from a commercial supplier, the question arises to whether the two different strategies might also be influenced by ontogenetic factors. In the light birds, a series of experiments with spinal temperature manipulations was performed during starvation. Pigeons had been prepared beforehand with spinal thermodes (Graf, 1980b). The experiments were performed in a respiration box positioned in a larger climatic chamber. The results corresponded well with the concept of an active downward regulation of Tb since metabolic response curves to spinal cooling were shifted dramatically to lower values (Fig. 4). Whereas a threshold shift of about 2°C is typical for the day-night adjustment in the fed state (Graf, 1980b), we found shifts of as much as 6°C in starving birds. These shifts are probably achieved by circadian rather than sleep influences on temperature control since differences in sleep state distribution have not been found to be of major importance in the starving bird (Walker et al., 1983). Our results correspond with studies on torpic hummingbirds which showed rather low Tb-thresholds for metabolic heat production so that Tb was regulated at about 15°C during the night (Wolf and Hainsworth, 1972).

FEVER

In a series of starvation experiments with intravenous injections of different substrates which will not be reported here it happened several times and in some cases after pure saline injections that the core tempe-

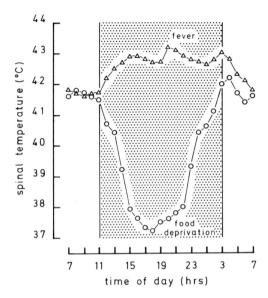

Fig. 5. Spinal temperature cycles at the fourth (food deprivation) and sixth (fever) day of starvation. The febril response was probably due to intravenous injections (normal saline) undertaken on the fifth day but occured during ongoing food deprivation.

rature rhythm was all of a sudden reversed (Fig. 5) despite preceding starvation periods of up to 5 days with the characteristic pattern of deep nocturnal temperature drops. The only reasonable explanation for this phenomenon was in our view that the birds developed fever. Kluger and D'alecy (1975) have reported on febrile responses in pigeons which were characterized by reversal of body temperature rhythms with an elevated (febrile) temperature during the night while day-time temperatures stayed at nonfebrile levels. Fever therefore seems to be another example for the flexibility of circadian body temperature control.

CONCLUSION

Pigeons exhibit a circadian variation of body temperature which is sensitive to alterations in environmental and/or internal conditions such as ambient temperature, light intensity, food deprivation, and fever resulting mainly in adjustments of the nocturnal temperature level. Studies of CNS temperature control show that the adaptations are achieved by active temperature regulation. Thus, our findings lead to the assumption that the adaptive capacity of the pigeons thermoregulatory system expresses circadian variability, with the primary capacity occuring nocturnally.

ACKNOWLEDGEMENT

We would like to thank Mrs. M. Meinecke and Mr. S. Sakaguchi for skillful technical assistance. This research was supported by National Institute of Neurological and Communicative Disorders and Stroke Grant NS-16317 and Deutsche Forschungsgemeinschaft Grants SFB 114 and Gr 724/1-1.

REFERENCES

Aschoff, J., 1970, Circadian rhythm of activity and of body temperature, in:"Physiological and behavioral temperature regulation", J.D. Hardy, A.P. Gagge, J.A.J. Stolwijk, eds., Springfield: Illinois.

Biebach, H., 1977, Reduktion des Energiestoffwechsels und der Körpertemperatur hungernder Amseln (Turdus merula). J Ornithologie., 118:294-300.

Binkley, S., Kluth, E., Menaker, M., 1971, Pineal function in sparrows: Circadian rhythms and body temperature. Science., 174:311-314.

Chossat, Ch., 1843, Recherches experimentales sur l'inanition. Des effects de l'inanition sur la chaleur animals. Ann Sci naturelles 2 Serie., 20:293-326

Dawson, W.R., Hudson, J.W., 1970, Birds, in:"Invertebrates and Nonmammalian vertebrates (comparative Physiology of Thermoregulation, I)," G.C. Whittow, ed., Academic Press: New York.

Gierse, A., 1842, Quaenium sit ratio cloris organici partium inflammatione laborantium febrium..., Diss., Halle.

Graf, R., 1980a, Diurnal changes of thermoregulatory functions in pigeons. I. Effector mechanisms. Pflügers Arch., 386:173-179.

Graf, R., 1980b, Diurnal changes of thermoregulatory fuctions in pigeons. II. Spinal thermosensitivity. Pflügers Arch., 386:181-185.

Graf, R., Necker, R., 1979, Cyclic and non-cyclic variations of spinal cord temperature related with temperature regulation in pigeons. Pflügers Arch., 380:215-220.

Heller, H.C., Graf, R., Rautenberg, W., 1983, Circadian and arousal state influences on thermoregulation in the pigeon. Am J Physiol., 245:R321-R328.

Kluger, M.J., D'alecy, L.G.,1975, Avian febrile response. J Physiol., 235:223-232.

Mac Millen, R.E., and Trost, C.H., 1967, Nocturnal hypothermia in the inca dove, Scarda fella inca. Comp Biochem Physiol.,23:243-253.

Peiponen, V.A., 1970, Body temperature fluctuations in the nightjav (Caprimulgus e. europaeus L.) in light conditions of southern Finland. Ann Zool Fennici., 7:239-250.

Reinertsen, R.E., 1983, Nocturnal hypothermia and its energetic significance for small birds living in the arctic and subartic regions. A review. Polar Res ns., 1:264-284.

Trost, C.H., 1972, Adaptations of horned larks (Eremophila alpestris) to hot environments. The Ank., 89:506-527.

van Kampen, M., 1974, Physical factors affecting energy expenditure, in: "Energy requirements of poultry," T.R. Morris, B.M. Freeman, eds., British Poultry Science Ltd.: Edingurgh.

Veghte, J.H., 1964, Thermal and metabolic responses of the grey jay to cold stress. Physiol Zool., 37:316-328.

Walker, L.E., Walker, J.M., Palea, J.W., Berger, R.J., 1983, A continuum of sleep and shallow torpor in fasting doves. Science., 221:195-196.

Wolf, L.L., Hainsworth, F.R., 1972, Environmental influence on regulated body temperature in torpid humming birds. Comp Biochem Physiol., 41:167-173.

THERMAL AND FEEDING REACTIONS OF PIGEONS DURING FOOD SCARCITY AND COLD

Michael E. Rashotte, Dori Henderson and Deborah L. Phillips

Psychobiology/Neuroscience Program
Department of Psychology, Florida State University
Tallahassee, Florida 32306-1051 USA

Birds living in the cold face energetic challenges from low ambient temperatures (Ta) as well as from food scarcity which often accompanies cold in nature. Birds react to low Ta with complex thermoregulatory and metabolic adjustments, and they also alter their foraging and feeding behavior rather dramatically when the availability of food becomes limited. In reaching a full understanding of adaptation to cold, it will be necessary to take into account the ways birds adapt to food scarcity as well as their thermal/metabolic adaptations to low Ta. This is a particularly interesting problem because of the interplay between thermoregulation and feeding. For example, when feeding strategies fail to produce sufficent food to maintain energy reserves, some birds reduce energy costs by lowering their body temperature (Tb). Also, exposure to cold not only activates the bird's thermoregulatory defenses, but alters feeding strategies in a way that results in increased food intake if sufficient food is available in the environment. Furthermore, food intake has short term thermogenic consequences which can be significant in certain circumstances.

We have chosen the pigeon (Columba livia) for further study in the laboratory of the relationship between thermoregulation and feeding. Pigeons' thermoregulatory responses have been well studied, and there is a sizable, but separate, body of research on their feeding behavior. The present paper describes some recent experiments from our laboratory which have begun to bring these two lines of research together. In this work, we hope to characterize the relationship between thermoregulation and feeding in a more complete fashion. With such information in hand, it may be possible to improve the understanding of adaptation to cold.

FOOD SCARCITY

Laboratory studies of the thermoregulatory response of birds to food scarcity have utilized a limited range of conditions of food availability. Fasting and the controlled presentation of a fixed daily ration are the two most commonly studied conditions. Recently, a procedure has been introduced which has the interesting property that it allows the birds to utilize various feeding strategies in combination with thermoregulatory strategies to cope

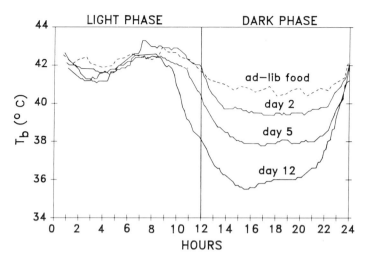

Fig. 1. Representative data from Phillips' (1989) experiment with fasting pigeons. A temperature-sensitive transmitter in the abdominal cavity measured core body temperature on a continuous basis. The daily cycle of Tb is shown for the final day of ad libitum feeding when the bird's body weight was 424g, and for selected days during the fasting period. The pigeon's body weight reached 80% of its free-feeding value on the 12th day.

with different levels of food availability. We have begun a series of experiments using this "closed economy" procedure, and we summarize one of them below. First, however, we characterize some aspects of the relationship we have found between different levels of food scarcity and the circadian pattern of body temperature in experiments where the traditional procedures of fasting and controlled feedings were employed.

Fasting

Under fasting conditions in the laboratory, small birds become hypothermic during the inactive part of the daily cycle, and some become torpid (e.g., Hudson, 1978; Walker, Walker, Palca and Berger, 1983). Pigeons are among the birds for which nocturnal hypothermia deepens progressively when the fast continues for several days, as is illustrated in Figure 1. The lowering of body temperature (T_b) can result in considerable energy conservation in a bird the size of a domestic pigeon (Chaplin, Diesel and Kasparie, 1984), and it appears that pigeons have this energetic strategy available to use when their energy reserves are being progressively depleted.

Single Daily Feedings: Effects of Variation in the Size and Timing of the Food Presentation

A condition of food scarcity that is less extreme than fasting is one in which a small amount of food is eaten each day during a single feeding bout. Phillips (1989) studied the effects of varying the size and the temporal placement within the daily cycle of this single feeding. Initially, the birds were fed at the beginning of the light phase. Later in the experiment, food was presented at the ninth hour of the light phase, at approximately the time when pigeons in the laboratory have their largest feeding bout under ad libitum feeding conditions (Zeigler, Green and Lehrer, 1971). As a point of reference, the amount of food eaten each day during ad libitum feeding was about 20g for each bird. On most days of the experiment,

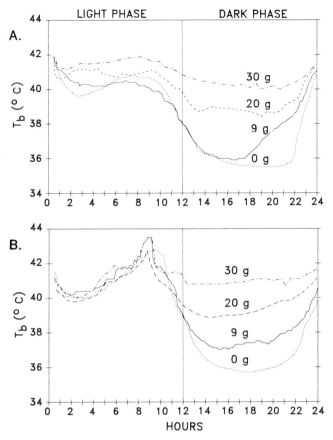

Fig. 2. Panel A shows, for a representative pigeon, the circadian cycle of Tb when food was given at the beginning of the light phase. Each record shows Tb across a single 24h period. Records are shown for days when 0g, 9g, 20g and 30g of food were given. Panel B shows the circadian cycle of Tb when the experiment was repeated with the daily feeding at the ninth hour of the light phase. In each panel, the 9g record is for a representative day after the Tb cycle had stabilized at the prevailing daily feeding time.

the birds were fed a 9g ration, which was sufficient to maintain their body weight at 80% of its value during ad libitum feeding.

When the birds were fed the 9g ration, the circadian cycle of Tb displayed a large day-night difference, with a pronounced lowering of Tb during the dark phase. The resting level of nocturnal Tb was about 1C higher after the 9g feeding was moved to late in the light phase. These characteristics of the data are illustrated by the 9g curves in Figure 2. Two additional points are of interest concerning the 9g data. First, rewarming from the low resting level of dark-phase Tb occurred across several hours, and, at the point of light onset, Tb was higher when feeding occurred at the beginning of the light phase. Possibly, this latter result indicates an anticipatory thermal effect of the impending feeding which is superimposed on the endogenous rewarming which appears to be under circadian control (cf. Figure 1). It is not possible to identify the mechanism of rewarming in this case, but the instrumentation of the experiment allowed Phillips to determine that the pigeon remained perched throughout the entire dark phase. This feature of the data reduces the likelihood that thermogenesis was related to gross motor activity during the dark phase. It is interesting that body weight loss increased progressively during the dark-phase hours, with about 80% of the total loss occuring in the final 6 hours. On the basis of further analysis, Phillips showed that most of this weight loss is accounted for by excretory activity. Although it is clear that excretory activity coincides with rewarming, the present data do not speak to the possibility that there is a causal relation between these two activities.

The second point to note is that, after the time of feeding was shifted to the ninth hour of the light phase, anticipation of the feeding was indicated by the progressive increase in Tb across several hours preceding the feeding. This effect developed across the first five days after the shift in feeding time. It is likely that this increase is related to the thermogenic effects of gross motor movements, although we cannot confirm this on the basis of the data available.

Phillips also studied the effect of varying the amount of food eaten in the daily feeding. For single days only, she presented a 0g, 20g, or 30g portion of food, separating such days by several on which the animals were fed their usual 9g ration. This procedure made it possible to determine the time-course over which the Tb cycle returns to the 9g baseline after a single change in the amount of food eaten. In all cases, the food was completely eaten within a few minutes of being presented. The data indicate that the circadian pattern of Tb is very responsive to the amount of food eaten in the single daily feeding bout. Figure 2 shows that nocturnal Tb is affected in a similar way by the amount eaten, whether the feeding occurs 12h or only 3h prior to dark. The minimum level of nocturnal Tb was strongly correlated with the body weight at light offset ($r > 0.91$ in all cases). Surprisingly, for a given amount of food eaten, the absolute level of body weight at light offset was virtually the same whether the feeding occurred 12h or 3h earlier. This aspect of the data raises questions about the transit time of food through the digestive system of chronically underfed pigeons. Even when water intake is taken into account, the pigeons appeared to retain the largest amount of the food eaten during the day until the dark phase began. Finally, the time course of recovery of Tb after a large amount of food was eaten extended over several days. The largest effect occurred after a 30g feeding, when the circadian cycle of Tb returned to baseline gradually across a three or four day period. When the 30g feeding occurred late in the light phase, there was a pronounced elevation of Tb throughout the next day's light phase.

Phillips' experiment demonstrates that the pigeon adopts a repeatable level of nocturnal hypothermia when a single 9g meal is eaten daily for several months. Apparently, this steady-state level of hypothermia contributes to the energetic balance necessary for survival during a chronic period of low food availability. However, the level of nocturnal hypothermia is very responsive to variation in the size of the daily meal. When the bird finds no food in the light phase, it displays deeper and more prolonged hypothermia the following night, apparently utilizing more fully the energy saving aspects of hypothermia. When the bird finds a larger than usual meal, however, its nocturnal Tb rises in proportion to the size of the meal eaten. In these cases, the pigeon apparently relinquishes much of the energy savings from hypothermia, despite the fact that its energy reserves remain depleted from the extended exposure to reduced food availablity. It will be interesting to investigate other properties of these transient, feeding-induced changes in nocturnal Tb. In particular, it should be determined whether these changes in the level of nocturnal Tb are regulated, as is shown in the case of fasting-induced nocturnal hypothermia (Rashotte, Rautenberg, Henderson and Ostheim, 1986). Also, the characteristics of sleep during the feeding-induced changes in nocturnal hypothermia should be determined (Walker et al., 1983).

Thermoregulation and Feeding Strategies During Food Scarcity

It is possible to vary the level of food availability while the bird completely controls how often and how much it eats each day, as well as the times in the day when it eats. In the "24h closed economy" procedure (Collier, 1983; Hursh, 1980, 1984), an animal lives in the experimental situation and is able to gain access to food at any time by paying a behavioral price. For example, a pigeon might be able to obtain a few seconds access to food in a hopper by performing a designated response (e.g. by pressing a foot-treadle) a specified number of times. Therefore, in the closed economy procedure, daily food intake is determined jointly by the number of times the pigeon gains access to the food hopper each day and by the amount it eats during each hopper presentation,

$$I = \Sigma \ A(n) \tag{1}$$

where I represents the daily food intake in grams, A represents the amount of food consumed per hopper presentation in grams, n represents the nth hopper presentation on that day, and Σ indicates summation of the amounts eaten from the first through the last hopper presentation each day.

A question of interest in feeding research has been how variations in the behavioral price of obtaining food affect the amount and temporal patterning of daily intake (e.g., Collier and Rovee-Collier, 1981; Rashotte and O'Connell, 1986). Rashotte and Henderson (in press) examined this question in an experiment that simultaneously investigated the pigeon's use of nocturnal hypothermia during rising food prices. Continuous measurement of subcutaneous temperature was made when the pigeons could obtain a 10-sec presentation of the food hopper by treadle pressing. The behavioral price of each hopper presentation was initially low (6 presses/hopper). Subsequently, the cost increased every six days in a series (60, 120, 240, 480, 960, and 1920 presses/hopper) that was terminated when the pigeon's body weight fell to 80% of its value at the initial low cost. As typically found in closed economy experiments, the pigeons changed their feeding strategies as the price rose: the number of times they worked to produce the hopper each day decreased, and the amount of food they ate from the hopper during each 10-sec presentation increased. These changes in feeding seem geared to improve efficiency by increasing the amount of food obtained for a given behavioral price. However, these strategies did not succeed in defending the level of food intake which the pigeons maintained when the price was lowest. Instead, the total amount of food eaten per day decreased as the cost rose, and a price was eventually reached at which most pigeons completely refused to work for food, even though food could always be obtained at the prevailing price.

The subcutaneous Tb data indicated that nocturnal hypothermia was utilized by individual pigeons at different points in the rising price series. However, in all cases there was a clear linkage between the depth of nocturnal hypothermia and the loss of body mass. An example is shown in Figure 3 for a pigeon that first used nocturnal hypothermia when the price rose to 960 presses/hopper. At this price, the pigeon's body weight had fallen to 87.4% of its baseline weight, and the pigeon earned 7.2g of food. When the price was 480 presses/hopper, the pigeon earned approximately the same amount of food (7g), but its body weight was 91.7% of baseline.

The occurrence of nocturnal hypothermia in Rashotte and Henderson's experiment indicates the involvement of an autonomically mediated thermoregulatory response which is related to a progressive reduction in energy reserves, as indexed by body weight loss. In this respect, the outcome is similar to that reported when fasting or controlled feedings are used to vary food availability. However, in the present case the loss of energy reserves is a conse-

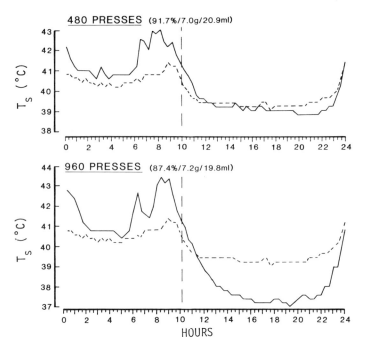

Fig. 3. Circadian pattern of subcutaneous temperature in a pigeon paying behavioral prices of 480 or 960 treadle-presses per 10-sec hopper presentation. The dashed line in each panel is the baseline temperature of this pigeon when the price of food was 6 presses/hopper. Each temperature curve is from the last of six consecutive days at each price. The light-dark transition at hour 10 is marked by a vertical line. The three numbers in parentheses in each panel show: body weight as a percentage of weight in the 6-press baseline (%), the day's total food intake (g), and the day's total water intake (ml). Total daily food intake in the 6-press condition averaged 21.1g for this bird.

quence of the pigeon's refusal to work for food that is available. Accordingly, it is tempting to view nocturnal hypothermia in the closed economy as one of the thermal/metabolic strategies that is engaged once the feeding strategies fail to maintain sufficient energy reserves. Undoubtedly, several thermal strategies are activated during periods of exposure to high-cost food in the closed economy. For example, Ostheim & Rautenberg (this volume) have shown that behavioral control of ambient temperature by pigeons is differentially affected as the cost of food rises. Also, it is likely that changes in feather posture occur during reduced food intake which lower heat loss (McFarland & Budgell, 1970). A knowledge of thermal/metabolic strategies, the interactions among them, and the hierarchy (if there is one) of their appearance and disappearance as the price of food varies, will provide a useful perspective on the ways birds adapt to stressful environmental conditions.

FEEDING STRATEGIES IN THE COLD

Exposure to cold ambient temperatures activates a number of adaptation mechanisms in birds and mammals, including changes in the level of daily energy intake (e.g., Kraly and Blass, 1976; Savory, 1986). It is well known that birds living in the laboratory with free access to food increase their daily food intake as a result of being exposed to cold ambient temperatures (e.g. Kendeigh, Kontogiannis, Mazac and Roth, 1969; Steen, 1957). However,

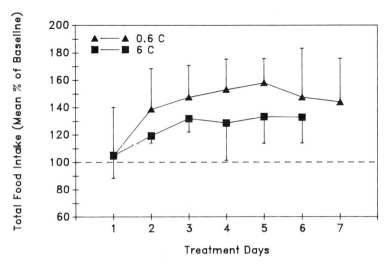

Fig. 4. Mean change in daily food intake during exposure to 6C and to 0.6C for four pigeons living in the closed economy. The change scores are relative to the baseline food intake at 21C immediately before each cold exposure began. The cost of 10-sec access to the food hopper was held constant at a low value (12 keypecks). Error bars show the standard deviation for the means.

in many experiments on the physiology of adaptation to cold, this aspect of the adaptation process is not emphasized. The closed economy procedure makes it possible to quantify the behavioral strategies which underlie increased energy intake during cold exposure and, ultimately, to investigate the relationship between thermoregulatory and feeding strategies in the cold.

Henderson (1988) has exposed pigeons to cold ambient temperatures in a closed economy. In preliminary training, four pigeons were adapted to living on a 12h:12h L:D cycle, with Ta at 21C, and with a low behavioral cost (12 keypecks) of obtaining a 10-sec period of access to food in a hopper. The apparatus was instrumented so that the amount of food eaten during each 10-sec access period was measured. With this type of data, it is possible to partition the total daily food intake into its individual components, in the manner described by Equation 1. The experimental protocol required that each pigeon achieve a stable baseline of food intake at 21C, and then receive six consecutive days of exposure to 6C. Subsequently, each bird was returned to 21C until food intake stablized, and then an exposure to 0.6C occurred for seven consecutive days.

Figure 4 shows the time course of change in the amount of food eaten daily by the pigeons at both levels of Ta. In general, food intake increased gradually across days. By the third day, all pigeons had achieved clearly higher intake than on the 21C baseline. As the error bars suggest, variability between animals was such that intake was not always greater in 0.6C than in 6C. Despite the sizable increase in daily food intake, the body weights of the birds were little affected (range: - 0.42% to + 1.1% of baseline body weight). Apparently, the additional energy obtained from the higher level of food intake in the cold was largely utilized to offset the higher costs of homeothermy in cold.

Figure 5 illustrates two basic feeding strategies that resulted in increased food intake in this situation (see Equation 1). The figure shows that, relative to the 21C baseline, P5 produced the hopper more frequently at 6C, but did not alter the amount it ate from the hopper on each presentation. On the other hand, P7 increased the amount of food it ate from the

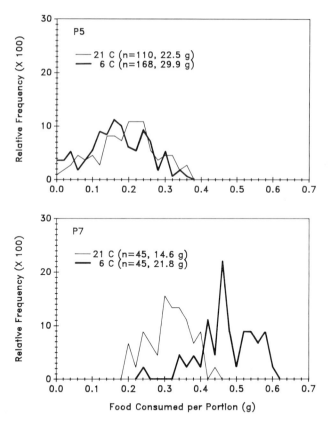

Fig. 5. Frequency distributions of the amounts of food eaten during each 10-sec hopper presentation for two pigeons in Henderson's (1988) experiment. A distribution is shown for the final baseline day when Ta=21C and for the final day of exposure to 6C. In all cases, the cost of gaining access to the food hopper was 12 keypecks. The legend in each panel shows the number of hopper presentations produced by the pigeon (n) and the total amount of food eaten (in grams) each day. The value of n is the number of observations in the frequency distribution.

hopper during each presentation, but it did not alter the number of times it produced the hopper each day. Of course, the most effective strategy for increasing food intake would be to combine the strategies used by P5 and P7, but none of the birds mixed the strategies in this way during exposure to 6C. In the 0.6C exposure, however, all birds ate larger amounts of food per hopper presentation, and one bird combined this strategy with increased number of hopper presentations. Possibly, these different strategies for increasing daily food intake represent the influence of thermal variables. For example, a pigeon that utilizes feather posture and body posture to conserve energy during cold (e.g., McFarland and Budgell, 1970), may find that the keypecking activity required to gain access to food results in heat loss due to disturbance of its postural adjustments. In such cases, the pigeons might be more likely to keypeck less often and to eat larger amounts when the hopper is available. It is also possible that the different feeding strategies are related to different thermoregulatory strategies as reflected in the daily cycle of Tb. The possible interactions between feeding and thermal variables in the closed economy remain to be explored.

The changes in feeding behavior summarized in Figure 5 emerged during the first two or three days of cold exposure. The gradual appearance of these changes in feeding strategy indicates that the effects of cold exposure were not mediated by the immediate sensory effects of cold (c.f. Kraly and Blass, 1976), but arose from processes involved in adaptation to cold which have a relatively long time course.

Henderson (1988) observed that the feeding activities of pigeons in the closed economy is distributed across the light phase in a way that seems quite resistant to change. She reported that every three hours, the birds have approximately five discrete feeding bouts (each separated by at least 10 minutes without feeding). The size of the feeding bouts (in terms of amount of food eaten) is largest in the third and fourth quarter of the light period. This pattern of eating corresponds to the familiar pattern when pigeons eat food ad libitum in the laboratory (e.g. Zeigler et al., 1971). Even during cold exposure, the birds persisted in eating in this diurnal pattern. They did not, for example, eat a very large meal at the beginning of the light phase, as might be expected if they had greatly depleted their energy reserves overnight.

Finally, Henderson (1988) explored the consequences of raising the price of food from 12 to 60 keypecks per hopper presentation during exposure to cold. After the final day of exposure to 6C and to 0.6C, she raised the price of food for a single day. As noted earlier, the time course of the effects of a change in the price of food appear within a few attempts to obtain food at the new price, and certainly are evident within a single day. Henderson found that the effects of an increase in price are quite comparable at 21C, 6C and 0.6C. In all cases, pigeons are less willing to work for food, thereby producing the food hopper less frequently each day. And, with one exception, the pigeons eat more food when the hopper is presented. The exception seems to arise in pigeons who are already eating near the maximum amount possible as a result of their exposure to 0.6C.

CONCLUSION

Food scarcity and cold are environmental conditions that affect thermoregulation and feeding behavior in birds. The understanding of relationships between thermoregulation and feeding can be improved by combining experimental procedures for the study of feeding and for the study of thermoregulation. The outcome of such studies is likely to improve the understanding of adaptation to cold.

ACKNOWLEDGEMENT

We thank Ross Henderson, biomedical engineer on the support staff of the Psychobiology/Neuroscience program, for valuable discussions about the research and for instrumentation related to the experiments summarized in this chapter. We also thank Professor Werner Rautenberg and Joachim Ostheim for advice and assistance in conducting some of the experiments. Their participation was made possible by a NATO grant to Rautenberg and Rashotte for international collaboration in research.

REFERENCES

Collier, G. H., 1983, Life in a closed economy: The ecology of learning and motivation, in: "Advances in Analysis of Behavior, Vol. 3," M. D. Zeiler and P. Harzem, eds., Wiley: New York.

Collier, G. H., and Rovee-Collier, C. K., 1981, A comparative analysis of optimal foraging behavior: Laboratory simulations, in: "Foraging Behavior: Ecological, Ethological and Psychological Approaches," A .C. Kamil and T. D. Sargent, eds., Garland Press: New York.

Henderson, D., 1988, "Feeding in the Cold: Pigeons (Columba livia) in a Closed Economy," Masters thesis. Florida State University, Tallahassee, Florida.

Hudson, J. W., 1978, Shallow daily torpor: A thermoregulatory adaptation. in: "Strategies in Cold: Natural Torpidity and Thermogenesis," L. C.H. Wang and J. W. Hudson, eds., Academic Press: New York.

Hursh, S. R., 1980, Economic concepts for the analysis of behavior, J Exp Anal Behav., 37:219.

Hursh, S. R., 1984, Behavioral economics, J Exp Anal Behav., 42:435.

Kendeigh, S. C., Kontogiannis, J. E., Mazac, A., and Roth, R. R., 1969, Environmental regulation of food intake by birds, Comp Biochem and Physiol., 31:941.

Kraly, F. S., and Blass, E. M., 1976, Mechanisms for enhanced feeding in the cold in rats, J Comp & Physiol Psychol, 90:714.

McFarland, D., and Budgell, P., 1970, The thermoregulatory role of feather movements in the Barbary Dove (Streptopelia risoria), Physiol Behav., 5:763.

Phillips, D. L., 1989, "Thermal Effects of Feeding during Chronic Food Scarcity in the Pigeon (Columba livia)," Masters thesis. Florida State University, Tallahassee, Florida.

Rashotte, M. E., and Henderson, D., in press, Coping with rising food costs in a closed economy: Feeding behavior and nocturnal hypothermia in pigeons (Columba livia), J Exp Anal Behav.

Rashotte, M. E., and O'Connell, J. M., 1986, Pigeons' reactivity to food and to Pavlovian signals for food in a closed economy: Effects of feeding time and signal reliability. J Exp Psychol (Animal Behav Proc)., 12:235.

Rashotte, M. E., Rautenberg, W., Henderson, D. and Ostheim, J., 1986, "Thermal, metabolic and feeding reactions of pigeons when food is scarce," Paper read at the 27th annual meeting of the Psychonomic Society, New Orleans, Louisiana.

Savory, C. J., 1986, Influence of ambient temperature on feeding activity parameters and digestive function in domestic fowls, Physiol Behav., 38:353.

Steen, J., 1957, Food intake and oxygen consumption in pigeons at low temperatures, Acta Physiol Scand., 39:22.

Zeigler, H. P., Green, H. L. and Lehrer, R., 1971, Patterns of feeding behavior in the pigeon, J Comp & Physiol Psychol., 76:468.

METABOLISM AND BODY TEMPERATURE DURING CIRCADIAN SLEEP AND TORPOR IN THE FED AND FASTING PIGEON

Nathan H. Phillips and Ralph J. Berger

Department of Biology
University of California
Santa Cruz, CA 95064 USA

INTRODUCTION

Shallow torpor is utilized by many small-to-medium sized birds and mammals to extend their limited energy reserves by minimizing energy expenditure when food is scarce and/or ambient temperatures are low. In some of these animals circadian torpor also occurs as an endogenous rhythm under constant conditions. In rodents, the duration of circadian torpor increases proportionally to decreases in body weight as energy reserves decline (Tucker, 1966; Brown and Bartholomew, 1969; Wolff and Bateman, 1978) while depth of torpor remains relatively constant (Walker *et al.*, 1979; Harris *et al.*, 1984). However, in birds the depth of circadian torpor is inversely proportional to their body weight while the duration of torpor is generally constant and restricted to the dark portion of the light-dark cycle. A characteristic pattern of progressively larger nocturnal decreases in body temperature (T_b) and metabolic rate (MR) and a daily return to euthermic levels accompanies body weight loss in the tiny willow tit (Reinertsen and Haftorn, 1984), white-crowned sparrow (Ketterson and King, 1977) and larger kestrel (Shapiro and Weathers, 1981). Declining energy reserves in these diurnally active birds are thus followed by increasing nocturnal energy conservation during the inactive portion of the circadian rhythm, which in ringed turtle doves was shown to be characterized predominantly by slow wave sleep (SWS; Walker *et al.*, 1983).

At sleep onset, a decrease in hypothalamic thermosensitivity leads to a decreased thermogenic responsiveness and consequently a reduction in T_b and oxygen consumption ($\dot{V}O_2$) regardless of the phase of the circadian T_b rhythm at which it occurs (Heller and Glotzbach, 1977; Parmeggiani, 1980). This down-regulation of thermosensitivity is extended further during entrance into hibernation leading to even greater energy savings through further reductions in T_b and $\dot{V}O_2$. Despite this reduced responsiveness, proportional control of T_b and MR persists throughout sleep and hibernation, although of lower gain than during wakefulness (Heller and Glotzbach, 1977). Thus, electrophysiological and thermoregulatory changes associated with the onset of SWS and subsequent entrance into circadian torpor and hibernation indicate that sleep, circadian torpor and hibernation are homologous energy conserving processes based on a continuum of decreasing MR and T_b (Heller, 1979; Walker and Berger, 1980; Berger, 1984).

To test the hypothesis that nocturnal T_b and MR are regulated at levels proportional to energy reserves, T_b, $\dot{V}O_2$, carbon dioxide production ($\dot{V}CO_2$) and electrophysiological measures were taken on pigeons subjected to two fasting conditions. Energy reserves during each fast were manipulated by diurnal infusions of either glucose or saline solutions. To determine whether these daily changes in T_b and MR are endogenous circadian rhythms and their relationship to sleep, several pigeons were recorded during constant lighting and ambient temperature (T_a) conditions.

METHODS

Surgery

Implantations of electrodes, temperature telemetry capsule and catheterizations required two separate surgical operations, in which Halothane anesthetic was used. In the first operation, pigeons (*Columba livia*) were chronically implanted with stainless steel screw electrodes to record EEG and EOG, and with teflon coated stainless steel wire electrodes for the neck EMG. The electrode contacts were inserted into an Amphenol pin strip connector which was permanently affixed to the skull with acrylic dental cement. A Minimitter temperature telemetry capsule (Model XM) was also inserted into the abdominal cavity.

In the second operation, following a recovery period of at least one week, catheters were implanted in the right brachial vein and artery. Each catheter consisted of a 20-cm length of polyethylene tubing (PE-50) inserted 1 cm into a 5-cm length of silastic tubing (.020" i.d. x .037" o.d.). The length of the silastic tubing was inserted into the vessel and the catheter securely sutured in placed at the point of insertion. The PE tubing was threaded under the skin to the strip connector on the head and anchored to it with dental cement. The entire length of tubing was then filled with heparin solution and sealed.

Apparatus

The recording chamber consisted of an electrically shielded incubator placed within a sound attenuated (45dB) chamber. Pigeons were housed in the incubator in a 6.5 l Plexiglas metabolic chamber. Water was provided by a 35 ml drinking tube mounted on the inner wall of the chamber. Recording cables and catheter extensions were led out of the metabolic chamber through a sealed port and then to the outside of the recording chamber so that infusions and blood samples could be performed without disturbing the pigeon.

T_a was measured with a Bailey (BAT-8) digital thermometer and copper-constantan thermocouple attached to the inside wall of the metabolic chamber. The pulses transmitted from the Minimitter were received by an antenna built into the floor of the incubator and amplified by an AM receiver. The pulse rate of this signal, which was proportional to the temperature of the minimitter, was used to latch and reset the count from a 10 kHz clock. This clock count was then fed to the computer via a parallel port.

Oxygen and carbon dioxide concentrations were measured in an open flow system. Air was drawn through the metabolic chamber at approximately 1.2 l/min (adjusted so that CO_2 did not exceed 1%) dried (Drierite) and then metered (Tylan mass flow meter, Model FM-360). A 100 ml/min sample of this air stream was drawn through O_2 and CO_2 analyzers (Applied Electrochemistry Models S-3A and CD-3A, respectively) by a flow controller (Applied Electrochemistry, Model R-1). $\dot{V}O_2$ and $\dot{V}CO_2$ were calculated using the equations of Withers (1977).

The analog outputs of the O_2 and CO_2 analyzers, and the digital thermometer were connected to a 6809 based microcomputer (Tandy) through a 12-bit analog to digital converter (Technical Hardware Inc.). Temperature and metabolic measures were sampled at 1 sec intervals and 2 min means calculated and stored on floppy disk. Electrophysiological measures were recorded on an 8 channel polygraph (Beckman Model R-411) at a paper speed of 6 mm/sec and on a seven channel FM tape recorder (Sangamo, Model 3500).

Procedure

Following at least one week for recovery from the second surgery, each pigeon was connected to the recording cables and catheter extensions and placed in the metabolic chamber to begin the first of two 4-day fasts which were separated by at least one week of *ad lib* food and water in their home cage. On each day of one fasting period, 12-hr infusions (3 ml/hr) of saline (0.9% NaCl, 308 mOsm/l) were given during the light phase of the 12:12 hr light-dark cycle (light on at 0800 hr). On each day of the other fast, approximately 15% of the overall mean caloric expenditure was infused in the form of glucose (5.6%, 314 mOsm/l) over 12 hrs at the same rate beginning at the same time of day. The order of saline and glucose infusion was alternated in each successive bird. On days 2-4 of the fast, electrophysiological, temperature and metabolic measures were taken continuously except for approximately one hour per day beginning at 1100 hr. During this period, the pigeon was removed from the metabolic chamber, weighed and the chamber cleaned. T_a was maintained at 20°C ±0.5°C.

Figure 1 shows characteristic electromyograms (EMG), electroencephalograms (EEG) and electrooculograms (EOG) of four arousal states. Electrophysiological records were visually scored without knowledge of treatment condition in 25-sec epochs for wakefulness (W: low voltage, mixed frequency EEG, eye or nictitating membrane movements in the EOG less than 1 sec apart, phasic EMG), drowsiness (D: high voltage slow EEG, EOG events greater than 1 sec but less than 5 sec apart, absence of EMG phasic activity) and SWS (S: high voltage slow EEG, EOG events greater than 5 sec apart, absence of EMG phasic activity). Rapid eye movement sleep (R: low voltage, high frequency EEG, eye movement activity in EOG, decrease or no change in EMG activity from preceding SWS epoch) was scored in 5-sec epochs.

In a separate study of the effects of constant light and T_a conditions on sleep/wake and T_b rhythms, two pigeons (one from the metabolic study above) were connected to the recording cables and maintained in a 22 x 36 x 40 cm cage with a grid floor inside the incubator. Each pigeon was subjected to extended periods (up to 50 days) in three different photic conditions: 12:12 light-dark cycle (LD), constant bright light (LL: 40 W incandescent bulb, ~200 lux at cage floor) and constant dim red light (DD: 7.5 W incandescent bulb, ~3 lux at cage floor). Food and water were available *ad lib* except during a 4-day fast in each of the three conditions when only water was unrestricted. T_b was recorded continuously and 24-hr samples of electrophysiological measures were taken in each condition in the fed state and on Day 4 of fasting.

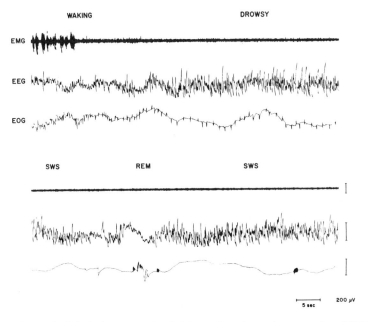

Fig. 1. Electrophysiological patterns of wakefulness, drowsiness, slow wave sleep (SWS) and rapid eye movement (REM) sleep in a pigeon.

Data Analysis

Statistical analyses were performed on the mean 12-hr light and mean 12-hr dark values of each variable of interest. Analysis of variance for repeated measures was used to test for overall significance, and two tailed t-tests were used for post-hoc comparisons. Reported *r* values are Pearson product-moment correlations pooled across pigeons. The significance level was set at 0.05.

RESULTS

LD Fast

The pattern of T_b in an individual bird over days of fasting under both saline and glucose conditions is shown in Figure 2. Mean diurnal and nocturnal T_b for 5 pigeons in both conditions is shown in Figure 4. Mean diurnal T_b did not change significantly across days of fasting in either the saline or glucose condition and did not differ significantly between conditions. Mean nocturnal T_b on Day 3 and Day 4 was significantly lower than on Day 2 of the fast in both conditions (saline, $P < 0.05$; glucose, $P < 0.01$) and was significantly lower under saline than under glucose infusions on Day 3 ($P < 0.05$) and Day 4 ($P < 0.05$) of the fast.

There was a strong positive correlation between T_b and $\dot{V}O_2$ in both saline (pooled $r = 0.70, P < 0.001$) and glucose (pooled $r = 0.65, P < 0.001$) conditions. Figure 3 shows the concordance between the patterns of T_b and $\dot{V}O_2$ in a pigeon. Unlike T_b, mean diurnal $\dot{V}O_2$ decreased significantly over days of fasting in both conditions (Fig. 5; saline, $P < 0.01$; glucose, $P < 0.01$) as did nocturnal $\dot{V}O_2$ in the saline condition ($P <$

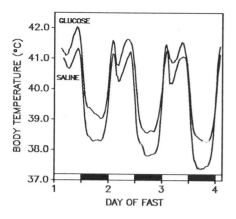

Fig. 2. 2-hr moving averages of body temperature in a pigeon on days 2-4 of fasting under 12-hr diurnal infusions of saline or glucose. Dark periods of 12:12 light-dark cycle indicated by black bars on abscissa.

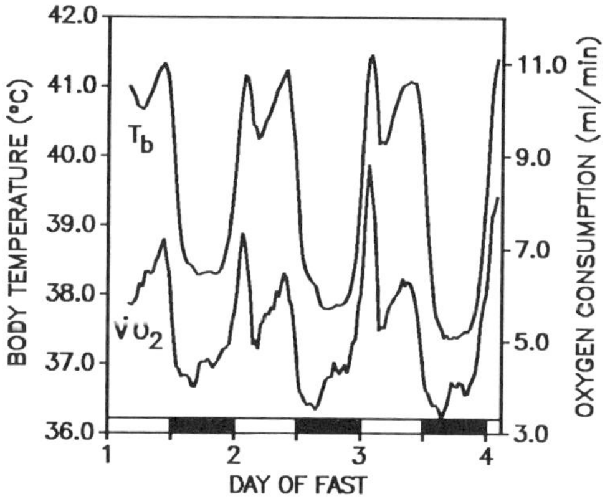

Fig. 3. 2-hr moving averages of body temperature (T_b) and oxygen consumption ($\dot{V}O_2$) on days 2-4 of fasting under 12-hr diurnal infusions of saline in the same pigeon as Fig. 2. Dark periods of 12:12 light-dark cycle indicated by black bars on abscissa.

0.05). Moreover, mean $\dot{V}O_2$ did not significantly differ between the saline and glucose conditions on any day of the fast.

Mean diurnal respiratory quotient (RQ) was significantly higher during glucose infusions than during saline infusions on all of the last three days of fasting (Fig. 6; $P < 0.01$). Mean nocturnal RQ was significantly higher in the glucose condition than in the saline condition only on Day 3 ($P < 0.01$). Mean RQ did not differ significantly across days of fast in either condition at night, but rose significantly by day on Day 3 in the glucose condition ($P < 0.05$).

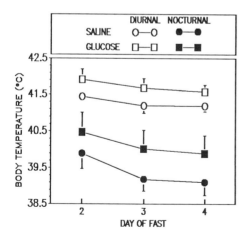

Fig. 4. 12-hr mean diurnal and nocturnal body temperature on days 2-4 of fasting with 12-hr diurnal infusions of saline or glucose. Pigeons maintained on 12:12 light-dark cycle. Vertical lines denote SEM (n = 5).

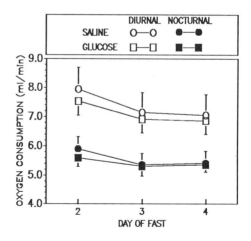

Fig. 5. 12-hr mean diurnal and nocturnal oxygen consumption on days 2-4 of fasting with 12-hr diurnal infusions of saline or glucose. Pigeons maintained on 12:12 light-dark cycle. Vertical lines denote SEM (n = 5).

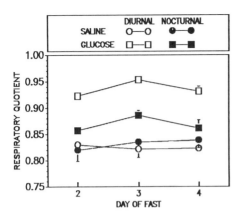
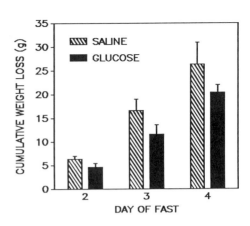

Fig. 6. 12-hr mean diurnal and nocturnal respiratory quotient on days 2-4 of fasting under 12-hr diurnal infusions of saline or glucose. Pigeons maintained on 12:12 LD cycle. Vertical lines denote SEM (n = 5).

Fig. 7. Mean cumulative body weight loss over days 2-4 of fasting under 12-hr diurnal saline or glucose infusions. Vertical lines denote SEM (n = 5).

Mean body weight fell significantly over days of fast in both conditions (Fig. 7; $P < 0.001$), and was positively correlated with both mean T_b and mean $\dot{V}O_2$ for the light period (T_b: pooled $r = 0.51, P < 0.05$; $\dot{V}O_2$: $r = 0.70, P < 0.001$) and the dark period (T_b: pooled $r = 0.59, P < 0.01$; $\dot{V}O_2$: $r = 0.62, P < 0.01$). The cumulative weight loss was significantly greater in the saline condition than in the glucose condition on each of the last three days of fasting ($P < 0.05, P < 0.01, P < 0.05$, respectively). This differential rate of weight loss was 23 percent less under glucose than under saline.

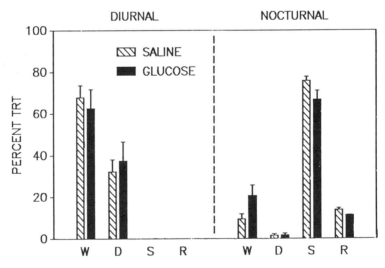

Fig. 8. Mean diurnal and nocturnal percent total recording time (TRT) in waking (W), drowsiness (D), SWS (S), and REM (R) on day 4 of fasting under 12-hr dirunal infusions of saline or glucose. Vertical lines denote SEM (n = 3).

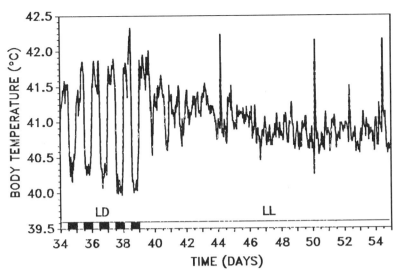

Fig. 9. Body temperature in a pigeon 5 days prior to and 16 days following transition from a 12:12 light-dark cycle (LD) to constant bright light (LL). Black bars on abscissa indicate dark portion of LD cycle. The spikes of high body temperature on days 44, 50 and 54 are the result of experimenter intrusions to replenish food and water and clean the chamber.

Analysis of the sleep parameters for three pigeons on Day 4 of fasting indicates no significant differences between the two conditions in percent of total recording time (TRT) of W, D, S or R (Fig. 8). However, there was a trend for a higher percent S ($P < 0.09$) and a lower percent W ($P < 0.08$) during the nocturnal portion of the LD cycle in the saline condition than in the glucose condition.

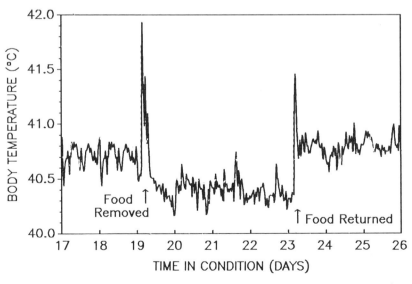

Fig. 10. Body temperature of a pigeon in constant bright light. A 4-day fast began on day 19 and ended on day 23.

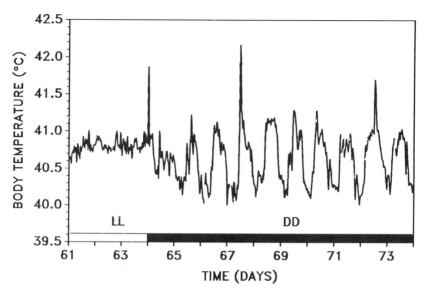

Fig. 11. Body temperature of a pigeon 3 days prior to and 10 days following a transition from constant bright light (LL) to constant dim red light (DD) on day 64.

Constant Conditions

Circadian T_b rhythms in two birds consistently were sharply attenuated and subsequently disappeared following transition from LD to LL in the fed condition (Fig. 9). T_b remained arrhythmic during fasting but generally fell to a slightly lower level (Fig. 10). In DD, a circadian rhythm of T_b with a free-running period of slightly less than 24 hrs persisted following transitions from LD to DD and developed soon after LL to DD transitions (Fig. 11). Fasting in DD led to an increase in the amplitude of the T_b rhythm similar to fasting patterns in LD.

Analysis of electrophysiological data from one pigeon indicates that S and R were eliminated immediately after transition from LD to LL, and replaced by W alternating with relatively brief periods of D throughout the 24-hr period in both fed and fasted conditions. Fasting in LL led to a 38% increase in W (fed = 68% TRT, fasted = 93% TRT) and a concomitant 79% decrease in D (fed = 32% TRT, fasted = 7% TRT). The circadian T_b rhythm in DD was synchronized with a circadian rhythm of sleep/waking both of which persisted in the fed and fasted conditions. The largest changes in arousal states due to fasting in DD were a 41% increase in S (fed = 33% TRT, fasted = 46% TRT) and a decrease in D (fed = 17% TRT, fasted = 7% TRT).

DISCUSSION

Reductions in spinal thermosensitivity in the pigeon are associated independently with both sleep and the dark phase of the LD cycle (Heller et al., 1983). The present finding of a significant progressive decline in nocturnal T_b over days of fasting confirms previous findings in the white-crowned sparrow (Ketterson and King, 1977), ringed turtle dove (Walker et al., 1983) and american kestrel (Shapiro and Weathers, 1981) during fasting, and is in accordance with a progressive reduction in spinal thermosensitivity during each successive night of fasting in the pigeon (Heller, 1988). Energy reserves, as indicated by body weight, were positively correlated with T_b and $\dot{V}O_2$. Reducing the rate of energy reserve utilization during the fast, through diurnal infusions of glucose, caused significantly higher nocturnal T_bs although they still declined over the fast. These results indicate that nocturnal T_b in the pigeon is regulated at levels proportional to its energy reserves.

Although the expected strong correlation between T_b and $\dot{V}O_2$ was evident, the overall pattern of $\dot{V}O_2$ differed slightly from that of T_b. $\dot{V}O_2$ declined both by day and night whereas T_b declined significantly over the fasting period only at night. One possible explanation of this difference is that maintenance of a relatively high and stable T_b is necessary to sustain a high level of vigilance during the light portion of the LD cycle when the pigeon is normally active and ecologically at the greatest risk of predation. MR can be divided into basal (maintenance) and activity components. Since $\dot{V}O_2$ has a larger activity component during the day than at night, relatively large reductions in daytime energy expenditure could be achieved through a decrease in behavioral activity without any change in T_b, whereas at night, when activity is normally minimal, substantial reductions in energy expenditure could only be derived from reductions in basal MR and associated changes in T_b. Therefore, it seems likely that during a fast T_b is the primary regulated variable and that during sleep MR is at least partially determined by the T_b set point.

A number of physiological and behavioral factors may contribute to the absence of significant differences in $\dot{V}O_2$ between glucose and saline conditions. Greater depletion of body weight in the saline condition results in an increase in the surface-to-volume ratio which would contribute to greater heat loss. Significantly higher RQs during the glucose condition, decreased weight loss and the absence of glucose in the cloacal fluids indicate that the infused glucose was metabolized. Heat production per mole of O_2 varies with the substrate being oxidized (Kleiber, 1975), that of glucose (5.0 kcal) being higher than that of fat (4.7 kcal). Therefore, higher $\dot{V}O_2$ would be required in the saline condition to generate the same amount of heat as in the glucose condition.

Fasting in LD condition, lead to significant changes in T_b only during the dark phase when sleep predominates. The results from the DD condition indicate that the pattern of T_b is under endogenous circadian control and is tightly linked with the sleep/wake cycle. Fasting in DD led to progressively greater circadian declines in T_b similar to those in LD and which maintained the same phase relation to the sleep cycle. The present T_b data from pigeons in DD conform to the T_b patterns reported by MacMillen and Trost (1967) in the inca dove fasted in constant darkness for 49 hrs. LL suppressed both the circadian T_b rhythm and SWS. The slight decline in T_b in response to fasting in LL did not appear to be related to sleep.

Conclusion

The results of the present study indicate that circadian rhythms of T_b and $\dot{V}O_2$ in the pigeon are regulated during sleep at levels proportional to its energy reserves. The largest changes in T_b occurred during the dark phase in LD conditions, and continued to be associated with the circadian sleep period in DD. In LL, SWS and circadian T_b rhythms were suppressed, and the fast-induced decline in T_b was greatly attenuated and continued to show no circadian rhythm.

ACKNOWLEDGEMENTS

We thank Dr. Rudolf Graf for technical advice during the development of the catheterization procedure, and Peter Knopf, Troy Fallon, Michael Seffinger, Keith Bachman and Aribert Bauerfeind for their assistance.
Supported by NIH grant DK 38402.

REFERENCES

Berger, R. J., 1984, Slow wave sleep, shallow torpor and hibernation: Homologous states of diminished metabolism and body temperature, *Biol. Psychol.*, 19:305.

Brown, J. H., and Bartholomew, G. A., 1969, Periodicity and energetics of torpor in the kangaroo mouse, *Microdipodops pallidus*, *Ecology*, 50:705.

Harris, D. V., Walker, J. M., and Berger, R. J., 1984, A continuum of slow-wave sleep and shallow torpor in the pocket mouse *Perognathus longimembris*, *Physiol. Zool.*, 57: 428.

Heller, H. C., 1979, Hibernation: Neural aspects, *Annual Rev. Physiol.*, 41:305.

Heller, H. C., 1988, Sleep and hypometabolism, *Can. J. Zool.*, 66: 61.

Heller, H. C., and Glotzbach, S. F., 1977, Thermoregulation during sleep and hibernation, *Internat. Rev. Physiol.*, 15: 147.

Heller, H. C., Graf, R., and Rautenberg, W., 1983, Circadian and arousal state influences on thermoregulation in the pigeon, *Am. J. Physiol.*, 245: R321.

Ketterson, E. D., and King, J. R., 1977, Metabolic and behavioral responses to fasting in the white-crowned sparrow (*Zonotrichia leucophrys* Gambelii), *Physiol. Zool.*, 50: 115.

Kleiber, M., 1975, "The Fire of Life," Krieger, Huntington, N.Y.

MacMillen, R. E., and Trost, C. H., 1967, Nocturnal hypothermia in the inca dove, *Scardafella inca*, *Comp. Biochem. Physiol.*, 23:243.

Parmeggiani, P. L., 1980, Temperature regulation during sleep: a study in homeostasis, in: "Physiology in Sleep," J. Orem and C. D. Barnes, ed., Academic Press, New York.

Reinertsen, R. E., and Haftorn, S., 1984, The effect of short-term fasting on metabolism and nocturnal hypothermia in the Willow Tit *Parus montanus*, *J. Comp. Physiol.*, 154(B): 23.

Shapiro, C. J., and Weathers, W. W., 1981, Metabolic and behavioral responses of American kestrels to food deprivation, *Comp. Biochem. Physiol.*, 68(A):111.

Tucker, V. A., 1966, Diurnal torpor and its relation to food consumption and weight changes in the California pocket mouse, *Perognathus californicus*, *Ecology*, 47:245.

Walker, J. M., and Berger, R. J., 1980, Sleep as an adaptation for energy conservation functionally related to hibernation and shallow torpor, *Prog. Brain Res.*, 53:255.

Walker, J. M., Garber, A., Berger, R. J., and Heller, H. C., 1979, Sleep and estivation (shallow torpor): Continuous processes of energy conservation, *Science*, 204: 1098.

Walker, L. E., Walker, J. M., Palca, J. W., and Berger, R. J., 1983, A continuum of sleep and shallow torpor in fasting doves, *Science*, 221: 194.

Withers, P. C., 1977, Measurement of $\dot{V}O_2$, $\dot{V}CO_2$, and evaporative water loss with a flow-through mask, *J. Appl. Physiol.*, 42:120.

Wolff, J. O., and Bateman, G. C., 1978, Effects of food availability and ambient temperature on torpor cycles of *Perognathus flavus* (Heteromyidae), *J. Mamm.*, 59: 707.

AUTONOMIC AND BEHAVIORAL TEMPERATURE REGULATION AS A PART OF THE RESPONSE COMPLEX TO FOOD SCARCITY IN THE PIGEON

Joachim Ostheim and Werner Rautenberg

Dept. Zoophysiol., Ruhr-Universität Bochum
Postbox 102148
D-4630 Bochum 1, FRG

INTRODUCTION

When faced with periods of food scarcity, an endothermic organism can use a variety of mechanisms for achieving a positive energy balance, or at least a steady state, in which energy intake compensates metabolic costs of survival. In general these mechanisms can be described as followed:
1) Changes in foraging behavior should increase the efficiency of searching for food and/or the utilization of detected sources.
2) Decreased deep body temperature (T_b) and decreased heat production lower energy output, especially the costs of endothermia, in mammals and birds.
3) Birds can decrease their heat loss to the environment by erection of plumage or by resting at warm places.

In addition most animals have an energy storage consisting of deposited fat. This energy storage permits survival for a fixed time without any food intake; but it should not be depleted when there is any opportunity to achieve a positive or at least steady state energy balance.

The present study provides evidence about the way these mechanisms interact during adaptation to increasing levels of food scarcity in the pigeon.

MATERIAL AND METHOD

Apparatus

The experiment was done in a 24-h "closed economy" behavioral procedure, which is a laboratory simulation of natural foraging circumstances (Collier & Rovee-Collier: 1981, Hursh: 1984). The procedure was modified to enable the birds to choose between warm (30°C) and cold (10°C) ambient temperature (T_a). The pigeons lived about six months in the foraging simulation, where they could earn access to small portions of food or water by pecking on seperate keys. Pressing the proper key a specified times led to operation of the food or water dispenser. It was possible to increase the "behavioral cost" of the food portions by increasing the number of keypecks required to get a 9 s access to food.

In the middle of the box a wooden perch connected to an electronic load-beam weighing circuit was installed. The body weight (BW) of the bird was measured continously when the animal was sitting on the perch. Infrared light beams ran lengthwise along the right and left side of the box (2.5 cm from wall). The bird could reach them by moving from one side of the perch to the other. Breaking a light beam started a fan to blow air through a radiator on the same side of the box, whereas the opposite fan was stopped. Once started, the fan blew in temperature-controlled air until the opposite light beam was broken. One radiator was continuously flushed with water of 10°C, the other with water of 30°C. The pigeons were able to change Ta of the box between these extreme temperatures gradually by shuttling from one side of the cage to the other. Ta was measured by a copper-constantan thermocouple located below the perch. The recorded Ta provided an estimation of temperature regulatory behavior.
Tb was obtained from a radio transmitter (Mini-Mitter X-Series) implanted into the intestinal cavity under general anesthesia (Halothan). Its signals (resolution = 0.1°C) were received by an antenna located within the perch.
During the light-phase of the 12/12 h L/D cycle, the box was illuminated at about 160 lx by two fluorescent tubes on the top of the box and by light from the two pecking keys. At the beginning of the dark-phase, all lights were switched off rapidly. Pecking the keys during the dark-phase gave no access to food or water. Data (Tb, Ta, BW, key pecking for food and water, presentations of food and water) were sampled once a minute by a PC, and means across 10 min were stored.

Animal and procedure

The experiment was done with experimentally naive, domestic pigeons (Columba livia), which were fed a commercial pellet diet (altromin). The keypeck requirement to operate the food and water dispensers was first set at a Fixed Ratio of 1 peck (FR 1) on the appropriate key. Within a few days the birds learned to operate the equipment. At this point the pecking requirement was increased to FR 6 for food and water. The required number of pecks for producing food was increased when the BW did not decrease for about one week. As the performance criterion allowed, the FR for food was increased through the series: 6, 12, 60, 120, 240, 480, and 960 pecks/hopper presentation. After the pigeons completed the FR 960, the requirement was reset to 12 in order to test whether the effects of high food costs were reversible. The pecking requirement of getting water remained at FR 6 throughout the entire experiment.

RESULTS

At increasing FR the diurnal and nocturnal BW both decreased (Fig. 1). At FR 480 and 960 the amplitude of BW-oscillation, which was rather constant at FRs up to 240, became smaller, reaching a minimum at FR 960. The final value of BW at FR 960 was about 70 % of the initial baseline value and the food intake was reduced to about 30% of its initial value (Tab. 1).
When the FR was raised from 12 to 240, the number of trials/day decreased; but the amount of food eaten within a trial increased gradually. This improvement reached a saturation level at FR 240, which consisted through the higher FRs (Fig. 1). Up to the point where feeding behavior reached its saturation level, there was

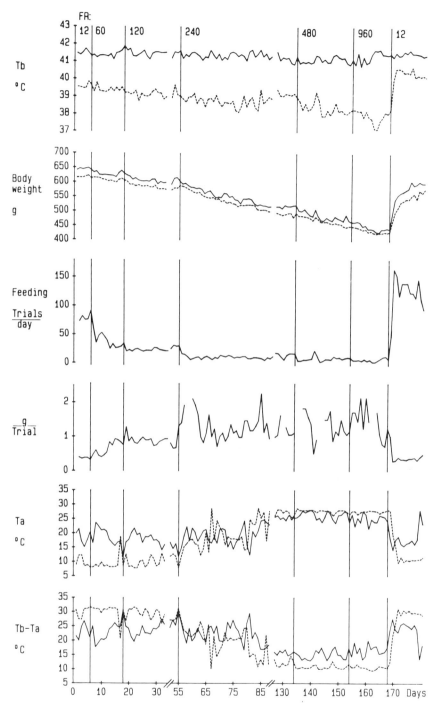

Fig. 1. Time courses of the recorded parameters for the entire period of a typical experiment. In order to approximate mean min. and max. values (means of 180 measurements) of the daily oscillation in each parameter (solid lines light-, broken lines dark-phases), data were averaged at different times: T_b, T_a, ($T_b - T_a$) from 11:30 to 14:30 h (light), 23:30 to 2:30 h (dark); BW from 17:30 to 20:30 h (light), 5:30 to 8:30 h (dark). See text for details.

Table 1. Mean steady state-values of the six final days at each FR. BW- and Tb-data were averaged over the same intervals as in Fig. 1. The food intake/day was calculated from weighing the food dispenser every 3 to 4 days.

FR	Food intake g/day	min. BW g	max. BW g	oscill. BW g	min. Tb °C	max. Tb °C	oscill. Tb °C
12	28,8	617 (3.5)	645 (2.8)	28.5 (2.9)	39,6 (0.12)	41,5 (0.13)	1.92 (0.19)
60	25,5	604 (3.3) **	625 (8.3) **	21.3 (8.0)	39.4 (0.11) *	41,4 (0.18)	2.00 (0.14)
120	20,3	585 (5.0) **++	609 (5.2) **+	24.0 (9.5)	39,1 (0.43) *	41,4 (0.21)	2.25 (0.43)
240	17,9	491 (6.0) **++	519 (4.6) **++	28.8 (8.3)	39,1 (0.05) **	41,2 (0.20) *	2.02 (0.17)
480	10,5	455 (5.8) **++	474 (8.8) **++	18.8 (10.3)	38.3 (0.24) **++	41.1 (0.21) *	2.87 (0.16) **++
960	8.6	429 (2.5) **++	438 (6.1) **++	8.5 (5.2) **	37.9 (0.40) **	41.5 (0.15) +	3.63 (0.52) **+
12	45.6	569 (12.6) **++	609 (13.1) **++	40.5 (13.7) ++	40.2 (0.05) **++	41.5 (0.14)	1.23 (0.15) **++

only a moderate decrease in BW. During the pigeon's adjustment to FR 240, maximum and minimum BW in the light- and dark-phases both decreased by about 15%. The adjustment to FR 240 required a longer period of adaptation than that of the other FRs. During the extended time of FR 240, food intake increased from initial low values (12.1 g/day in week 1; 9.9 g/day in week 2) to relatively high values (18.8 g/day, 17.9 g/day in the final 2 weeks). FR 240 seemed to be an important transition point. Ignoring the oscillation between diurnal and nocturnal Tas, the pigeon selected relatively low Tas at FRs up to 120. At FR 240, however, the bird began to prefer relatively high Tas which persisted through FR 480 and FR 960. Following the final return to FR 12, the pigeon reverted to select relatively low Tas. Perhaps the most interesting aspect of the Ta data is the phase-shift in Ta oscillation which occured during FR 240. The Ta panel in Fig. 1 shows that the pigeon preferred higher Tas in the light- than in the dark-phase at FRs below 240. But during the last third of FR 240 and at higher FRs it selected warmer Tas in the dark-phase. The change in thermoregulatory behavior occured gradually from the beginning of FR 240 and reached a steady state within the last third of this ratio. At the same time the BW stabilized as well. Except for the phase-shift during FR 240, temperature preference was relatively constant at the other ratios.
At FRs lower 240, the (Tb - Ta) difference, which is a parameter of thermal conductance, was relatively large (Fig. 1). This indicates a higher heat loss to the environment than at conditions of

greater food scarcity (FRs beyond 240). As discussed above, the pigeon varied Ta across a wide temperature range. Much of the change in the (Tb - Ta) gradient can be attributed to this source of variation.

In addition to changes in the thermoregulatory behavior there were changes in the autonomic parameter Tb. At increasing FR nocturnal Tb decreased (Fig. 1). This nocturnal Tb decrease was only a moderate one at FRs up to 240; but nocturnal hypothermia became more severe at higher FRs. At FR 240 an initial drop in Tb recovered to comparatively higher values, when Ta preference behavior and course of BW both had reached a steady state. The temporal conformity of these tendencies led to the conclusion,

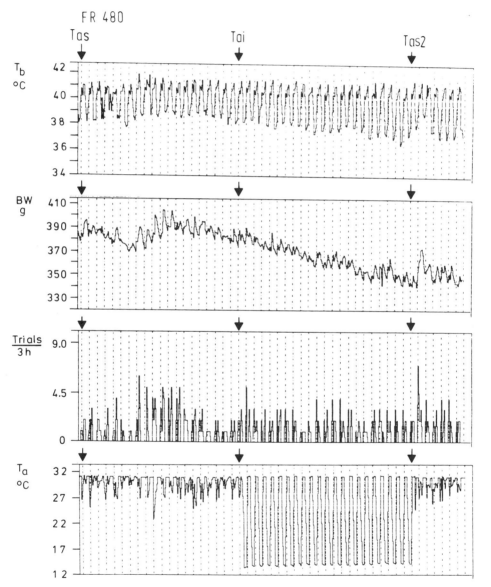

Fig. 2. Influence of Ta-selection on Tb and BW. From Tai to Tas2 a 10°C night was imposed. Broken vertical lines indicate light on at 7:00 h. The synchronism in feeding activity, BW, and Tb occured at all FRs. See text for details.

that the mechanisms of behavioral and autonomic temperature regulation are connected. This hypothesis was tested in a second experiment with a different pigeon. Until the bird selected a high Ta during the dark-phase, the procedure and results of this experiment were much the same as described above. The only difference between the two pigeons was, that the second one selected a cold night up to FR 480. When the preference of a warm night had reached a steady state condition, a fixed night temperature of 10°C, comparable to the selected Ta of lower FRs, was imposed to the bird (Tai). This procedure caused a BW- and Tb decrease, which both recovered, when the pigeon could influence Ta again (Tas2, Fig. 2).

In addition to nocturnal hypothermia there was a small drop in diurnal Tb at FR 240 and 480 which did not occure at the end of FR 960 (Fig. 1, Tab. 1). Data for diurnal Tb were averaged close to the time at which the pigeon was most likely to keypeck for food. Therefore the relatively high diurnal Tb at FR 960 was caused by heat production related to feeding activity, which increased at this state of the experiment (Fig. 1). The continued decline in BW may have reached a level where food intake was stimulated. The increase in the number of trials/day resulted in a BW plateau and increasing nocturnal Tb during the last third of FR 960.

When the FR was reinstated to FR 12, all recorded parameters returned towards their initial levels within a few days. The food intake reached its maximum values (Table 1), resulting in a BW-recovery and elevated nocturnal Tb by high anabolic metabolism.

DISCUSSION

The results lead to the conclusion that pigeons use the mechanisms available to solve the problem of food limited conditions. The most immediate adaptation to an increasing FR consisted in an improvement of feeding behavior. Within one hopper presentation the birds ate as much as possible, when the costs of the presentation was high. This finding does agree to results of Rashotte & Henderson (in prep.), who demonstrated, that pigeons increase their rate of eating dependent from the costs of feeding. When the costs rose to a level where foraging behavior failed to maintain energy reserves, the birds showed nocturnal hypothermia. But in our experiment the pigeons had the additional ability to influence Ta. Up to now it is not quite clear, why pigeons prefer a cold Ta during dark-phase, but they do, when the FR is low. Selecting a high Ta during night decreases the temperature gradient (Tb - Ta), which is an energy preserving mechanism. When a bird was prevented from this adjustment of behavioral temperature regulation, the nocturnal Tb decreased, as in the experiment of Rashotte & Hendersen. This points out that the response to a level of food scarcity is determined by the sum of the energy saving mechanisms available.

Within this experiment the pigeons also reduced Tb during light time. Energy saving by reduced Tb is not only dependent from the lower metabolism at 0.3°C to 0.4°C lower Tb (Q10); but the Tb decrease itself is a result of decreased activity. Movement induced heat production usually is dissipated by effectors of the thermoregulatory system, and Tb remains constant. Therefore the amount of saved energy by decreased activity is underestimated by a Tb-measurement. In previous studies (Rashotte et al., 86) a 50% decrease of diurnal metabolic rate was observed, which does not reflect in an adequate Tb decrease.

Up to the last third of FR 960 the BW panel of Fig. 1 shows a decrease at all FRs. It should be considered that a decrease in body mass itself lowers the costs of endothermia. But a decrease in body mass is limited by a critical minimal BW, at which energy storages are close to depletion. This seems to be reached at the BW plateau of FR 960. The begin of the BW plateau was correlated with an increase in food intake. Until now decreased food intake was only discussed as a variable influenced by the costs of feeding, which was compensated by autonomically and behaviorally mediated reactions. At this final state of FR 960 there seems to be a feedback mechanism of BW and/or T_b influencing their own input, which is the food intake. This means that there are two different strategies to overcome food scarcity: The "passive strategy" which saves energy and the "active strategy" which invests some energy to increase food intake. The "passive strategy" seems to dominate during the most parts of the adaptation to increasing levels of food scarcity, and the "active strategy" occurs to be of importance, when the energy sources of a pigeon are nearly depleted.

ACKNOWLEDGEMENT

The study was supported by DFG grant 815/86

REFERENCES

Collier, G.H. & C.K. Rovee-Collier, 1981, A comparative analysis of optimal foraging behavior: Laboratory simulations, in: "Foraging behavior: Ecological, ethological, and psychological approaches", A.C. Kamel & T. Sargent, eds., Garland STPM Press, New York.
Hursh, S.R., 1984, Behavioral economics, J. Exp. Anal. Behav., 42: 435.
Rashotte, M.E., W. Rautenberg, D. Henderson, and J. Ostheim, 1986, Thermal, metabolic, and feeding reactions of pigeons when food is scare, delivered at 27th annual meeting of the Psychonomic Society, New Orleans LA.
Rashotte, M.E. & D. Henderson, in prep., Coping with rising food costs in foraging simulations: Foraging behavior and nocturnal hypothermia in the pigeon (Columba livia).

BODY MASS, FOOD HABITS, AND THE USE OF TORPOR IN BIRDS

Brian K. McNab

Department of Zoology
University of Florida
Gainesville, Florida 32611, U.S.A.

INTRODUCTION

The sensitivity of birds to cold temperatures varies greatly. Some cannot tolerate temperatures below 0°C, while others can tolerate temperatures between -20 and -40°C (see Steen, 1958), even at high wind velocities. Various factors, including body size, rate of metabolism, and insulation, directly influence this sensitivity. Response to low temperatures, however, also includes their evasion by migration and by entrance into torpor. The selection of a response, itself, is influenced by many factors, including body size, food habits, and rate of metabolism. The object of this paper is to examine the physiological and ecological conditions that determine which response to cold is used by birds.

BODY MASS, FOOD HABITS, AND BASAL RATE OF METABOLISM

The conventional view of the factors setting the level of avian basal rates was established by Brody (1945), Lasiewski and Dawson (1967), and Aschoff and Pohl (1970). These authors collectively considered two factors, body mass and phylogenetic affiliation. They showed that 1) basal rate scales to body mass in a manner similar to that found in mammals (i.e., ca. $m^{0.75}$), 2) basal rate is higher in birds than in mammals, and 3) basal rate is higher in passerines than in other birds. Prinzinger and Hanssler (1980), however, concluded that much of the difference between passerines and other birds was related to a difference in body mass; when birds of similar size are compared, passerines and other birds have similar basal rates. Surely, large passerines and small procellariiforms, charadriiforms, and psittaciforms have high basal rates (McNab in press).

Factors other than body mass and phylogeny also have been suggested to influence avian basal rates. Thus, Scholander et al. (1950) hinted that the basal rates of birds might be correlated with climate, a later conclusion of Weathers (1979), Hails (1983), Ellis (1984), and Bennett and Harvey (1987). Avian basal rates appear to be correlated, at least in regions with high solar radiation, with plumage color (Weathers 1979; Ellis 1980, 1984). Ellis (1984) suggested that basal rate was correlated with feeding behavior, a conclusion that can be easily read into the data of Hayes and Gessaman (1980) and Wasser (1986). Clearly, basal rate of metabolism in birds is determined in a very complex manner.

McNab (in press) re-examined the variation among birds in basal rate of metabolism after body mass is accounted for. Much of the residual variation is correlated with the food habits of birds (Figure 1). A comparison of the association of basal rate with food habits in birds with that described in eutherian mammals (McNab, 1986) is of special interest because the endothermy of birds without doubt evolved independently of mammalian endothermy.

Much of the difference in basal rate described between birds and mammals is lost when food habits and body mass are considered simultaneously. For example, birds that feed on seeds or on grass generally have basal rates that are similar to those found in seed- or grass-eating mammals (high by the Kleiber standard), whereas feeding on flying insects and on fruits generally is associated with low basal rates in both groups, especially at large masses. The insectivorous birds include caprimulgids and some falconiforms. Fruit-eaters with low basal rates include trogons, colies, and trumpeters. Among passerines, tropical bulbuls have low basal rates. Yet, some differences between the vertebrate classes remain: at intermediate masses invertebrate-eating birds generally have much higher basal rates than invertebrate-eating mammals. The low basal rates of these mammals may reflect their extensive ingestion of soil particles (McNab, 1984), whereas equally-sized, invertebrate-eating birds principally eat aquatic invertebrates, which can be easily filtered. Furthermore, although most mammals and many birds that consume vertebrates have high basal rates, some birds with these habits have low basal rates. Vertebrate-eating birds with low basal rates also tend to soar while searching for prey. Both

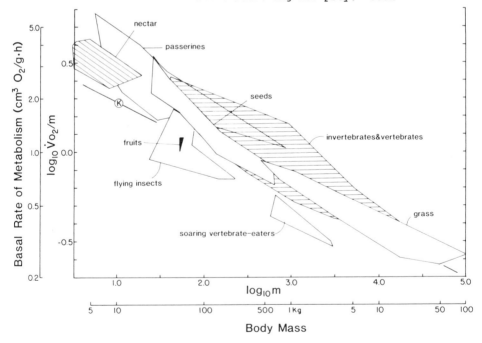

Fig. 1. $\log_{10}$ mass-specific basal rate of metabolism in birds as a function of $\log_{10}$ body mass. Passerines and various food habit categories are indicated for other birds. Data were derived from McNab (in press). The Kleiber (1961) metabolism-mass curve is indicated by K.

characteristics, soaring flight and a low basal rate, reduce energy expenditure in predators that have a low probability of finding prey.

Some birds have higher basal rates than is expected from a combination of body mass and food habits. They include large passerines and some smaller procellariiforms, charadriiforms, and psittaciforms (McNab in press). High basal rates in passerines are correlated with high growth rates, and have been suggested to be the basis for the widespread success of passerines in temperate terrestrial environments (McNab in press). In contrast to passerines, however, high basal rates in charadriiforms and procellariiforms are not associated with high growth rates.

BODY MASS, BASAL RATE, AND THE USE OF TORPOR

Many small mammals that live in seasonally cold climates evade the low temperatures and reduced food resources of winter by entering torpor. Relatively few birds enter torpor, and the majority that do enter torpor only on a daily basis. The propensity to enter torpor in mammals and birds is associated with body mass and the level at which energy is expended. Thus, McNab (1983) described in endotherms a ("boundary") curve between the basal rate of metabolism and body mass that separates those species that enter daily torpor from those species that do not. The use of torpor, therefore, is not just the consequence of a low rate of metabolism, but reflects the combination of a low rate of metabolism and an appropriately small body size. This combination is shown in Figure 2 in birds known for entrance into torpor, including hummingbirds (Trochilidae), colies (Coliidae), manakins (Pipridae), a dove (Columbidae), and some nightjars (Caprimulgidae). Seasonal torpor is definitively known only in the North American poorwill (<u>Phalaenoptilus nuttallii</u>), but may also be present in some swifts and swallows (McAtee, 1947).

The exact quantitative description of the boundary curve at present is uncertain. When first described (McNab, 1983), total rate of metabolism was said to be proportional to $\underline{m}^{0.33}$, i.e., mass-specific rate was proportional to $\underline{m}^{-0.67}$, but a recent analysis of scaling basal rate in mammals (McNab, 1988a) indicated that total rates in the boundary curve may be proportional to $\underline{m}^{0.50}$–$\underline{m}^{0.60}$, which would make mass-specific rates proportional to ca. $\underline{m}^{-0.45}$. The first estimate was derived from a very broad mass and taxonomic comparison, the second was derived from a narrower, mammalian analysis. Both boundary curves are indicated in Figure 2, and at first glance the earlier estimate appears to be most appropriate, but some evidence suggests that the second estimate may be more useful.

This evidence is derived from examining the response of birds to the boundary curve. To test the possibility that this boundary is a graded zone, rather than a sharp line, the capacity to maintain a temperature differential with the environment is examined in birds relative to the boundary curve. This requires a measure of a bird's position relative to the boundary curve. The best estimate of position is the coefficient in a power function having the same power as the boundary curve; it can be obtained by equating the power function to the measured basal rate in each species evaluated. For example, if <u>BMR</u> is the mass-specific basal rate, then the species' position relative to the boundary curve on a metabolism-mass graph is given by $\underline{c} = \underline{BMR}/m^{b-1} = \underline{BMR}\ m^{1-b}$ where b is the power of the boundary curve. This method simply is a means of calculating a set of curves parallel to the boundary curve and determining which of these curves corresponds to the basal rate of a particular species at its mean mass. These curves differ only in their coefficient $\underline{c}$.

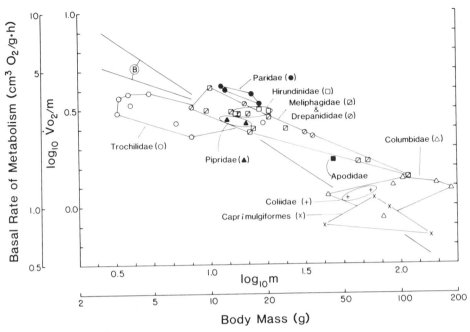

Fig. 2. $\log_{10}$ mass-specific basal rate of metabolism in small birds as a function of $\log_{10}$ body mass. Data were derived from Bartholomew et al. (1962), Lasiewski (1963), Lasiewski and Dawson (1964), Lasiewski and Lasiewski (1967), Lasiewski et al. (1967), MacMillen and Trost (1967a), Dawson and Fisher (1969), Bartholomew and Trost (1970), Lasiewski et al. (1970), MacMillen (1974), Chapin (1976), Collins et al. (1980), Prinzinger et al. (1981), Weathers and Riper (1982), Bartholomew et al. (1983), and Reinertsen and Haftorn (1986). The two "boundary" curves, labelled B, were taken from McNab (1983) and McNab (1988).

Such a calculation requires a known or assumed power for the boundary curve. When the original estimate for the boundary curve, i.e., proportional on a mass-specific basis to $\underline{m}^{-0.67}$, is used, the birds indicated in Figure 2 show a significantly positive correlation of the minimal temperature differential with the coefficient ($r = 0.687$, $n = 21$). But when the mass-specific boundary curve is set proportional to $\underline{m}^{-0.45}$, a tighter ($r = 0.891$, $n = 21$) relationship is shown between the lowest temperature differential maintained with the environment and the coefficient (Figure 3), which suggests that -0.45 is the better mass-specific estimate of the power of the boundary curve.

Whatever the power of the boundary curve, the response of birds to the boundary curve clearly is gradual, so that at a fixed mass it is best considered to be a zone of transition from rigid endothermy (at high basal rates) to a marked propensity to enter torpor (at low basal rates). If data were available, a similar interspecific pattern might well indicate at a fixed small mass that the frequency of entrance into torpor would increase as basal rate fell.

One fact must be remembered. The boundary curve is an attempt to describe which endotherms enter torpor. It does not knowingly describe why this association exists, except that as body mass falls (and storage time for heat and energy decreases), rate of metabolism must be proportionally higher for torpor to be evaded. In other words, a small species can avoid torpor only if it has sufficiently high rates of food

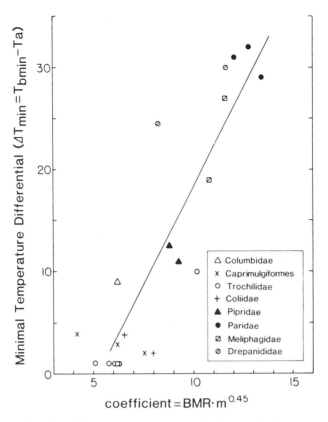

Fig. 3. Minimal temperature differentials maintained by the birds in Fig. 2 as a function of the coefficient derived for each species by setting its basal rate equal to the boundary curve whose mass-specific power equals -0.45 (see text). Data were taken from the sources in Fig. 2 and MacMillen and Trost (1967b).

availability to reduce its dependence on food storage. If for any reason a small species is limited in its rate of food acquisition and if this limit is reflected in a reduction in its basal rate of energy expenditure, it will upon occasion enter daily torpor, the depth (and frequency?) of which depends on the degree to which basal rate has been lowered. At the smallest masses energy storage is so small relative to demand that high food availability in the environment does not deter entrance into torpor unless the endotherm (like certain shrews, but unlike hummingbirds and insectivorous bats) is nearly continuously active in the procurement of food.

A DISTINCTIVE RESPONSE TO COLD EXPOSURE

In response to an exposure to low temperatures, an endotherm usually either maintains a high, regulated body temperature, which requires a high rate of metabolism, or it enters torpor, which permits a great reduction in rate of metabolism. Another distinctive pattern among these variables has been recently described in some tropical, arboreal mammals (McNab, 1988b): these species maintain a high body temperature at low temperatures with a minimal expenditure of energy, often as low as, or lower than, the "standard" rate. This pattern requires a marked reduction in peripheral circulation. It has been found in palm-civets, lemurs, a tree-kangaroo, and the red panda.

Preliminary measurements on rate of metabolism and body temperature in the gray-winged trumpeter (Psophia crepitans) show a pattern similar to that found in the arboreal mammals (Figure 4). Notice that as environmental temperature falls below 33°C, rate of metabolism increases, but at temperatures below 12°C, rate of metabolism decreases even though cloacal temperature remains constant at about 40°C. Evidence of a reduction in peripheral circulation was seen in the very cold legs of the trumpeters at ambient temperatures between 6 and 10°C.

This response to low temperatures may well be widespread in endotherms of intermediate mass. Among mammals, it was found in species that weighed from 2 to 15 kg. This trumpeter weighed one kilogram. Recent measurements (P. Gray, pers. comm.) on the sandhill crane (Grus canadensis), which weighed 4.1 kg, showed a similar reduction in rate of metabolism at environmental temperatures below 20°C without any reduction in core temperature.

COLD TOLERANCE, MIGRATION, AND TORPOR

The differential use of various solutions to a problem depends on many factors. The responses to an extended cold exposure include tolerance, migration, and entrance into torpor.

Cold Tolerance

No species can tolerate cold unless sufficient food is available: starvation for appreciable periods is only compatible with temperature

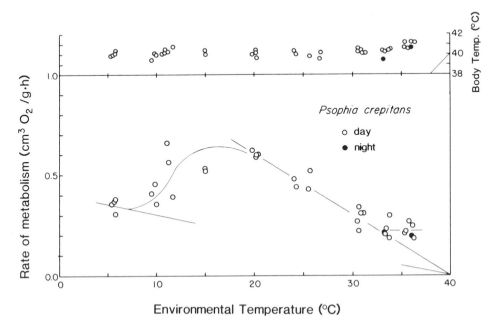

Fig. 4. Rate of metabolism and body temperature in two gray-winged trumpeters (Psophia crepitans) as a function of environmental temperature.

regulation in large species. Species of intermediate size can increase their tolerance to a short-term cold exposure by reducing energy expenditure because they have a sufficient mass to permit a separating of core and shell temperatures. Even the presence of food, however, may not be sufficient for temperature tolerance in the smallest species. For example, most birds cannot forage in continuous darkness. And as a matter of fact, few birds overwinter in continuous darkness: Irving (1960) showed that Arctic passerines, like Acanthis and Plectophenax, migrate at least 900 km south into regions that have winter daylight. He nevertheless noted that a few birds spend the winter above the Arctic Circle, including the rock ptarmigan (Lagopus mutus), snowy owl (Nyctea scandiaca), downy woodpecker (Picoides pubescens), black-capped chickadee (Parus atricapillus), and pine grosbeak (Pinicola enucleator). Birds can tolerate low temperatures only if they acquire energy equal to, or greater than, the amount required on a daily basis, regardless of the length of the light period. Tolerance to cold exposure, then, depends on food habits, i.e., on food availability. Cold tolerance is most prevalent in terrestrial birds that feed on seeds, insect larvae, and vertebrates. Aquatic birds obviously require open water.

Migration

Migration, requiring as it does the capacity for long-distance movement, is widely used by birds. The use of migration is correlated with food habits. Among terrestrial species it is most prevalent in species that eat adult insects and in those (few) facultative frugivores found in temperate environments. Aquatic species that live in areas that are cold enough for large bodies of water to freeze also must migrate. Many vertebrate-eating birds migrate, apparently in relation to a reduced prey base: this is most clear in species that prey on small birds (most of which migrate) and small mammals (some of which are protected by snow cover and others of which enter seasonal torpor).

Torpor

Torpor generally is reciprocally associated with the occurrence of migration. This, then, is a great difference between temperate mammals and birds. Most small mammals cannot evade a cold climate by migration (because of their limited capacity to move a distance) and some compensate by entering torpor. In contrast, most small birds seasonally migrate, which is why few use torpor. Because torpor is correlated with low basal rates, the relative infrequency of torpor in birds exaggerates the difference in basal rate between these classes. Again, torpor in birds is most prevalent in small species and in those that feed on flying insects.

CONCLUSIONS

This paper explores the responses of birds to a long-term or seasonal exposure to low temperatures. Several solutions are used, including tolerance, migration, and torpor. The response used is influenced by the interactions among body size, food habits, and the basal rate of metabolism. These factors are associated with each other: basal rate is correlated with body mass, food habits with mass, basal rate with food habits, torpor with basal rate, and torpor with body mass. Such an interlocking set of interactions makes an analysis of the cause and effect of the response of birds to low temperatures very difficult. Nevertheless, the response of birds to cold clearly is related to their other characteristics, which accounts for the diversity among birds in their response to life in a cold environment.

ACKNOWLEDGMENTS

I thank Drs. Claus Bech and Randi Reinertsen for the invitation to participate in this symposium and Paul Gray for the use of unpublished data.

REFERENCES

Aschoff, J., and Pohl, H., 1970, Der Ruheumsatz von Vögeln als Funktion der Tageszeit und der Körpergrosse, J. Ornith., 111:38-47.
Bartholomew, G. A., and Trost, C. H., 1970, Temperature regulation in the speckled mousebird, Colius striatus, Condor, 72:141-146.
Bartholomew, G. A., Hudson, J. W., and Howell, T. R., 1962, Body temperature, oxygen consumption, evaporative water loss, and heart rate in the poor-will, Condor, 64:117-125.
Bartholomew, G. A., Vleck, C. M., and Bucher, T. L., 1983, Energy metabolism and nocturnal hypothermia in two tropical passerine frugivores, Manacus vitellinus and Pipra mentalis, Physiol. Zool., 56:370-379.
Bennett, P. M., and Harvey, P. H., 1987, Active and resting metabolism in birds: allometry, phylogeny and ecology, J. Zool., Lond., 213:327-363.
Brody, S., 1945, "Bioenergetics and Growth," Reinhold Publ. Corp., New York.
Chapin, S. B., 1976, The physiology of hypothermia in the Black-capped Chickadee, Parus atricapillus, J. Comp. Physiol., 112B:335-344.
Collins, B. G., Cary, G., and Payne, S., 1980, Metabolism, thermoregulation and evaporative water loss in two species of Australian nectar-feeding birds (Family Meliphagidae), Comp. Biochem. Physiol., 67A:629-635.
Dawson, W. R., and Fisher, C. D., 1969, Responses to temperature by the spotted nightjar (Eurostopodus guttatus), Condor, 71:49-53.
Ellis, H. I., 1980, Metabolism and solar radiation in dark and white herons in hot climates, Physiol. Zool., 53:358-372.
Ellis, H. I., 1984, Energetics of free-ranging seabirds, pages 203-234 in: "Seabird Energetics," G. C. Whittow and H. Rahn, eds., Plenum Press, New York.
Hails, C. J., 1983, The metabolic rate of tropical birds, Condor, 85:61-65.
Hayes, S. R., and Gessaman, J. A., 1980, The combined effects of air temperature, wind and radiation on the resting metabolism of avian raptors, J. Therm. Biol., 5:119-125.
Irving, L., 1960, "Birds of Anaktuvuk Pass, Kobuk, and Old Crow", Bull. 217, U. S. National Museum, Smithsonian Instit., Washington, D.C.
Lasiewski, R. C., 1963, Oxygen consumption of torpid, resting, active, and flying hummingbirds, Physiol. Zool., 36:122-140.
Lasiewski, R. C., and Dawson, W. R., 1964, Physiological responses to temperature in the common nighthawk, Condor, 66:477-490.
Lasiewski, R. C., and Dawson, W. R., 1967, A re-examination of the relation between standard metabolic rate and body weight in birds, Condor, 69:13-23.
Lasiewski, R. C., and Lasiewski, R. J., 1967, Physiological responses of the Blue-throated and Rivoli's Hummingbirds, Auk, 84:34-48.
Lasiewski, R. C., Weathers, W. W., and Bernstein, M. H., 1967, Physiological responses of the giant hummingbird, Patagona gigas, Comp. Biochem. Physiol., 23:797-813.
Lasiewski, R. C., Dawson, W. R., and Bartholomew, G. A., 1970, Temperature regulation in the Little Papuan Frogmouth, Podargus ocellatus, Condor, 72:332-338.

MacMillen, R. E., 1974, Bioenergetics of Hawaiian honeycreepers: the Amakihi (Loxops virens) and the Anianiau (L. parva), Condor, 76:62-69.

MacMillen, R. E., and Trost, C. H., 1967a, Thermoregulation and water loss in the Inca dove, Comp. Biochem. Physiol., 20:263-273.

MacMillen, R. E., and Trost, C. H., 1967b, Nocturnal hypothermia in the Inca dove, Scardafella inca, Comp. Biochem. Physiol., 23:243-253.

McAtee, W. L., 1947, Torpidity, Amer. Midl. Nat., 38:191-206.

McNab, B. K., 1983, Energetics, body size, and the limits to endothermy, J. Zool., Lond., 199:1-29.

McNab, B. K., 1984, Physiological convergence amongst ant-eating and termite-eating mammals, J. Zool., Lond., 203:485-510.

McNab, B. K., 1986, The influence of food habits on the energetics of eutherian mammals, Ecol. Monogr., 56:1-19.

McNab, B. K., 1988a, Complications inherent in scaling the basal rate of metabolism in mammals, Quart. Rev. Biol., 63:25-54.

McNab, B. K., 1988b, Energy conservation in a tree-kangaroo (Dendrolagus matschiei) and the red panda (Ailurus fulgens), Physiol. Zool., 61:

McNab, B. K., in press, Food habits and basal rate of metabolism in birds, Oecologia, Berl.

Prinzinger, R., Goppel, R., Lorenz, A., and Kulzer, E., 1986, Body temperature and metabolism in the red-backed mousebird (Colius castanotus) during fasting and torpor, Comp. Biochem. Physiol., 69A:689-692.

Prinzinger, R., and Hanssler, I., 1980, Metabolism-weight relationship in some small nonpasserine birds, Experientia, 36:1299-1300.

Reinertsen, R. E., and Haftorn, S., 1986, Different metabolic strategies of northern birds for nocturnal survival, J. Comp. Physiol. B, 156:655-663.

Scholander, P. F., Hock, R., Walters, V., and Irving, L., 1950, Adaptation to cold in arctic and tropical mammals and birds in relation to body temperature, insulation, and basal rate of metabolism, Biol. Bull., 99:259-271.

Steen, J., 1958, Climatic adaptation in some small northern birds, Ecology, 39:625-629.

Wasser, J. S., 1986, The relationship of energetics of falconiform birds to body mass and climate, Condor, 88:57-62.

Weathers, W. W., 1979, Climatic adaptation in avian standard metabolic rate, Oecologia, Berl., 42:81-89.

Weathers, W. W., and Riper III, C. V., 1982, Temperature regulation in two endangered Hawaiian honeycreepers: the Palila (Psittirostra bailleui) and the Laysan finch (Psittirostra cantans), Auk, 99:667-674.

ENERGY SAVING DURING BREEDING AND MOLT IN BIRDS

Jean-Patrice Robin, Yves Handrich, Yves Cherel and Yvon Le Maho[*]

Laboratoire d'Etude des Régulations Physiologiques
(associé à l'Université Louis Pasteur)
Centre National de la Recherche Scientifique
23 rue Becquerel, 67087 Strasbourg, France

INTRODUCTION

Success of birds in breeding and molt obviously implies adequate availability in either food or body fuel reserves. However, due to conflicting requirements, a limit in these energetic conditions for success may be reached. For example, following an impairment of climatic conditions, availability of food may be reduced while the decrease in the body fuel reserves of the breeders is accelerated. In the same time, increased foraging by the breeders may be limited due to protection of eggs or chicks from cold or predation (Calder and Booser, 1973; Milne, 1976; Aldrich and Raveling, 1983). Molt may also bring birds into a near marginal energetic condition, particularly in those species which simultaneously replace all their feathers. Ability for foraging may be impaired while there are higher energetic demands due to decreased thermal insulation and building of new plumage (Groscolas, 1978).

The present review deals with the different ways that birds facing a limited food availability may save energy while breeding or molting, and with how a premature depletion in their body reserves may induce failure in the activity in which they are engaged.

I. INCREASE OF ENERGY REQUIREMENTS AND LIMITATION OF FOOD INTAKE

1. Increased Energy Requirements Associated with Molt and Breeding

<u>Molt</u>. Metabolic rate of birds increases during molt, from about 9% to about 35% above the premolt levels. This increase is due to a rise in basal metabolic rate that is associated with synthesis of new feathers. It may also result from a cold-induced thermogenesis as a consequence of reduced thermal insulation (for review, see King, 1980). However, the consequence of molting on total energy daily requirement could be attenuated by correlative adjustments of the locomotor activity (decrease of activity and/or of its cost).

<u>Breeding.</u> Birds are mostly inactive when incubating their eggs. Since small species with body masses below 350-400 g must shiver when inactive

---
[*] To whom correspondence should be addressed.

at most environmental temperatures normally encountered in temperate, subarctic and arctic regions (West, 1965), an interesting question is therefore whether birds also shiver, i.e. have a cold induced thermogenesis, when incubating.

Would data for the American kestrel, *Falco sparverius* (Gessaman and Findell, 1979), for the Laysan albatross, *Diomedea immutabilis* and Bonin petrel, *Pterodroma hypoleuca* (Grant and Whittow, 1983) and for the canary, *Serinus canaria* (Weathers, 1985) be generalized for other birds, incubation can be accomplished at about the level of adult resting metabolism, i.e. without an additional energy cost for maintaining eggs at a high ambient temperature. However, in these species that periodically leave the nest, there is a very high energy demand for rewarming the eggs when they have cooled off during the absences of the incubators (Tøien et al., 1986).

After hatching, the parents have increased energy requirements of foraging, in order to provide the food for the chicks.

2. Added Energy Requirements Due to Impairment of Climatic Conditions

The higher energy demand due to cold has been for example demonstrated in the marsh tit, *Parus palustris*. In this species, the male feeds the incubating female until hatching. To support her higher requirements, the male's provisioning rate increases when ambient temperature decreases. Unusual cold may however put such a higher demand that the clutch can be jeopardized (Nilsson and Smith, 1988). Of course, molt is also a process which can be influenced by impairment of weather conditions, particularly when there is a significant reduction in thermal insulation.

3. Limitation of Food Intake

Despite the increased energy requirements due to breeding or molt, birds could maintain their body mass simply by eating more. The available information in the literature however gives various examples indicating that, instead, birds partly or totally rely on their energy reserves even with food readily available.

A first example is that of the Eurasian kestrel, *Falco tinnunculus*, for which the molt limits the possibility of increased feeding as this process reduces the success of flight hunting (Masman et al., 1988b). Another example is those females of ducks, geese and ptarmigans which assume the entire task of the incubation and moreover have to rely on their own in order to feed. Prolonged foraging therefore means longer absences on the nest, higher risk of predation and higher cost of incubation due to the rewarming of eggs. This altogether explains that these females, despite food may be available in the vicinity, usually have a progressive decrease in their energy reserves during the course of the incubation (Ankney and MacInnes, 1978; Milne, 1976; Mainguy and Thomas, 1985; Gabrielsen and Unander, 1987).

Birds may also be unable to feed under usual conditions of breeding or molting. It is so, for example, when those species that feed only at sea, stay ashore to breed or molt. An extreme situation is that of the emperor penguin, *Aptenodytes forsteri*. It breeds on winter sea-ice, at a distance which may be as much as 120 km from the open sea where it feeds (Budd, 1962). There are records of emperor tracks at least 300 km from the nearest known open-water (Sladen and Ostenso, 1960). The largest of the fasts associated with breeding lasts for 4 months in the males, which entirely assume the task of the incubation. They each year overall fast for as much as 6 months, including not only the successive fasts associated with breeding but also the one month long fast which coincides with the summer molt (Prévost, 1961; Groscolas, 1982).

Food intake of birds may also be restricted due to unusual climatic conditions. For example, tiny hummingbirds, which nest in the chilling nocturnal climate of the Rocky mountains may become in a difficult energetic situation when feeding opportunity is reduced by inclement weather (Calder and Booser, 1973).

Finally, as there is an anorexia associated with the beginning of molt in domestic geese (Le Maho, unpublished data), it might be that the biochemical adjustments associated with this process necessitate a limitation in food intake.

In this context, two major questions therefore arise. First, how can breeding or molting birds save energy, when exposed to cold while food is limited? Second, what is the limit in the depletion of energy reserves that breeding or molting birds may tolerate and what occurs when this limit has prematurely been reached?

II. WAYS OF SPARING ENERGY

Although there is now detailed information on how birds may spare energy and therefore adapt to long-term starvation (for review see Le Maho, 1983a; Le Maho et al., 1987; Cherel et al., 1988b), there still is a restricted number of species for which this information is available in the context of breeding or molting under natural conditions.

1. Adequate Timing

Timing according to seasons. The first way that breeding or molt of a bird may occur at the lowest energy cost is obviously through an adequate timing in relation to environmental conditions. Since food availability is usually the highest in the warm season, breeding or molting at that time offers both advantages of cheaper foraging in association with lower cost of temperature regulation.

For example, the molting fast of the emperor penguin coincides with the antarctic summer, while ambient temperature is around 0°C and sun radiation is the highest. Still - due to the decreased thermal insulation associated with the loss of old feathers and the building of the new plumage - the per cent decrease in body mass of a molting bird is, in one month only, as large as it is for a non-molting bird during four winter months with ambient temperatures between -10 and -30°C (Groscolas, 1978). A molting bird already shivers - thus increasing its metabolic rate - below -5°C, versus -10°C for a non molting bird (Le Maho et al., 1976). Would the molt of the emperor penguin occur in the winter, its fuel reserves would be insufficient to cope with the higher energy requirements.

Similarly, the molt of the male kestrel occurs when the environmental conditions are the best and food availability is the highest. This therefore compensates for both the increased cost of thermoregulation and the reduced effectiveness for hunting that are associated with the molt. Importantly, breeding is over at that time and the male has therefore no more need to provide for the increased energy requirements of the chicks (Masman et al., 1988a,b).

Adequate timing through task differentiation. In the kestrel, the female has a timing for molt which differs from that of the male, since it does occur during incubation. The reduced effectiveness of flight hunting does not create an energetic problem, as the female does not leave the nest when incubating and is fed by the male (see Masman et al., 1988a,b).

2. Reduction of Cost of Maintenance

The cost of maintenance of a bird is defined here as the rate of energy expenditure when it does not move, that is the sum of basal metabolic rate (metabolic rate at thermoneutrality) and metabolic rate related to temperature regulation (cold-induced thermogenesis).

Spontaneous reduction in food intake (anorexia). An effective way to reduce the cost of maintenance is through a spontaneous reduction in food intake.
The spontaneous character of the food suppression which occurs in association with molt in penguins is showed by the fact that captive birds, when molt occurs, do not feed or at least take very little food (Gillespie, 1932).

Another example is that of the incubating hen of the junglefowl *Gallus gallus*, which has a food intake only about 25% of the usual value. Importantly, she does not eat more if the food is directly available on the nest. This again demonstrates that the reduced food intake of the bird has a spontaneous character. Due to the importance of this food restriction associated with anorexia, the body mass of the hen regularly decreases. Food intake can be totally suppressed for several days without that the bird leaves the nest. When food is again available, the hen feeds in order to regain not her body mass at food suppression but the lower value that she would have reached if continuing the spontaneous limited intake. Similar data, according with the concept that body mass is regulated in relation with a changing set-point (Mrosovsky and Powley, 1977), were obtained following food manipulation during hibernation in the ground squirrel, *Spermophilus lateralis* (Mrosovsky and Sherry, 1980) and before the winter in the Svalbard ptarmigan, *Lagopus lagopus hyperboreus* (Mortensen and Blix, 1985). This is therefore a general process, i.e. that is not restricted to the anorexia associated with breeding and is common at least to mammals and birds.

The way that anorexia allows a reduction of the cost of maintenance of a bird is through fasting-induced physiological adjustments. First, food restriction rapidly induces a reduction in basal metabolic rate which is independent of body mass and reflects a rapid involvement of physiological processes which allow minimizing heat loss. In addition to a possible slight decrease in body core temperature, there should be a decrease in shell temperatures and/or a reduction in the size of the body core (Le Maho, 1983a). Second, basal metabolic rate progressively decreases with fasting-induced reduction of body mass (Le Maho, 1983a; Cherel et al., 1988b). Finally, in addition to these energy saving consequences, anorexia has the advantages of allowing the single incubating bird to reduce its abandonment of nest and associated cooling of eggs as well as of reducing the risk of predation.

More sleep. In addition to the body mass related decrease in the cost of maintenance, there might be an energy saving due to fasting-induced changes in sleep. Basal metabolic rate is usually the lowest during episodes of slow-wave sleep. Indeed, time spent asleep might increase in the anorexic incubating bird as was found to occur during an experimental fast in domestic geese (Dewasmes et al., 1984) as well as when artificially fed emperor penguins had no more food (Dewasmes et al., 1989).

Deeper sleep (torpor). An effective way to reduce both basal metabolic rate and cold induced thermogenesis is by a profound decrease of the body core temperature. Torpor has indeed been described in various species of small birds (Krüger et al., 1982; Lyman, 1982; Reinertsen and Haftorn, 1986). Most of this information pertains to torpor experimentally induced by cold or food restriction. However, using temperature sensors embedded in synthetic eggs, torpor has been shown to occur in the broad-tailed hummingbird *Selasphorus platycercus* breeding under natural conditions. This torpor has been correlated with weather conditions that pose chilling demands upon marginal energy reserves (Calder and Booser, 1973).

Avoiding cold-induced thermogenesis. When bird-breeding is not timed in such a way that it coincides with thermoneutral conditions, heat loss may be minimized through behavioral adjustments, thereby enabling reduction of thermogenesis. This is illustrated by the emperor penguin. The males huddle together during most of the 2 month-long incubation, which allow them to maintain at basal metabolic rate and, therefore, to keep at minimal cost of maintenance (Le Maho, 1983a,b).

The cold-induced thermogenesis that occur at most ambient temperatures in small passerine birds when inactive (West, 1965) may possibly be avoided during incubation, due to the structure of the nest. This requires further investigation.

3. Locomotor Activity

With adequate timing of breeding and molt, a further way to minimize energy expenditure is by decreasing the level of activity, as well as the energy cost of a given activity.

Reduction in level of activity. The importance of the energetic consequences of a reduction in locomotion is remarkably illustrated by data on the Svalbard ptarmigan: an autumnal decrease in locomotor activity enables an increase in body mass even though food intake spontaneously decreases (Stokkan et al., 1986).

The most favorable energetic situation for an incubating bird is then when it does not leave the nest and is fed by the other mate. This is shown in many species of Passeriforms, for example in the female blue tit, *Parus caeruleus* (Nilsson and Smith, 1988) and also of raptors (Newton, 1979). In single sex incubators, however, locomotor activity is usually minimized but not totally suppressed. This is for example illustrated by the female of Canada geese, *Branta canadensis*, which has a limited recess time and subsequent low activity level during most of the incubation (Aldrich and Raveling, 1983). This is again enabled by an anorexia. That a low food intake and an accordingly limited activity level still persist is presumably due to physiological constraints.

The behavioral repertoire of the breeding emperor penguin is characterized by limited locomotor activity. Moreover, the different postures which constitute this repertoire are essentially shown during pair formation and egg laying. Behavioral postures are minimum during the long fast of the male (Jouventin, 1978). Similarly, while the mean distance emperor penguins daily walk is only 150 m at the time the pairs are formed, it is even lower (30 m per day) at the time the males incubate (Geloen and Jouventin, in Le Maho, 1983b). That these males do not totally suppress their locomotor activity can be explained by the social thermoregulatory behavior of huddling. The huddle continuously moves, as the males in the periphery walk very slowly in order to get inside it (Jouventin, 1971).

In some molting captive birds, perch-hopping activity decreases by as much as 90% as compared with control periods, mimicking the quiescience of free-living birds, when molting. This drastic decrease in locomotor activity minimizes the concurrent increase in metabolic rate induced by molt (review by King, 1980).

<u>Reduction in cost of activity</u>. An effective way of reducing the energy cost of movement is through moving at a reduced body mass. Thus, the decrease in body mass which is associated with the anorexia of breeding or molting again allows energy saving when the bird moves.

For example, male emperor penguins weigh on average 39 kg at their arrival in the colony when they start their breeding cycle. At that time the open sea is usually very close to the colony and there is therefore a short distance to walk on sea-ice. At 39 kg, the metabolic rate of an emperor penguin is 340 W when it walks at its usual speed, 1.4 km·h^{-1}; this is as much as 4.5 times the resting metabolic rate. By contrast, the walking metabolic rate of an emperor penguin is only 220 W (3 times the actual resting metabolic rate) when the bird weighs only 23 kg, that is the mean body mass of the males when they leave their colony. By departing at 23 kg, emperor penguins still have some leeway in the fuel reserves that are necessary to cover the usually long distance on sea-ice between the colony and the sea (Dewasmes et al., 1980). Would both males and females alternate in assuming the task of the incubation, they could go back and forth at a much larger body mass between the colony and the open-sea. The safety margin in energy reserves would be higher, but the part of the energy used in transportation and, accordingly, the energy requirements at sea would be considerably larger. Indeed, food availability at sea is minimum in winter time, thus while emperors breed.

Another example, which illustrates the advantage of a reduced body mass for limiting energy expenditure is again that of the kestrel. Flight in this bird is the most expensive action. At the time of breeding, the male does not feed before the evening. During the day, he stores in a cache the preys that have not been captured for the female or the chicks and his body mass is therefore maintained at low value, which minimizes the cost of flying (Masman, 1986). A similar example to the diurnal kestrel is illustrated by the nocturnal barn owl, *Tyto alba*. The male, which feeds the incubating female, hunts at a low body mass during the whole night and only starts to eat and therefore increase in mass at dawn (Handrich, unpublished data).

Finally, in the line of a lower cost of bird flight in relation with a lower body mass, it is also interesting to point out the sexual dimorphism in body mass which may be associated with differentiation of tasks. It is particularly well illustrated in birds of prey. In those species for which the body mass of the female is much larger than that of the male, the cost of flight is accordingly much smaller in the smaller male. Indeed, he is usually hunting while the female does not leave the nest as she incubates. This may explain, for example, that female Tengmalm's owls, *Aegolius funereus*, choose light males; presumably these males hunt more effectively and economically (Korpimäki, 1986). In other raptors, where there is not such a marked difference in body mass between the mates, both male and female alternate in incubating and hunting (Newton, 1979).

III. CHANGES IN FUEL UTILIZATION

When investigating the way energy may be spared in a bird fasting in association with breeding or molt, it is not sufficient to consider the overall energy stored, i.e. the sum of the energy represented by the different tissue reserves. Also important is the selective mobilization versus sparing of these different tissues, because their energetic value differs as well as their amount which can be depleted.

Considering the energetic value of body fuels, there is in birds, as for mammals, an insignificant contribution of glycogen to energy expenditure. The utilization rate of body proteins rapidly decreases after food suppression and there is a shift to the preferential use of lipids. Without such an early shift, which characterizes the so-called phase I of fasting (Le Maho, 1983a; Cherel et al., 1988b), birds would be unable to tolerate prolonged food restriction. Body proteins yield only half the energy value provided by lipids and, moreover, no more than a 50 % decrease in body proteins reserves can be tolerated (Garrow et al., 1965). In contrast, lipids can harmlessly disappear almost completely (Pond, 1978).

At the end of phase I, that is after several days in large birds, the proportion of the energy which originates from lipids has reached a high fixed value and, accordingly, that from proteins a low fixed value. Such a steady-state (phase II) can thereafter be maintained for as long as 3-4 months in an emperor penguin (Robin et al., 1988b).

A key factor, in setting during phase I the proportion of the energy which respectively originates from lipids and proteins during phase II, is the importance of adiposity at the onset of the fast: a higher initial adiposity induces a higher proportion of the energy derived from lipids, throughout phase II, and an accordingly lower proportion from proteins. The huge lipid reserves that various species of birds store, before spontaneously fasting in association with breeding, therefore, enable not only to provide a fuel of a high energetic value (lipids), but also to delay the time when a critical value in body protein depletion has been reached (Robin et al., 1988b; Le Maho et al., 1988). For this reason, the reversed dimorphism in body mass, associated with distribution of tasks in various species of raptors, seems again of advantage: presumably, the larger female is more effective in sparing her body proteins when partly relying on her energy reserves while incubating.

IV. WHEN REFEEDING BECOMES A PRIORITY

There is a time during a prolonged fast, in the emperor penguin, where there is an increase in the contribution of body proteins to energy expenditure and where the proportion of the energy which originates from lipids accordingly decreases (phase III). This end of protein sparing is not the consequence of lipid exhaustion, as significant fat reserves are still available (Robin et al., 1988b).

Data in the literature suggest that the existence of this new metabolic situation, marked by an increase in the utilization of proteins, can be generalized to a wide variety of birds undergoing long-term food restriction (Cherel et al., 1987). Prolongement of this phase would bring the birds into a critical stage of protein depletion, and death would be the consequence.

1. An Internal Signal Induces Refeeding

We already know that male emperor penguins whose females are delayed in returning to the colony can prolong their fast. Although they are then fasting for about 4 months, they can feed their newly hatched chick on milk secreted by the oesophagus (Prévost and Vilter, 1962). However, at some stage the males do not extend their fast further and they abandon their chick to refeed at sea. This explains that for 30 years of observation there was never found an emperor penguin starving to death in the breeding colony of Pointe Géologie (Adélie Land). Accordingly, an internal signal must trigger refeeding before a critical level in energy reserves has been reached (Le Maho et al., 1988).

In fact, the average body mass of the emperor penguins that depart from their colony corresponds to the end of the phase of protein sparing (Groscolas, 1986; Robin et al., 1988b). A complication, obviously, is that it is impossible to distinguish between the birds departing because having being relieved by their mate and those departing on their own and, therefore, abandoning their chick. Presumably, the males departing at low body masses, i.e. about 21 kg, are birds whose departure has been triggered by an internal signal (Groscolas, 1986). By leaving their colony at 21 kg, emperor penguins are already into the stage where protein utilization has increased, which suggests that the internal signal for refeeding is associated with this metabolic state. Why the hypothesized signal for refeeding anticipates total depletion of lipid reserves is explained by the fact, as indicated above, that significant lipid reserves are still available when protein utilization increases. Moreover, only 20% of body protein have been depleted at this stage, thus still far from the critical proportion of 50% (Robin et al., 1988b). It has been calculated that emperor penguins leaving their colony at 21 kg weigh about 18 kg when reaching the open sea. Based on refeeding of penguins having reached this 18 kg mass, it still corresponds to a reversible metabolic situation (Groscolas, 1982).

The concept of a signal for refeeding that is associated with increased protein utilization accords with data for other breeding birds. For example, female eiders that abandon their eggs before the term of the incubation (Tinbergen, 1958; Korschgen, 1977) have probably prematurely entered the stage of increased protein utilization that usually occurs (Milne, 1976) at the end of incubation. Decreased attentiveness on the nest in late incubation in Canada geese (Aldrich and Raveling, 1983) might be related to the depletion of lipids and the catabolism of body proteins which occur at that stage (Ankney and MacInnes, 1978. Nesting white-crowned sparrows (*Zonotrichia leucophrys*) which abandon their nests when subjected to a catastrophic storm, have an increased blood level of corticosterone (Wingfield, 1988) and such a rise is indeed associated with the fasting-induced increase in protein utilization (Cherel et al., 1988c). A rise in corticosterone moreover fits with the idea that refeeding is induced, since a consequence of this rise is an increase in foraging (Wingfield, 1988).

How then evolves the rate of energy expenditure of a bird in search of food, after refeeding has been induced by an internal signal?

2. A Cost of Maintenance which Still Remains Reduced

The basal metabolic rate of a penguin still continues to decrease with body mass when the bird is into the stage of increased protein utilization (see Cherel et al., 1988b). Thus, after its departure from the breeding colony has been induced by an internal signal, the penguin undergoes a long treck on sea-ice at a low cost of maintenance. There is no further way of reducing the cost of maintenance due to increased sleep, as slow-wave sleep, in contrast to the previous stage of protein sparing, decreases when protein utilization rises (Dewasmes et al., 1984). Indeed, such a decrease in sleep might enable a bird in search of food to find it more rapidly when refeeding has become a priority.

3. A Higher Level of Locomotor Activity but at a Lower Cost

The long walk of a male emperor penguin before it reaches the sea at the end of incubation or the intense foraging activity of a crowned-sparrow that has failed in breeding are obviously associated with a high energy expenditure due to locomotor activity. However, that both these birds have a low body mass at this stage minimizes the cost of moving. Even more, when an emperor penguin leaves its colony while it is in the stage of increasing protein utilization it is also into a stage where the cost of transport decreases more than simply due to the decreasing body mass. This is because its depletion of body tissues is such that its gait for walking then changes (Dewasmes et al., 1980).

4. A high effectiveness in refueling

When birds with depleted reserves start to refeed, the restoration of their muscular mass is presumably accelerated due to an increased efficiency in protein assimilation, as recently found in domestic geese after a prolonged fast (Robin et al., 1988a). Within the first week of refeeding, the geese were able to eat as much as 2 to 3 times more food than during the prefast period. At that time, 5 g of nitrogen was daily retained, due to the rise in food intake and also to a rise in the nitrogen assimilation rate. This nitrogen retention suggests a daily build-up of 100-150 g of muscular tissue.

V. REGULATION OF BODY MASS

From the present review, it is clear that a reduction in body mass, resulting from an anorexia, often plays a major role for energy saving when birds breed and molt. The existence of a seasonal cycle in body mass is essential since energy reserves are usually increased prior to the anorexia induced energy debt. As illustrated above for the spontaneous reduction in body mass which may occur whereas food is readily available, this body mass cycle is not the mere consequence of natural variations in food availability. It is a regulated process. Still, unusual cold may compromise the success in breeding or the proper renewal of plumage by limiting the anticipatory accumulation in energy reserves and/or inducing a premature depletion in these reserves.

REFERENCES

Aldrich, T.W., and Raveling, D.G., 1983, Effects of experience and body weight on incubation behavior of Canada geese, Auk, 100:670.

Ankney, C.D., and MacInnes, C.D., 1978, Nutrient reserves and reproductive performance of female lesser snow geese, Auk, 95:459.

Budd, G.M., 1962, Population studies in rookeries of the emperor penguin Aptenodytes forsteri, Proc. Zool. Soc. London, 139:365.

Calder, W.A., and Booser, J., 1973, Hypothermia of broad-tailed hummingbirds during incubation in nature with ecological correlations, Science, 180:751.

Cherel, Y., Stahl, J.-C., and Le Maho, Y., 1987, Ecology and physiology of fasting in king penguin chicks, Auk, 104:254.

Cherel, Y., Leloup, J., and Le Maho, Y., 1988a, Fasting in king penguin. II. Hormonal and metabolic changes during molt, Am. J. Physiol. 254 (Regulatory Integrative Comp. Physiol. 23):R178.

Cherel, Y., Robin, J.-P., and Le Maho, Y., 1988b, Physiology and biochemistry of long-term fasting in birds, Can. J. Zool., 66:159.

Cherel, Y., Robin, J.-P., Walch, O., Karmann, H., Netchitailo, P. and Le Maho, Y., 1988c, Fasting in king penguin. I. Hormonal and metabolic changes during breeding, Am. J. Physiol. 254 (Regulatory Integrative Comp. Physiol. 23):R170.

Dewasmes, G., Buchet, C. and Le Maho, Y., 1989, Sleep changes during fasting in emperor penguins, Am. J. Physiol. (Regulatory Integrative Comp. Physiol.), in press.

Dewasmes, G., Cohen-Adad, F., Koubi, H. and Le Maho, Y., 1984, Sleep changes in long-term fasting geese in relation to lipid and protein metabolism, Am. J. Physiol. 247 (Regulatory Integrative Comp. Physiol. 16):R663.

Dewasmes, G., Le Maho, Y., Cornet, A., and Groscolas, R., 1980, Resting metabolic rate and cost of locomotion in long-term fasting emperor penguins. J. Appl. Physiol.: Respirat. Environ. Exercise Physiol., 49:888.

Gabrielsen, G.W. and Unander, S., 1987, Energy costs during incubation in Svalbard and Willow Ptarmigan hens, Polar Res., 5n.s.:59-69.

Garrow, J.S., Fletcher, K., and Halliday, D., 1965, Body composition in severe infantile malnutrition, J. Clin. Invest., 44:417.

Gessaman, J.A. and Findell, P.R., 1979, Energy cost of incubation in the American kestrel, Comp. Biochem. Physiol., 63A:57.

Gillespie, T.H., 1932, "A Book of King Penguins", Herbert Jenkins Limited, London.

Grant, G.S. and Whittow G.C., 1983, Metabolic cost of incubation in the Laysan albatross and Bonin petrel, Comp. Biochem. Physiol., 74A:77-82.

Groscolas, R., 1978, Study of molt fasting followed by an experimental forced fasting in the emperor penguin Aptenodytes forsteri: Relationship between feather growth, body weight loss, body temperature and plasma fuel levels, Comp. Biochem. Phys., 61A:287.

Groscolas, R., 1982, Modifications métaboliques et hormonales en relation avec le jeûne prolongé, la reproduction et la mue chez le Manchot empereur, Thesis, University of Dijon, France.

Groscolas, R., 1986, Changes in body mass, body temperature and plasma fuel levels during the natural breeding fast in male and female emperor penguins Aptenodytes forsteri, J. Comp. Physiol. B, 156: 521.

Jouventin, P., 1971, Incubation et élevage itinérants chez les manchots empereurs de Pointe Géologie (Terre Adélie), Rev. Comp. Animal, 5:189.

Jouventin, P., 1978, Ethologie comparée des Sphéniscidés, Thesis, University of Montpellier, France.

King, J.R., 1980, Energetics of avian moult, in: "Acta XVII Congressus Internationalis Ornithologici", R. Nöhring, ed., Deutsche Ornithologen-Gesellschaft, Berlin.

Korpimäki, E., 1986, Reversed size dimorphism in birds of prey, especially in Tengmalm's owl Aegolius funereus: a test of the "starvation hypothesis", Ornis Scand., 17:326.

Korschgen, C.E., 1977, Breeding stress of female eiders in Maine, J. Wildl. Mgmt., 41:360.

Krüger, K., Prinzinger, R., and Schuchmann, K.-L., 1982, Torpor and metabolism in hummingbirds, Comp. Biochem. Physiol., 73A:679.

Le Maho, Y., 1983a, Metabolic adaptations to long-term fasting in antarctic penguins and domestic geese, J. therm. Biol., 8:91.

Le Maho, Y., 1983b, Le manchot empereur: une stratégie basée sur l'économie d'énergie, Le Courrier du CNRS, 50:15.

Le Maho, Y., Delclitte, P., and Chatonnet, J., 1976, Thermoregulation in fasting emperor penguins under natural conditions, Am. J. Physiol., 231:913.

Le Maho, Y., Robin, J.-P., and Cherel, Y., 1987, The metabolic features of starvation, in: "Comparative Physiology of Environmental Adaptations", vol. 2, P. Dejours, ed., Karger, Basel.

Le Maho, Y., Robin, J.-P., and Cherel, Y., 1988, Starvation as a treatment for obesity: The need to conserve body protein, NIPS, 3:21.

Lyman, C.P., 1982, Who is who among the hibernators, in: "Hibernation and Torpor in Mammals and Birds", C.P. Lyman, J.S. Willis, A. Malan, and L.C.H. Wang, ed., Academic Press, New York.

Mainguy, S.K., and Thomas, V.G., 1985, Comparisons of body reserve buildup and use in several groups of Canada geese, Can. J. Zool., 63:1765.

Masman, D., 1986, The annual cycle of the kestrel *Falco tinnunculus*. A study in behavioral energetics. Ph.D. Thesis, Rijksuniversiteit Groningen, Drukkerij Van Denderen, B.V. Groningen.

Masman, D., Daan, S., and Beldhuis, H.J.A., 1988a, Ecological energetics of the kestrel: daily energy expenditure throughout the year based on time-energy budget, food intake and doubly labeled water methods, Ardea, 76:64-81.

Masman, D., Daan, S., and Dijkstra, C., 1988b, Time allocation in the kestrel *Falco tinnunculus*, and the principle of energy minimization, J. Anim. Ecol., 57:411-432.

Milne, H., 1976, Body weights and carcass composition of the common eider, Wildfowl, 27:115.

Mortensen, A., and Blix, A.S., 1985, Seasonal changes in the effects of starvation on metabolic rate and regulation of body weight in Svalbard ptarmigan, Ornis Scand., 16:20.

Mrosovsky, N., and Powley, T.L., 1977, Set points for body weight and fat, Behav. Biol., 20:205.

Mrosovsky, N., and Sherry, D.F., 1980, Animal anorexias, Science, 207:837.

Newton, I., 1979, "Population Ecology of Raptors", Berkhamsted, Poyser.

Nilsson, J.-Å., and Smith, H.G., 1988, Incubation feeding as a male tactic for early hatching, Anim. Behav., 36:641.

Pond, C.M., 1978, Morphological aspects and the ecological and mechanical consequences of fat deposition in wild vertebrates, Ann. Rev. Ecol. Syst., 9:519.

Prévost, J., 1961, "Ecologie du Manchot Empereur", Hermann, Paris.

Prévost, J., and Vilter, V., 1962, Histologie de la sécrétion oesophagienne du Manchot empereur, in: "Proceedings, 13th International Ornithological Congress", American Ornithologists' Union, Baton Rouge.

Reinertsen, R.E., and Haftorn, S., 1986, Different metabolic strategies of northern birds for nocturnal survival, J. Comp. Physiol. B, 156:655.

Robin, J.-P., Cherel, Y., Girard, H., Chaban, C., and Le Maho, Y., 1988a, Augmentation du rendement azoté et hyperphagie associées à la réalimentation après un jeûne prolongé chez l'oie domestique, C.R. Acad. Sci. Paris, 306, série III:375.

Robin, J.-P., Frain, M., Sardet, C., Groscolas, R. and Le Maho, Y., 1988b, Protein and lipid utilization during long-term fasting in emperor penguins, Am. J. Physiol. 254 (Regulatory Integrative Comp. Physiol. 23):R61.

Sladen, W.J.L., and Ostenso, N.A., 1960, Penguin tracks far inland in the Antarctic, Auk, 77:466.

Stokkan, K.-A., Mortensen, A., and Blix, A.S., 1986, Food intake, feeding rhythm, and body mass regulation in Svalbard rock ptarmigan, Am. J. Physiol. 251 (Regulatory Integrative Comp. Physiol. 20):R264.

Tinbergen, N., 1958, "Curious Naturalists", Countrylife Ltd., London.

Tøien, Ø., Aulie, A., and Steen, J.B., 1986, Thermoregulatory responses to egg cooling in incubating bantam hens, J. Comp. Physiol. B, 156:303.

Weathers, W.W., 1985, Energy cost of incubation in the canary, Comp. Biochem. Physiol. 81A:411-413.

West, G.C., 1965, Shivering and heat production in wild birds, Physiol. Zoöl., 38:111.

Wingfield, J.C., 1988, Changes in reproductive function of free-living birds in direct response to environmental perturbations, in: "Processing of Environmental Information in Vertebrates", M.H. Stetson, ed., Springer-Verlag, Heidelberg.

EFFECT OF CLUTCH SIZE ON EFFICIENCY OF HEAT TRANSFER

TO COLD EGGS IN INCUBATING BANTAM HENS

>
> Øivind Tøien
>
> Div. of General Physiology
> Dept. of Biology, University of Oslo
> Box 1051 Blindern, N-0316 Oslo 3, Norway

INTRODUCTION

The brood patch works as an important organ of heat transfer between the incubating bird and the eggs by increasing circulation when exposed to cold. (Midtgård et al., 1985). The heat loss of the bird to the eggs has to be compensated by increased heat production.

Increased O_2 consumption of birds incubating cold eggs has been demonstrated by Biebach (1979), Vleck (1981), Tøien et al.(1986) and Gabrielsen et al. (1987). Biebach (1986) found, however, little correlation between the total energy required for egg rewarming and clutch size. Rewarming two eggs seemed to be more expensive than expected from the rise in the heat content of the eggs. The efficiency of rewarming appeared to be higher with a clutch size of eight. It should thus be energetically more favorable to incubate a large clutch than a small one with respect to the egg rewarming. As suggested, this result may be due to the lack of control with the heat content of the bird's body. A pronounced drop in body temperature was observed in the Great Tit immediately after egg rewarming started (Haftorn and Reinertsen 1982).

Tøien et al. (1986) found that the O_2 consumption of bantam hens exposed to cold water circulated eggs first increased and then stabilized at a plateau. In this work the steady state situation obtained with water circulated eggs has been made use of to measure the amount of heat removed by eggs, and metabolic response of bantam hens to 2, 4 and 8 cold eggs.

EXPERIMENTAL DESIGN

Bantam hens (Gallus domesticus, 520-820 g) had free access to water and food until the experiments started. They were then placed on a test nest consisting of a plastic nestbowl with 4-8 eggs circulated with pretempered water or coolant fluid (fig. 1). The eggs were bantam eggshells reinforced with epoxy resin. They were fixed to the nest bottom by the entrance tubes, positioned on an ellipse according to how hens placed their own eggs in the nest. Hens tended to be restless if offered only two eggs. Thus, to obtain only two cold eggs with a clutch size of four, the other two eggs were temperated at approx 40°C. Thermocouples were mounted in the center of each egg (T_e). One thermocouple was taped to the upper surface of one of the eggs

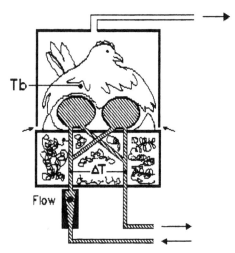

Fig. 1. Schematic diagram of the test nest. The tubes to the eggs were connected to temperature controlled reservoirs. Fluid flow trough the eggs was measured with a rotameter and temperature difference of incurrent and excurrent fluid (ΔT) with a thermopile. Air was drawn through the metabolic chamber by a pump and passed to an oxygen analysis system.

at a position likely to be in contact with the brood patch. The temperature difference of incoming and excurrent fluid was measured by a thermopile connected to the tubes just before they branched to the eggs. It was made from 25 serially connected thermocouples. The room beneath the nest was filled with insulating cotton waste.

A flow of 0.3-1 l/min through the eggs was used and was measured with a rotameter calibrated at each temperature. The heat capacity of the coolant fluid was measured by constant electric heating in a calorimeter, comparing it to water which was assumed to have a heat capacity of 4.184 $J \cdot cm^{-3} \cdot °C^{-1}$ (1.00 $cal \cdot cm^{-3} \cdot °C^{-1}$).

The test nest was situated in a 18.5 l plexiglass chamber through which air was drawn at a rate of 4-5 l/min by a membrane pump. CO_2 and water vapor was removed with soda lime and silicagel before the gas was passed through a flowmeter (Fisher & Porter Triflat). A sample of 100 ml/min was directed through a paramagnetic oxygen analyzer (Servomex OA-1100 with pressure compensation transducer). Back pressure was measured with a water manometer, and barometric pressure with a mercury barometer. The main flowmeter was calibrated against a calibrated spirometer in the pressure range used during the experiments. The O_2 analyzer was calibrated with pure N_2 and air. Correction was made for a sample delay of one minute. O_2 consumption was calculated according to the formula of Bartholomew et al. (1981). An energy equivalent of 5.58 W per l O_2 (STPD)/h was used (Schmidt-Nielsen 1979).

Temperatures were measured with 36 ga copper constantan thermocouples connected to thermocouple amplifiers (AD595). Cloacal temperature (T_b) was measured 40-50 mm into the cloaca. Ambient temperature (T_a) was measured close to the air outlet of the respiratory chamber. The signal from the thermopile was amplified 100X with an instrumentation amplifier (AD524). All signals were fed into an A/D-converter (Scientific Solutions Labmaster) connected to an AT-compatible computer controlled by software written in Turbo Pascal. Accuracy of the T_b measurements was ±0.1°C; for the other temperatures ±0.3°C because of a wider measuring range. The thermopile had an accuracy of ±0.01°C temperature difference. The hen was monitored with a video-camera.

During natural incubation, heat transfer from the incubating bird to the eggs includes heat needed to rewarm the eggs and to keep the eggs warm. The use of water circulated eggs in this study was aiming at only measuring heat exchange between hen and eggs ($\dot{Q}$) If ambient temperature is higher or lower than the egg temperature there will be some heat exchange between tubes to the eggs, the eggs and the environment. This was compensated by subtracting the background heat exchange measured with no hen on the nest from heat exchange measured when the hen was incubating:

$$\dot{Q} = \dot{V} \cdot c \cdot (\Delta T - \Delta T_0)$$

where $\dot{V}$ is flow is in ml/sec, c is the specific heat capacity of the circulating fluid (J/ml·°C), ΔT is the temperature difference of the thermopile with hen on the nest (°C) and ΔT_0 - without hen on the nest. Usually the eggs were warmed to approx 40°C before the hen was placed on the test nest. After she had calmed down, fluid with the desired temperature was shunted through the eggs. O_2 consumption and body temperature was then allowed stabilize before a measuring period of approx 10 min. Following, warm water

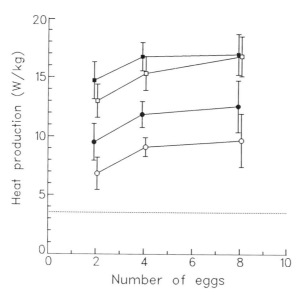

Fig. 2. Heat production of bantam hens vs number of eggs cooled to 10°C (■), 20°C (□), 30°C (●) and 35°C (○). Each data point represent a mean value of 5-27 observations, on 5 hens. The error bars indicate SD, and the stippled horizontal line is the resting metabolism of bantam hens (Tøien et al. 1986).

was shunted through the eggs and O_2 consumption allowed to approach resting level. The hen was then taken off the nest and ΔT_0 was then measured at the end of an identical period of egg cooling. In some experiments the eggs were cooled throughout the experiment. ΔT_0 was then measured shortly after the hen was removed from the nest.

Because the hen also insulates the eggs when she sits on them, the background heat exchange may be overestimated when T_e is deviating much from T_a (24-28°C). Some of the data on heat transfer has thus been restricted to a T_e range close T_a. Efficiency of heat transfer to the eggs was defined as the amount of heat transferred to the eggs (W) divided by the increase in the heat production of the hen above resting level (3.52 W/kg, Tøien et al. 1986).

Difference between means was tested with a Wilcoxon-Mann-Whitney test. A significance level of 5% was used. Regression analysis was performed with least squares method.

RESULTS

The metabolic response of bantam hens to cold eggs was not linearly related to clutch size (fig. 2), but rather reached a plateau at a clutch size of 4-8. Fig. 3 shows the same results plotted against egg temperature. With a clutch size of 8, heat production increased linearly to 2.8 times the

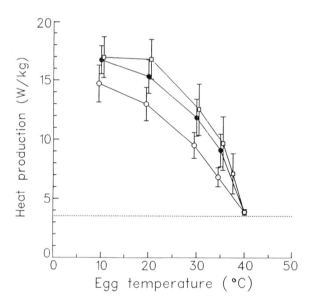

Fig. 3. Heat production of bantam hens exposed to 2- (O), 4- (●) and 8 (□) cold eggs vs temperature of the eggs. Data points are means ±SD (n= 5-27, 5 hens). Observations at 40°C are was obtained with 4-8 eggs and the stippled horizontal line is the resting metabolism of bantam hens (Tøien et al., 1986).

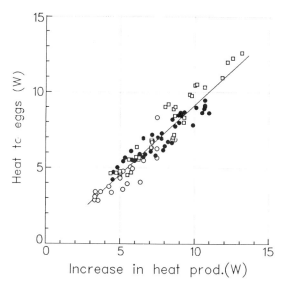

Fig. 4. Heat transferred to 2- (○), 4- (●) and 8 (□) eggs by 3 bantam hens vs the increase in heat production above resting level; data with egg temperatures between 10 and 30°C. At higher egg temperatures the background heat exchange between the eggs and the environment was considerable and may have been overestimated. The correlation coefficient of the regression line was 0.957.

resting level at 35°C. Further egg cooling gave a nonlinear increase to a maximum heat production of 4.8 times the resting level at 20°C. At 10°C the heat production was 17.0 ±1.7 W/kg (n=10) and not significantly different from the one at 20°C. The heat production with 4 cold eggs tended to be slightly lower than with 8 eggs, but was only significantly different at 20°C (P<0.05). At 10°C it reached the same level of heat production as with 8 eggs. Two cold eggs gave a linear increase in heat production to 2.7 times the resting level at 30°C. Below this temperature the response seemed to be nonlinear, but never reached a plateau. Heat production was significantly lower than with both 4 and 8 eggs at all egg temperatures (P<0.01).

There was a considerable variation in the metabolic response at a given egg temperature and clutch size. Some of this variation may be ascribed to differences in body size and tendency to incubate of the individual birds. When two identical exposures was following each other in the same bird, however, a considerable variation in the response could persist. In some cases it was observed that the hen adjusted herself differently to the eggs.

The amount of heat transferred to the eggs by the hen was closely correlated to the increase in heat production (fig. 4, Y:= -0.13+0.934·X, R= 0.957). As shown in fig. 5 the efficiency of heat transfer was quite high and dispersed at an egg temperature of 35°C. Towards lower egg temperatures it tended to decrease and was significantly lower at 20°C than at 30°C for 4 eggs (p<0.01), and at 10°C than at 20°C for 8 and 2 eggs (p<0.05).

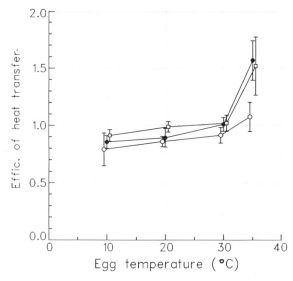

Fig. 5. Efficiency of heat transfer to 2- (○), 4- (●) and 8 (□) eggs vs egg temperature. The results are from 3 hens and are given as means ±SD (n=3-16).

At 10°C there was often signs of condensation of water on the eggs when the hen was taken off the nest. Some condensation also formed during measurement of background heat loss. For these reasons the efficiency of heat transfer graphed against clutch size in fig. 6 has been confined to a T_e between 20°C and 30°C which was close to T_a. With 2 cold eggs the efficiency of heat transfer was 0.89 ± 0.06 (n=18). Although the efficiency tended to increase with 4 eggs (0.93 ± 0.09, n=28), the difference was not significantly different (p<0.1). A clutch size of 8 gave an efficiency of heat transfer of 0.98 ± 0.08 (n=23) which was significantly higher than with 2 and 4 eggs (p<0.05).

DISCUSSION

The nonlinear dependence of metabolic response on clutch size in incubating bantam hens was not caused by differences in efficiency of heat transfer to the eggs. At egg temperatures close to ambient only minor changes in efficiency was observed with clutch size. Almost all the heat produced was transferred to the eggs.

This contrasts Biebach's (1986) energetic study on egg rewarming in the starling. Rewarming of 2 eggs seemed to require about the same amount of heat produced by the bird as rewarming 8 eggs. Since one would expect a small clutch to be rewarmed faster according to the present study, it is possible that Biebach missed the end of the rewarming period with the larger clutch sizes. Temperature was only measured in one egg and the central eggs are likely to be heated faster than peripheral eggs.

Since the metabolic response to 8 eggs was almost the same as with 4 eggs, the bantam hens in this study seemed to have a limited area of brood patch to bring in contact with the eggs. Midtgård (1987) found that bantam

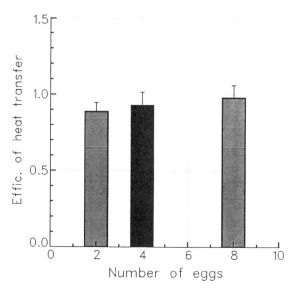

Fig. 6. Efficiency of heat transfer to eggs of 20-30°C vs clutch size. Data from different egg temperatures are pooled. The results are mean values with SD indicated as positive error bars (n=18-28).

hens offered 1-4 eggs regulated the total amount of weight exerted on the eggs according to clutch size. If the weight applied per egg is proportional to the area of the brood patch in contact with eggs, this indicates that the hen tries to keep as close contact to each egg as possible without breaking it or compressing the blood vessels in the brood patch. Fixed eggs in the nest, as used in the present study, may however represent a problem to the bird. Bantam hens incubating their own eggs were often observed to keep some eggs under the wings. This is not possible with eggs fixed to the nest. The capacity of the brood patch of bantam hens is thus probably higher than 4-8 eggs. A normal clutch size in bantam hens is about 15 eggs (Steen and Parker, 1981).

Both in this study, and in the study by Tøien et al. (1986) the metabolic response was nonlinearly related to the amount of egg cooling at lower egg temperatures. Biebach (1979) only tested the starling with slight egg cooling where our bantam hens showed a linear response. With clutch sizes of 4 and 8 the heat production of the bantam hens seems to level out at a plateau of 4.6-4.8 times the resting level. This is the heat production expected at a cold exposure of -60 to -70°C (Tøien et al, 1986). It is probably close to the maximum capacity for prolonged heat production of the bantam hens. As efficiency of heat transfer tended to decrease rather than increase at these egg temperatures, the amount of heat transferred to the eggs also levels off at intense egg cooling. The decrease in T_b of the hen upon cold egg exposure was also to small to explain this effect.

Thus, the total conductance of the brood patch is decreasing at lower egg temperatures. This does not seem consistent with the local vasodilatation found when the brood patch of bantam hens was exposed to 5-10°C (Midtgård et al., 1985). Though, if the area of brood patch in contact with the eggs decreased at low temperatures the total conductance of the brood patch would also decrease. But in the study by Midtgård et al. (1987), the weight applied to the eggs by bantam hens was independent of egg temperature. Unless the hen by other means is able regulate the area of brood patch in contact with the eggs, for instance by controlling muscles around contour feathers, it seems that the response of the brood patch may rather be an intense vasodilatation at medium cooling and then a gradual vasoconstriction at strong cooling. Both potent vasodilator (VIP-immunoreactive and AchE-positive) and vasoconstrictor (adrenegic) nerve fibres are present in the arteriovenous anastomoses of the brood patch (Midtgård, 1988). The slight vasoconstriction present in areas of the brood patch adjacent to the cooled ones (Midtgård et al. 1985) may indicate a general vasoconstrictory tone on top of the local vasodilator response.

Heat is not only transferred from the brood patch to the eggs by direct contact. When the hen half rose from the nest there was still a certain heat transfer indicating that eggs are also warmed by the nest air, which in turn are warmed by the brood patch. Some nest air may be lost to the environment or cooled by the nest material. Considering the low insulation of the test nest used in this study, the efficiency of heat transfer was remarkably high. Although the hen increased her insulation by feather erection, and could possibly channel some of her resting heat production through the eggs, the experiments were performed close to the lower critical temperature of a bantam hen (24°C, Tøien et al. 1986). Further change in conductance would be very small. The explanation of the higher efficiency with a clutch size of 8 may be that there are a larger surface area of eggs exposed to the nest air.

The present study did not include measurement of the amount of heat needed to keep the eggs warm. In the Blue Tit, the eggs seemed to act as an extension of the body leading to an increase in conductance of heat from the bird-egg complex proportional to clutch size (Haftorn and Reinertsen, 1985).

During egg-rewarming, increased conduction of heat from shivering muscles to the environment could possibly explain a decrease in efficiency towards lower egg temperatures. In bantam hens, shivering occurs mostly in muscles in the hind-quarter (Aulie and Tøien 1988). Situated remote from the brood patch, a rise in temperature of skin external to these muscles would not benefit heat transfer to the eggs. Ventilatory heat loss may also increase at low egg temperatures due to higher ventilatory requirements. Finally, the background measurements may have over-estimated the amount of heat added to cold eggs from the environment, because the hen insulates them. This would result in a too low efficiency.

The heat transfer to eggs thus seems to be optimized. At a T_e close to ambient temperature the deviations of the efficiency of heat transfer from unity are very small. The size of the brood patch seems to be well adapted to the normal clutch size of the species. For energetic calculations it may be a good approximation to calculate the theoretical increase in the heat content of the eggs, and then add the energy required to keep the eggs warm.

Acknowledgements

I wish to thank dr. Arnfinn Aulie and dr. Johan B. Steen for comments on the manuscript. The work was supported by the Nansen Foundation.

REFERENCES

Aulie, A. and Tøien, Ø., 1988, Threshold for shivering in aerobic and anaerobic muscles in bantam cocks and incubating hens., J. Comp. Physiol.: In Press.

Bartholomew, G. A., Vleck, D. and Vleck, C.M., 1981, Instantaneous measurement of oxygen consumption during pre-flight warm-up and post-flight cooling in Sphingid and Saturniid moths, J. Exp. Biol., 90:17-32.

Biebach, H., 1979, Energetik des brütens beim Star (Sturnus vulgaris), J. Orn., 120:121-138.

Biebach, H., 1986, Energetics of rewarming a clutch in starlings (Sturnus vulgaris), Physiol. Zool., 59:69-75.

Gabrielsen, G.W. and Unander, S., 1987, Energy cost during incubation in Svalbard and Willow Ptarmigan hens, Polar Res., 5:59-69.

Haftorn, S. and Reinertsen, R.E., 1982, Regulation of body temperature and heat transfer to eggs during incubation, Ornis Scand., 13:1-10.

Haftorn, S. and Reinertsen, R.E., 1985, The effect of temperature and clutch size on the energetic cost of incubation in a free-living blue tit (Parus caeruleus)., The Auk, 102(3):470-478.

Midtgård, U., 1987, How heavily does the hen sit on her eggs during incubation? Experientia, 43:1232-1233.

Midtgård, U., 1988, Innervation of arteriovenous anastomoses in the brood patch of the domestic fowl., Cell Tissue Res., 252:207-210.

Midtgård, U., Sejrsen, P. and Johansen, K., 1985, Blood flow in the brood patch of Bantam hens: evidence of cold vasodilatation, J. Comp. Physiol. B, 155:703-709.

Schmidt-Nielsen, K., 1979, Animal physiology: Adaptation and environment., Cambridge University Press, London.

Steen, J.B. and Parker, H., 1981, The egg-"numerostat". A new concept in the regulation of clutch- size, Ornis Scand., 12:109-110.

Tøien, Ø., Aulie, A. and Steen, J.B., 1986, Thermoregulatory responses to egg cooling in incubating bantam hens, J. Comp. Physiol. B, 156:303-307.

Vleck, C.M., 1981, Energetic cost of incubation in the zebra finch, Condor, 83:229-237.

EMU WINTER INCUBATION: THERMAL, WATER, AND ENERGY RELATIONS

William A. Buttemer and Terence J. Dawson

School of Zoology
The University of New South Wales
Kensington, Australia 2033

INTRODUCTION

The general pattern of avian life histories commonly displays minimal overlap of energy-demanding activities. For example, most birds reserve breeding and molting phases to times when their maintenance costs (e.g., thermoregulatory and foraging costs) are at their lowest annual levels. The emperor penguin (*Aptenodytes forsteri*) and emu (*Dromaius novaehollandiae*) appear to contradict this tendency in that both species are midwinter breeders and, for breeding males of each species, incubation is performed to the exclusion of feeding for about 9 weeks in penguins and 8 weeks in emus. Male emperor penguins possess many physiological and behavioral qualities which ameliorate their protracted exposure to the Antarctic winter (Le Maho, 1977), however, little is known about the energetics of emu incubation.

Although emus should benefit by reducing energy expenditure while incubating, the extent to which this occurs will be limited by the thermal requirements for successful embryo development. The latter concern includes a need to synchronize hatching in eggs that are incubated for disparate periods. Consequently, we investigated the thermal and energy relations of emus undergoing incubation and the effects of temperature on embryonic metabolism.

MATERIAL AND METHODS

Experimental Animals

We studied a colony of 10 adult emus (6 males; 4 females) which were maintained 50 km north of Sydney at the Cowan Field Station of the School of Zoology, University of New South Wales. All birds ranged freely in a 100 by 200 m enclosed section of dry sclerophyll forest and were provided pelleted commercial food to supplement their natural diet. Prior to incubation, we anesthetized 5 males with Ketamine HCl (Grubb, 1983) and inserted calibrated temperature transmitters (J. Stewart Enterprises) intraperitoneally. We subsequently implanted ECG transmitters (J. Stewart Ent.) subcutaneously into two of these birds. At the onset of laying, the enclosure was surveyed daily for eggs which, when found, were labelled with an indelible marker. We inspected clutches at least weekly during the first 7 weeks of incubation and at least daily thereafter until hatching.

Temperature Measurements

We evacuated three emu eggs, sawed them in two, and lined their inner surfaces with resin-coated fiberlass matting. A buoyant miniature temperature transmitter (J. Stewart Enterprises) was placed in each egg which was then joined with resin and filled with mineral oil. This permitted maintenance of each transmitter at the egg's upper surface despite egg turning by the incubating male. All temperature transmitters were calibrated against a certified mercury-in-glass thermometer at the beginning and end of the study. For two transmitters which displayed a slight divergnce in these calibrations, data were interpolated linearly over time betwen the calibration dates.

Transmitter signals were converted to a voltage output through a Telonics receiver and digital processor and each transmitter was recorded for 1 min at 10-min intervals on a Datataker 100F datalogger. Air temperature was similarly logged from a shaded 36 gauge copper-constantan thermocouple placed 0.5 m above ground at a site between the incubating emus.

Embryo Thermal Sensitivity

A detailed description of procedures used for these measurements is presented elsewhere (Hoyt, Vleck, and Vleck, 1978; Buttemer, Astheimer, and Dawson, 1988) and only a brief outline follows here. On a given day, we selected three eggs, placed them individually in open 4-l paint cans, and maintained them at 30, 34, or 38 °C for at least 2.5 h within a constant temperature cabinet. Following this equilibration period, we sealed each can and withdrew 100 cm^3 of chamber air with a syringe. A second air sample was withdrawn after sufficient time had elapsed to reduce the chamber's oxygen concentration to between 20.0 and 20.5%. We measured fractional oxygen concentrations of all samples using a Servomex oxygen analyzer (Model OA-272) after absorbing water vapor and CO_2 with Drierite ($CaSO_4$) and Ascarite (sodium asbestos anhydride), respectively. Using the formulation of Vleck (1987), we calculated oxygen consumption ($\dot{V}_{O_2}$) by:

$$\dot{V}_{O_2} = V (F_I - F_E)/(1.0 - F_E) t \tag{1}$$

where V is the chamber dry gas volume (STPD), F_I and F_E are the initial and final fractional oxygen concentrations in the chamber, and t is the elapsed time between initial and final gas sampling.

Using our measurements of each egg's $\dot{V}_{O_2}$ at the three temperatures on a given day, we evaluated embryonic metabolic thermal sensitivity by computing each egg's temperature coefficient (Q_{10}) for the temperature intervals 30-34 and 34-38 °C through:

$$Q_{10} = (R_2/R_1)^{(10.0/T_2 - T_1)} \tag{2}$$

where T_2 and T_1 are the higher and lower temperatures, respectively, at which the corresponding rates of oxygen consumption (R_2 and R_1) were measured.

Body Water Measurement

We measured water flux and total body water using tritiated water (HTO), following procedures recommended by Nagy and Costa (1980). Before injecting HTO, the bird was removed from its nest, placed in a large hessian bag, suspended from a clockface scale, and weighed to the nearest 100 g. A blood sample was then taken from a jugular vein to establish background radiation levels and then a measured quantity of HTO was injected intramuscularly (ca. 39 M Bq of tritium in 1 ml of isosmotic saline; each injection was weighed to the nearest 0.001 g). Following this, the bird was returned to its nest and an equilibrium blood sample was collected intravenously 4 h later. At the end of incubation, each emu was again weighed, a background sample of blood was collected, and its total body water re-evaluated through a final injection of tritiated water. The techniques and materials for evaluating tritium in emu blood are stated by Dawson, Herd, and Skadhauge (1983).

Heart Rate and Oxygen Consumption

The relation between heart and oxygen consumption rates was measured for two emus at the end of their incubation bout. For these measurements, each bird was placed in a gastight rectangular chamber (ca. 1.4 x 0.8 x 0.8 m) which had inlet and outlet gas ports at opposite ends. Dry air was pumped through the chamber at flow rates between 40 and 60 l min^{-1} as measured by a calibrated dry gas meter. Part of the effluent air was routed to a Beckman F3 oxygen analyzer after passing through a desiccant (Drierite) and CO_2 absorbent (Ascarite). Air temperatures within the metabolic chamber and gas meter were monitored using calibrated thermocouples connected to a Datataker 100F datalogger. Birds were placed in the chamber about midday and remained there for 16 to 22 h. During these measurements of $\dot{V}_{O_2}$, the output of each bird's ECG transmitter was simultaneously monitored using the same recording system as that employed for datalogging while the bird was incubating. All measurements of $\dot{V}_{O_2}$ were calculated using Equation 2 of Hill (1972) and were corrected to STPD. The entire respirometric system was calibrated under conditions which obtained for physiological measurements, except that the system was animal-free, using an iron-burning technique (Young, Fenton, and McLean, 1984). We assumed an energy equivalent of 20.1 kJ for each liter of oxygen consumed. Unless stated otherwise, values in this paper are presented as means plus or minus their standard error.

RESULTS

Incubation Temperature and Embryo Thermal Sensitivity

Mean daily temperatures of oil-filled eggs (T_{egg}) increased steadily from 32 to 34 °C during the first 10 days of incubation (Fig. 1). The range of T_{egg} values was much greater during the first week of incubation than for subsequent periods (Fig. 1). After day 10, T_{egg}'s remained near 34 °C followed by a rise to a plateau of 36 °C from day 35 onward (Fig. 1).

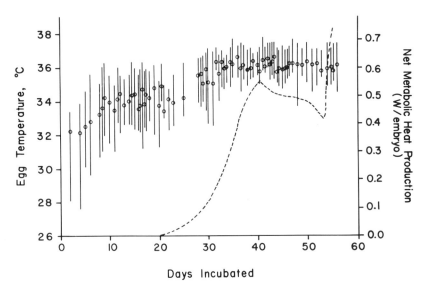

Fig. 1. Internal temperatures of oil-filled eggs over the course of natural incubation. Mean daily values from telemetered eggs placed in three different emu nests (open circles) are accompanied by their daily thermal ranges (vertical lines). The dashed line represents net embryonic heat production based on metabolic data of Vleck, Vleck, and Hoyt (1980).

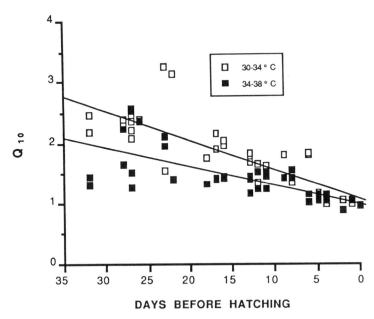

Fig. 2. Thermal sensitivity (Q_{10}) of emu embryonic oxygen consumption as a function of embryo age for the temperature intervals 30-34 °C (open squares; upper regression line) and 34-38 °C (filled squares; lower regression line).

The thermal sensitivity (Q_{10}) of embryonic $\dot{V}_{O_2}$ was greater at the lower temperature range (30-34 °C) than at the higher one (34-38 °C) and declined significantly as incubation proceeded (Fig. 2). The equation describing embryonic Q_{10} as a function of time to hatching for the lower temperature interval is,

$$Q_{10} = 1.03 + 0.053 \text{ (DBH)} \tag{3}$$

and for the upper temperature interval,

$$Q_{10} = 1.00 + 0.032 \text{ (DBH)} \tag{4}$$

where, DBH represents days before hatching. From analysis of covariance, the slopes and elevations of Equations 3 and 4 both differ significantly from one another ($P<0.05$ and $P<0.01$, respectively). Consequently, the thermal sensitivity of emu embryonic metabolism is inversely related to both temperature and embryonic age over the temperature interval 30 to 38 °C.

Adult Body Temperature

Mean daily body temperature (T_b) of incubating birds varied little for the two emus which successfully hatched chicks, averaging 37.7 ± 0.04 °C (n=32) and 37.9 ± 0.02 °C (n=37) for emus 5 and 11, respectively (Fig. 3). Individual variation in daily T_b over the course of incubation was related to neither average daily air temperature nor day-length ($P>0.3$ for both comparisons). Simultaneous T_b records, beginning at day 9 of incubation for one bird and 5 days before incubation for another, did not differ significantly over a 60-h period ($P>0.1$) although values for the latter emu had greater daily variation (Fig. 4). Generally, T_b's were lowest 1-2 h before dawn, rose sharply just before or at sunrise, and often continued rising until mid-afternoon (Fig. 4). Once eggs hatched, both brooding males had mean daily T_b's which were significantly higher than during incubation ($P<0.01$) and averaged 38.6 ± 0.09 °C for both emus.

Water Flux and Body Mass Change

Approximately 1 month before incubation, emus 5 and 11 weighed 40.1 and 42.8 kg, respectively. Neither bird was weighed during the first week of incubation to minimize disturbance during this period. Following this, emu 5 displayed a steady loss of body mass (M_b in kg) which is described by the relation:

$$M_b = 39.5 - 0.113 \text{ (Days Incubated); } (r=0.99; s_b=.009; s_{y \cdot x}=0.286) \tag{5}$$

Emu 11 proved less tractable and, consequently, was weighed only twice during its incubation cycle. From day 9 through day 56 of incubation, this bird lost mass at a rate of 115 g per day.

Water turnover was similar for the two incubating emus, with daily efflux averaging 210.4 ± 0.4 ml and daily influx 166.0 ± 2.0 ml. Based on the dilution of injected tritium, emu 11's total body water was 57.9% of its mass on day 14 of incubation and 60.2% on day 56, its final day of incubation. For emu 5, these values were 59.7 and 63.1% on days 17 and 56 of incubation, respectively. Net daily water loss for the two

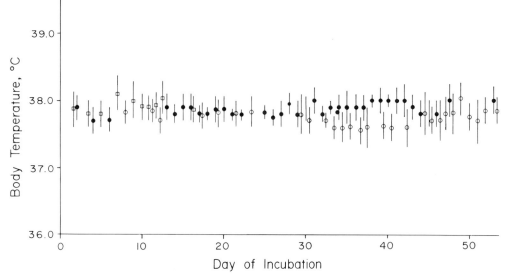

Fig. 3. Mean daily body temperatures and their standard deviation (vertical lines) for emus 5, 7, and 11 (open circles, open squares, and filled circles, respectively, as a function of day of incubation.

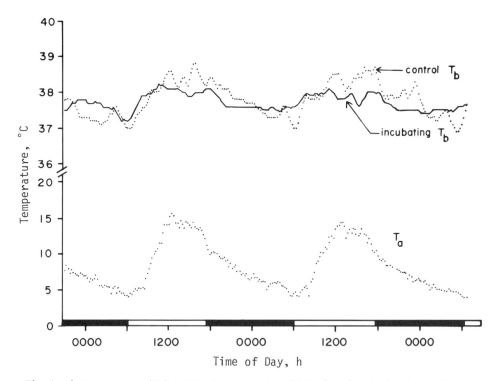

Fig. 4. Air temperature (T_a) and body temperature (T_b) of an incubating (emu 5) and non-incubating bird (control; emu 11) as functions of time of day. The light and dark bars on the bottom axis represent day and night, respectively.

incubating birds averaged 42 ml/d for emu 5 and 47 ml/d for emu 11.

Heart Rate and Oxygen Consumption

During incubation, the heart rate (HR) of emu 11 at times ranged between 17-20 beats/min. As such rates were beyond the 2.5 sec interpulse limit of our telemetry system, only emu 5 furnished consistently reliable HR data throughout incubation. For the latter bird, its HR correlated significantly with its rate of oxygen consumption ($\dot{V}_{O_2}$ in l/h) and is described by the relation:

$$\dot{V}_{O_2} = -0.426 + 0.214(HR); \quad (r=0.96; \; s_b=0.018; \; s_{y \cdot x}=0.708; \; n=15) \quad (6)$$

Between days 16 and 54 of incubation, daily mean HR averaged 31.5 ± 0.5 beats/min (n=21) and ranged between 28.0 and 35.6 beats/min. This latter variation correlated with neither daily mean air temperature nor daylength (P>0.4 for both comparisons) but correlated significantly with its daily mean body temperature (P<.01). Applying Equation 6 to our field measurements of emu 5's HR while incubating, we estimated the 95% confidence interval of its average daily energy expenditure to be 2700 to 3400 kJ/d.

DISCUSSION

Because emu chicks are precocial and become ambulatory within two days, the attending male must quickly shift duties from incubating eggs to brooding and shepharding the mobile chicks. For such species, fledging success will be importantly affected by the extent of intraclutch hatching synchronization. Birds could ensure synchronous hatching by either deferring incubation until the last egg is laid or by appropriately adjusting the rate of embryonic development in eggs incubated for disparate periods. The latter situation may apply to emus as both males in our study had eggs added to their 6-egg clutch at 1, 4, and 7 days following the start of incubation and yet all chicks of each resulting 9-egg clutch hatched within 48 h of one another.

Improved hatching synchrony would follow promotion of accelerated growth by embryos of "late" eggs. Because of the rise in egg temperature over the early stages of incubation (Fig. 1), eggs laid after the start of incubation will experience higher initial temperatures than their nestmates. Furthermore, as the thermal sensitivity of emu embryos is highest at this time (Fig. 2), such conditions will foster more rapid growth by these late additions. For example, if we assume that emu embryonic growth has a Q_{10} of 2.5 at the start of incubation and declines to 1.0 at hatching (Fig. 2; Buttemer et al., 1988), we can compare the growth rates attending differing T_{egg}'s for eggs added to a clutch 3, 6, and 9 days after the start of incubation. From this estimation, the higher relative T_{egg} of the "late" eggs reduces their required incubation time by 0.9, 1.8, and 2.7 days, respectively (Fig. 5). Although these differential growth rates do not account fully for our observation that all emu nestmates hatched within 48 h of one another, any growth rate enhancement for late additions will facilitate hatching synchrony. This follows from observations that accelerated hatching by precocial birds (e.g., Vince, 1966) requires that younger chicks are developed sufficiently to respond to acoustic stimuli from older chicks prior to their hatching (Woolf et al., 1976).

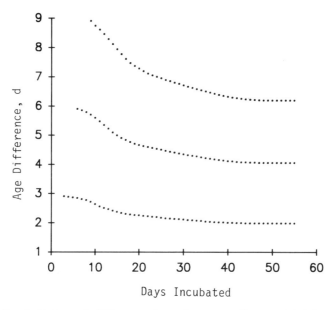

Fig. 5. Estimated differences in embryo age of eggs added 3, 6, and 9 days after the start of incubation from those of embryos incubated from its onset.

The suspension of activity by emus during their protracted incubation bout will reduce substantially their daily energy expenditure (DEE). Our estimate of emu 5's DEE is 63 to 79% of the value expected for an inactive non-passerine during the resting phase of its daily cycle (Aschoff and Pohl, 1970). Given the maintenance of near-normal body temperatures during incubation (Figs. 3 and 4), this level of metabolic expenditure seems suspicious, especially as it is based on extrapolation of physiological measurements for only one bird. Fortunately, our evaluation of change in body mass and body water content permits another way to estimate DEE of incubating emus. Since we measured body water content near the beginning and at the end of incubation for both birds, we can appraise the water-free component of their body mass decline. In estimating DEE, we assume that the latter values represent catabolized energy stores of 86% fat and 14% protein as in fasting emperor penguins (Groscolas and Clement, 1976) and that these substrates have energy contents of 39.3 and 17.8 kJ/g, respectively (Schmidt-Nielsen, 1975). Accordingly, the DEE for these emus during their incubation fast was 2580 and 2470 kJ/d for emus 5 and 11, respectively. Although these estimates average 58% of those expected for resting non-passerine birds (Aschoff and Pohl, 1970), they are nearly indistinguishable from values predicted from allometric appraisal of kiwi and ostrich resting metabolism (Withers, 1983; Fig. 6). Thus, through combination of phylogeny and inactivity, the energy costs of incubation are relatively lower for emus than reported for any other bird.

Compared to emus deprived of water for 9 days at the same season (Dawson et al., 1983), incubating males' rate of decrease in body water content was only 20% of that for the latter birds. Part of this difference owes to the fact that the water-deprived birds continued feeding during the initial phases of the experiment whereas there was no indication that incubating birds ever fed. Because fecal water loss is normally the major route of water loss in emus (Dawson et al., 1983), fasting would significantly reduce the water requirements of incubating birds. In addition, the curtailment of activity by incubating emus will reduce their respiratory water losses through attendant reduction in ventilatory frequency.

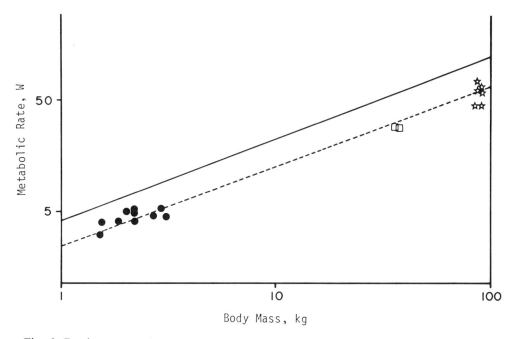

Fig. 6. Resting metabolic rate as a function of body mass in three species of kiwis (filled circles; Calder and Dawson, 1978) and the ostrich (stars; Withers, 1983). The dashed line represents the least-squares regression for the latter species (Withers, 1983) and the solid line is the resting metabolism predicted for non-passerine birds (Aschoff and Pohl, 1970). The metabolic rates estimated for the two incubating emus of this study are designated by the open squares.

ACKNOWLEDGEMENTS

We thank Lee Astheimer for contributions to this study and Shane Maloney, Jeff Vaughn, and Ray Williams for assisting with animal capture and maintenance. The Australian Research Grants Scheme and The University of New South Wales provided financial assistance. All emus were held under the provisions of a licence from the National Parks and Wildlife Service of New South Wales.

LITERATURE CITED

Aschoff, J., and Pohl, H., 1970, Der Ruheumsatz von Vögeln als Funktion der Tageszeit und der Körpergrösse, *J. Ornithol.*, 111:38.

Buttemer, W. A., Astheimer, L. B., and Dawson, T. J., 1988, Thermal and water relations of emu eggs during natural incubation, *Physiol. Zool.*, (in press).

Calder, W. A., and Dawson, T. J., 1978, Resting metabolic rates of ratite birds: the kiwis and the emu, *Comp. Biochem. Physiol.*, 60A:479.

Dawson, T. J., Herd, R. M., and Skadhauge, E., 1983, Water turnover and body water distribution during dehydration in a large arid-zone bird, the emu, *Dromaius novaehollandiae*, *J. Comp. Physiol.*, 153:235.

Groscolas, R., and Clément, C., 1976, Utilisation des réserves énergétiques au cours du jeûne de la reproduction chez le Manchot empereur, *Aptenodytes forsteri*, *C. R. Acad. Sci. Paris*, 282D:293.

Grubb, B., 1983, Use of ketamine to restrain and anesthetize emus, *Vet. Med./Sm. Anim. Clin.*, 78:247.

Hill, R. W., 1972, Determination of oxygen consumption by use of the paramagnetic oxygen analyzer, *J. Appl. Physiol.*, 33:261.

Hoyt, D. F., Vleck, D., and Vleck, C. M., 1978, Metabolism of avian embryos: ontogeny of temperature effects in the ostrich, *Condor*, 80:265.

Le Maho, Y., 1977, The emperor penguin: a strategy to live and breed in the cold, *Am. Scientist*, 65:680.

Nagy, K. A., and Costa, D. P., 1980, Water flux in animals: analysis of potential errors in the tritiated water method, *Am. J. Physiol.*, 238:R454.

Schmidt-Nielsen, K., 1979, "Animal Physiology: Adaptation and Environment", Cambridge University Press, Cambridge.

Vince, M. A., 1966, Artificial acceleration of hatching in quail embryos, *Anim. Behav.*, 14:389.

Vleck, D., 1987, Measurement of O_2 consumption, CO_2 production, and water vapor production in a closed system, *J. Appl. Physiol.*, 62:2103.

Vleck, D., Vleck, C. M., and Hoyt, D. F., 1980, Metabolism of avian embryos: ontogeny of oxygen consumption in the rhea and emu, *Physiol. Zool.*, 53:125.

Withers, P. C., 1983, Energy, water, and solute balance of the ostrich *Struthio camelus*, *Physiol. Zool.*, 56:568.

Woolf, N. K., Bixby, J. L., and Capranica, R. R., 1976, Prenatal experience and avian development: brief auditory stimulation accelerates the hatching of Japanese quail, *Science*, 194:959.

Young, B. A., Fenton, T. W., and McLean, J.A., 1984, Calibration methods in respiratory calorimetry, *J. Appl. Physiol.*, 56:1120.

ENERGY SAVING IN INCUBATING EIDERS

Geir Wing Gabrielsen

Department of Biology
Norwegian Polar Research Institute
1330 Oslo Lufthavn, Norway

INTRODUCTION

During the last decade many estimates have been made of the energy cost during incubation in free living birds. The two methods most often used to determine energy expenditure during incubation are loss of body mass (Prince et al. 1981, Croxall 1982, Croxall and Ricketts 1983, Grant and Whittow 1983) or measurements of oxygen consumption (Norton 1973, Biebach 1979, Gessaman and Findell 1979, Vleck 1981, Haftorn and Reinertsen 1985, Gabrielsen and Unander 1987). While most studies have been performed in the laboratory there are few on free-living birds on the nest site (Grant and Whittow 1983, Brown 1984, Brown and Adams 1984).

The eider (Somateria mollissima) is the most common duck in the Svalbard archipelago. The largest colonies are found on islands along the west coast. During the 26 days of incubation the female eider rarely leaves the nest. They abstain from feeding, resulting in a weight loss of 30-45 % during the time of incubation (Milne 1963, 1976, Cantin et al. 1974, Korschgen 1977).

The present paper reviews data on female eiders presented by Gabrielsen et al.(1989). We have measured incubation metabolic rate (IMR) of two wild incubating eiders. Resting metabolic rate (RMR) and thermal conductance (TC) were determined in 12 non-incubating eiders in the laboratory. Based on weight loss of eiders during the incubation period we have also been able to calculate daily energy expenditure (DEE). Since fasting is accompanied by increased levels of thyroxin (T_4) (Falconer 1971) and decrased levels of triiodothyronine (T_3) (May 1978) we studied plasma concentrations of these hormones both in incubating and non-incubating eiders. Le Maho et al. (1981), studying water turnover rates in geese, observed increased drinking after 21 days of fasting. In order to evaluate whether incubating birds drink during the last part of the fasting period we used tritium to study water influx rate.

MATERIAL AND METHODS

This study was carried out during June/July 1983-87 on female eiders breeding near the Research Station of the Norwegian Polar Research Institute at Ny-Ålesund, Svalbard (79 ^{0}N).

On the fifteenth day of incubation two eider nests with eggs and down were moved from the nest bowl and placed inside a plexiglass box at the original nest site. IMR was measured in an open system in which the plexiglass box functioned as the metabolic chamber. The chamber consisted of two parts, the upper part was lowered down while performing the metabolic measurements. Metabolic measurements both in the field and in the laboratory were carried out as described by Gabrielsen et al. 1988.

Body weight measurements of incubating female eiders were measured by capturing and weighing birds nesting on two islands in Kongsfjorden The number of days each bird had incubated was estimated by measuring the density of eggs at nest. Birds collected for body weight measurements were also used for studies of plasma concentrations of thyroid hormones. A blood sample (2-3 ml) was taken from the wing vein and analyzed after the method described by Andresen et al. 1980.

Tritium was used to determine water turnover rates in incubating eiders, close to hatching. Each bird was injected in the pectoral muscles with 0.75 ml tritium. Their body weight was determined and the birds were released. Two hours after the injection we recaptured the bird to collect the initial blood sample. Over 2-3 days we recaptured, weighted and sampled 6 birds one or two times. Blood samples were stored in microhematocrit capillary tubes and were vacuum-distilled to obtain pure water. Isotope levels in water were measured by liquid scintillation spectrometry as described by Wood et al. 1975.

RESULTS/DISCUSSION

Incubation metabolic rate (IMR), measured toward the end of incubation (day 15-20), was 0.80 ($\pm$ 0.02, number of measurements=13) ml O_2/g·h or 610 KJ/day. Both birds showed a decrease in mean daily energy expenditure with decreasing body mass, but there was no significant decrease in specific metabolic rate. Resting metabolic rate (RMR) in twelve non-incubating eiders was slightly, but significantly higher, 0.86 ml O_2/g·h or 675 KJ/day, i.e. 7.5 % higher than the IMR value of incubating birds. These results are in contrast to studies of smaller bird species indicating that IMR is 15-30 % higher than RMR (Biebach 1979, Vleck 1981, Haftorn and Reinertsen 1985, Gabrielsen and Unander 1987), but comparable to studies of larger birds which show that IMR is similar to or lower than RMR of non-incubating individuals (Grant and Whittow 1983, Brown 1984, Brown and Adams 1984). While smaller birds periodically leave the nest to forage and have limited tolerance to starvation, larger birds such as the eider can incubate continuously and survive relatively long periods of fasting.

Eider ducks show a large reduction in body weights during incubation. A loss of body weight by 30-45 % is observed in studies of incubating eiders in Scotland, Maine and Quebec (Milne 1963, 1976, Cantin et al. 1974, Korchgen 1977). The present study shows a 44 % decrease in body weight from the start of egglaying until hatching in Svalbard eiders. From the start of incubation until hatching there was a 35 % decrease in body weight. Based on loss of body weight during incubation we calculated daily energy expenditure (DEE) to 520 KJ/day. This is 15 and 23 % less than DEE based on metabolic measurements of incubating and non-incubating eiders, respectively.

Our metabolic measurements gave a respiratory quotient (RQ) value of 0.70, indicating that fat was the major metabolic fuel during the long period of fast in eiders. Le Maho et al. (1981) found a reduction of 39 % of body weight after 40 days of fasting and suggested that lipid accounted for 95 % of the energy expenditure in geese. In eiders, based on body composition analysis (Korchgen 1977, Parker and Holm, unpublished), it was calculated that lipid accounted for 97.5 % of the total energy expenditure during incubation.

Measurements of thyroxin (T_4) and triiodothyronine (T_3) indicate that while the level of T_4 was stable, the T_3 level increased through the incubation period. This is in contrast to the studies on chickens and penguins (May 1978, Cherel et al. 1988) which showed that the T_3 level was depressed after food deprivation. Compared to starved geese (Le Maho et al. 1981) our metabolic measurements of starved eiders are not reduced to the same level. While there is a 7-8 % difference between specific IMR and RMR in eiders there is a 17 % difference in RMR between starved and non-starved geese. Since the thyroid hormones are involved in metabolism, it may be argued that the metabolic level observed in incubating eiders may be due to the fact that incubating eiders did not lower their metabolism because constant heating of the eggs had to be maintained. Measurements of body temperature may support this. While the body temperature (40.1 ^{0}C) was stable in eiders, Le Maho et al. (1981) found a reduction in body temperature (40.1 to 39.3 ^{0}C) after 30 days of fasting in geese.

Studies of water content in starved geese showed that while the extracellular fluid volume was maintained throughout fast, there was an increase after 21 days of fasting (Le Maho et al. 1981). Studies of body composition of eiders (Korschgen 1977, Parker and Holm, unpublished) showed that the water content increased from 51 to 63 % from egg laying until hatching. It may be argued that this increase is due to depletion of body fat but our studies of water turnover rates at the end of incubation in eiders indicate that they also drink water. The water influx rate was 90 ml/day. Edema associated with starvation is well known in mammals (Keys et al. 1950). Birds may drink to maintain a steady body mass when approaching a critical weight level (stage III, as pointed out by Le Maho et al. 1981). This stage is characterized by increased protein utilization and a rapid drop in body weight. It is also suggested that drinking at the end of incubation is the main reason for underestimation of DEE using weight loss calculation. Caution should therefore be taken when using weight loss for calculating DEE in starved birds.

REFERENCES

Andresen, Ø., Amrud, J., Grønholt, L.E., Helland, G., Schie, K.A., and Sylliås G.A. 1980. Total thyroxine and free thyroxine index of dairy cows in relation to strength of heat. Acta Vet Scand, 21:108-112.
Biebach, H., 1979. Energetik des Brutens beim Star (Sturnus vulgaris). J Ornithologie, 120:121-138.
Brown, C.R., 1984. Resting metabolic rate and energetic cost of incubation in Marcaroni Penguins (Eudyptes chrysolophus) and Rockhopper Penguins (E. chrysocome). Comp Biochem Physiol, 77A:345-350.
Brown, C.R., and Adams, N.J., 1984. Basal metabolic rate and energy expenditure during incubation in the wandering albatross (Diomedea exulans). Condor, 86:182-186.

Cantin, M.J., Bedard, J., and Milne, H., 1974. The food and feeding of common eiders in the St.Lawrence estuary in summer. Can J Zool, 52:319-334

Cherel, Y., Robin, J.P., Walch, O., Karmann, H., Netchitailo, P. and Le Maho, Y., 1988. Fasting in king penguin. I. Hormonal and metabolic changes during breeding. Am J Physiol, 254:R170-R177.

Croxall, J.P., 1982. Energy cost of incubation and moult in petrels and penguins. J Anim Ecol, 51:177-194.

Croxall, J.P., and Ricketts, C., 1983. Energy cost of incubation in the wandering Albatross (Diomedea exulans). Ibis, 125:33-39.

Falconer, I.R., 1971. The thyroid gland. In: Bell, D.J., and Freeman, B.M., eds. Physiology and Biochemistry of the domestic fowl. Academic Press, New York, pp 459-472.

Gabrielsen, G.W., and Unander, S., 1987. Energy cost during incubation in Svalbard and Willow Ptarmigan hens. Polar Research, 5:59-69.

Gabrielsen, G.W., Mehlum, F., and Karlsen, H.E., 1988. Thermoregulation in four species of arctic seabirds. J Comp Physiol B, 157: 703-708.

Gabrielsen, G.W., Mehlum, F., Karlsen, H.E., Andresen, Ø., and Parker, H., 1989. Energy cost during incubation and thermoregulation in common eider ducks (Somateria mollissima). Norsk Polarinstitutt Skrifter (in press).

Gessaman, J.A., and Findell, P.R., 1979. Energy cost of incubation in the American Kestrel. Comp Biochem Physiol, 63A:57-62.

Grant, G.S., 1984. Energy cost of incubation to the parent seabird. In: Whittow, G.C., and Rahn, H., eds. Seabird energetics. Plenum Press, New York, London, pp.59-71.

Grant, G.S., and Whittow, G.C., 1983. Metabolic cost of incubation in the Laysan Albatross and Bonin Petrel. Comp Biochem Physiol, 74A: 77-82.

Haftorn, S., and Reinertsen, R.E., 1985. The effect of temperature and clutch size on the energetic cost of incubation in free living blue tit (Parus caeruleus). Auk, 102:470-478.

Keys, A., Brozek, J., Henschel, A., Mickelsen, O., and Taylor, H.L., 1950. In: The Biology of Human starvation. Univ. of Minnesota Press, Minneapolis, Minn., vol 1, ch.23.

Korchgen, C.E., 1977. Breeding stress of female eiders in Maine. J Wildl Manage, 41:360-373.

Le Maho, Y., Vu Van Kha, H., Koubi, H., Dewasmes, G., Girard, J., Ferre, P., and Cagnard, M., 1981. Body composition, energy expenditure, and plasma metabolites in long-term fasting geese. Am J Physiol, 241: E342-E354.

May, J.D., 1978. Effect of fasting on T_3 and T_4 concentrations in chicken serum. Gen Comp Endocrinol, 34:323-327.

Milne, H., 1963. Seasonal distribution and breeding biology of the eider, Somateria mollissima L., in the north-east Scotland. Ph.D. thesis, Aberdeen University, Scotland.

Milne, H., 1976. Body weight and carcass composition of the Common Eider. Wildfowl, 27:115-122.

Norton, D.W. 1973. Ecological energetics of Calidridine sandpipers breeding in northern Alaska. Ph.D. thesis, University of Alaska, USA.

Prince, P.A., Ricketts, C., and Thomas, G., 1981. Weight loss in, incubating albatrosses and its implications for their energy and food requirements. Condor, 83:238-242.

Vleck, C.M., 1981. Energetic cost of incubation in the Zebra Finch. Condor, 83: 229-237.

Wood, R.A., Nagy, K.A., McDonald, N.S., Wakakuwa, S.T., Beckman, R.J., and Kaaz H., 1975. Determination of oxygen-18 in water contained in biological samples by charged particle activation. Anal Chem, 47:656-660.

ADAPTATIONS TO COLD IN BIRD CHICKS

Robert E. Ricklefs

Department of Biology
University of Pennsylvania
Philadelphia, Pennsylvania 19104-6018 USA

INTRODUCTION

Adult birds maintain high body temperatures in cold environments by thick insulation (West 1972, Dawson and Carey 1976, Dawson et al. 1983), huddling and communal roosting (Knorr 1957, Mackenzie 1959), use of protected microsites (Kendeigh 1961, Mayer et al. 1982), and high capacity for thermogenesis. Several studies have shown that basal metabolic rate (BMR) increases with latitude (Weathers 1979, Ellis 1984), that temperature tolerances vary seasonally (Barnett 1970) and geographically (Blem 1973) owing to increased insulation as well as thermogenic capacity (Hart 1962, Dawson and Carey 1976, Dawson et al. 1983), particularly shivering thermogenesis of the flight muscles (Hart 1962, West 1965).

Nestlings and chicks have the same options for manipulating their thermal relationships with the environment, but these options are constrained by considerations that apply uniquely to young birds. First, chicks are small compared to their parents and therefore have less favorable surface/volume ratios for heat conservation. Second, the muscles of neonates, particularly the flight muscles, generally are small, poorly developed, or both, and contribute little to the heat production of the neonate (Aulie 1976, Ricklefs 1979a). Third, enhancement of thermogenic capacity exacts unique costs on developing birds, particularly by decreasing the potential growth rate (Ricklefs 1979b). Finally, because adults and chicks form a single unit within which energy is transferred among individuals, the thermal budget of the entire family group must be optimized. Depending on the circumstances, this may favor greater or lesser thermal dependence of the chick upon its parents.

Experimental studies have demonstrated the effect of environmental temperature on growth rate in precocial birds. In young domestic fowl, growth rate declines below an air temperature of 21°C, especially in birds raised singly (Osbaldiston 1966), although direct effects of hypothermia and indirect consequences of increased heat production were not distinguished. According to Jorgensen and Blix (1985), however, slow growth in willow ptarmigan at -2°C resulted from shifting the allocation of limited energy from growth to heat production. They suggested that food processing is the rate limiting step in metabolism, which is supported by force-feeding studies on domestic fowl (Nir et al. 1974, 1978).

Ectothermy in chicks does not preclude living in cold environments. Species of passerines (e.g., snow buntings <u>Plectrophenax nivalis</u>) and pelecaniformes (e.g., blue-eyed shags <u>Phalacrocorax atriceps</u>), whose chicks have altricial development (Nice 1962, Ricklefs 1983) extend their breeding distributions to polar latitudes, although Remmert (1980) noted a preponderance of precocial species with downy chicks at high latitudes (Charadriiformes and Anseriformes in the arctic; Procellariiformes in the subantarctic). At Palmer Station, on the Antarctic Peninsula, the four largest species of breeding birds vary in thermal dependence of the chick from the altricial blue-eyed shag, which requires continuous brooding through the early part of the development period, to the precocial southern giant fulmar (<u>Macronectes giganteus</u>) and South Polar skua (<u>Stercorcarius maccormicki</u>), whose chicks are not closely attended from an early age; the Adelie penguin (<u>Pygoscelis adeliae</u>) is intermediate (Ricklefs 1982). Mode of development of these species is consistent with related species at lower latitudes, indicating the basic taxonomic conservatism of development pattern and thermal relationship between parent and chick (Ricklefs 1983). Each of these species must make other adjustments in its thermal relationship to the environment.

HEAT BALANCE OF THE CHICK

In any homeothermic system, heat gain must balance heat loss. For endotherms, this means that

$$\text{heat production} = \text{thermal conductance} \times \text{surface area} \times \text{temperature gradient } (T_b - T_a). \qquad [\text{eq.1}]$$

Typically thermal conductance is expressed on a mass basis, in which case body mass is substituted for surface area. Temperature gradient includes conduction, convection, and radiation (Bakken 1976, 1980; Gates 1980), which are components of the environmental temperature. Clearly, as the cooling power of the environment increases (T_a decreases) individuals must adjust other terms to balance equation [1], generally they must increase metabolism, reduce thermal conductance, or reduce the temperature gradient by lowering body temperature (T_b). Each of these responses to cold stress has attendant costs and limitations. Surface area is generally determined by the body size and proportions of the species and the stage of development of the chick, although large neonate size (egg size) clearly would reduce thermal problems during the initial part of the growth period. Presumably, hypothermia may reduce metabolic capacity and growth rate when thare are no compensating physiological adaptations. Increased insulation requires allocation of structure to integument and also may interfere with the transfer of heat between the parent and the chick. Enhancement of thermogenic capacity requires more precocious growth and differentiation of the heat-generating tissues, which may prolong the development period.

<u>Microenvironment</u>

Several factors influence the temperature gradient between the body and the environment: season, nest placement and structure, and body temperature. Most birds breed during the warmest period of the year, so that even in polar latitudes, chicks rarely are exposed to temperatures below -5°C. In some cases, however, the season is so short compared to the length of the development period that hatching may be advanced to the spring, or even the winter in the case of the emperor penguin (Le Maho 1977). Nest placement may greatly influence convective and radiative avenues of heat loss (e.g., Horvath 1964; Ricklefs and Hainsworth 1969; Calder 1971, 1973; Austin 1974; Walsberg and King 1978; Webb and King

1983). In general, in progressively colder climates, nests are placed to avoid wind currents, reduce night-time radiation, and increase insolation, although availability of nesting sites and presence of predators may be overriding factors.

Remmert (1980) noted that the nests of passerines "in the vicinity of the forest limits are invariably much more elaborate in construction and far better insulated than those of the same species in regions with milder temperatures." Tiainen et al. (1983) related the northern limits of *Phylloscopus* warblers to the insulative properties of the nest. In songbirds, thermal conductances of nests are on the order of 4-6 W m^{-2} $°C^{-1}$ (Skowron and Kern 1980), which is on the order of the integuments of the chicks of several precocial birds (Chappel 1980, Taylor 1986).

Hypothermia

Adult birds frequently employ hypothermia to reduce energy expenditure during the night or the nonbreeding season (Bartholomew et al. 1957; Hainsworth and Wolf 1970, 1978; Haftorn 1972; Bucher and Worthington 1982) and may be more pronounced following fasting (Koskimies 1948, Bartholomew et al. 1983). Young chicks may also reduce body temperature after fasts (Boersma 1986), although manx shearwater (*Puffinus puffinus*) chicks may maintain T_b as low as 30°C in mild temperatures (T_a = 22°C; Bech et al. 1982) and arctic charadriiform chicks regulate body temperatures at 30-35°C while actively foraging (West and Norton 1975), and as low as 25°C in the snipe (*Gallinago gallinago*; Myhre and Steen 1979) without loss of locomotor ability. These species, whose environmental temperatures may drop to the freezing point and which require extensive parental brooding undoubtedly realize substantial energy savings through hypothermia, although the mechanisms and physiological consequences have not been determined.

Chicks of most species undoubtedly can selectively cool body appendages and perhaps even the skin of the body core. For example, a 10-day-old southern giant fulmar chick maintained at 20°C and then exposed to 0°C for one hour quickly dropped the temperature of the webs of its feet to a regulated level of about 16°C and that of the axial apterium from 38 to about 32°C. At -19°C, web temperature decreased to between 6 and 10°C and that of the axial apterium to 30°C. Esophageal temperature, maintained at 38°C at an ambient temperature of 20°C, also dropped under cold stress, to 35°C at 0°C and 33°C at -19°C (Ricklefs, unpubl.). Use of partial hypothermia to balance heat budgets may also vary developmentally. For example, in the southern giant fulmar, 3-4 day-old chicks increase metabolism in response to decreased air temperature, whereas the relationship between metabolism and temperature in 7-8 day-old chicks is flat. Adelie penguin chicks show a similar pattern. In the south polar skua, 5-8 day-old chicks are effectively homeothermic down to 0°C, whereas 10-20 day-old chicks rapidly drop their body temperatures under mild cold stress, by as much as 10°C when exposed to 0°C for an hour. It is possible that the adoption of a hypothermic response coincides with the general thermal independence of the chick in nature, although we did not observe depressed body temperatures in wild-caught skua chicks of 10-20 days of age (Ricklefs 1982).

Surface Area

The surface area across which the chick loses heat can be modified by changes in body mass, shape, brood size, and huddling. Except for the last, these are not changes effected by the chick; they are primarily adaptations of the adults and presumably more related to foraging than to considerations of breeding biology. The body mass of the neonate, hence its relative ability to conserve heat compared to its ability to generate heat, depends primarily upon the adult body size and relative egg size of the species. Several groups of birds are known for laying large eggs, particularly the Procellariiformes and the Charadriiformes, which gives the neonates a considerable thermal advantage and may be partly reponsible for the success of these groups at high latitudes. I have studied rates of body cooling in 10 species of waders at Churchill, Manitoba, ranging in size from the northern phalarope (Phalaropus lobatus; neonate mass = 4.0 g) to the whimbrel (Numenius phaeopus; 31.7 g). Maintenance of body temperature under mild cold stress (ca 15°C) varied in direct relationship to neonatal mass, reflecting both greater thermal inertia and greater heat generating capacity (unpubl.). In this case, small body size may preclude affordable endothermy, and so chicks of smaller species initially have poor thermogenenic capacities, rely more heavily on parental brooding, and potentially may grow more rapidly and with greater energetic efficiency.

Huddling is an effective way of reducing heat loss, especially for large broods (Mertens 1969, Yarbrough 1970, O'Connor 1975, Dunn 1976, Clark and Balda 1981, Clark 1982, Webb and King 1983). Because of this factor, broods of passerine species often achieve effective homeothermy at moderate ambient temperatures well before they develop a metabolic response to body cooling (e.g., Clark 1982). As Mertens (1969) and O'Connor (1975) have pointed out, the addition of a chick to the brood often does not increase the food requirements of the brood in direct proportion to number, owing to the thermoregulatory advantage of the larger brood mass.

Insulation

Both seasonal and latitudinal comparisons demonstrate that modification of the insulative value of the plumage is an effective adaptation of adult birds to cold (Scholander et al. 1950, West 1972, Dawson and Carey 1983). Although Koskimies and Lahti (1964) demonstrated that neonates of the arctic nesting common eider (Somateria molissima) had low conductances, there appears to be no general relationship between cold stress and conductance among neonates. Indeed, Beck et al. (1984) noted that the thermal conductance of kittiwakes (Rissa tridactyla) is rather high compared to other gulls, including temperate-nesting species; neonatal ptarmigan (Lagopus lagopus) have similarly high conductances (Aulie and Moen 1975, Aulie 1976).

Conductance affects not only loss of heat to the surroundings but also transfer of heat from the parent to the chick. It would seem therefore that conductance might reflect the parent-offspring relationship more closely than environmental temperature, and this is born out by the data in Table 1. Petrels and waterfowl, which are generally independent of parental brooding at an early age, have low conductances relative to galliform and charadriiform chicks, which rely on protracted brooding. Lower critical temperatures, which reflect both conductance and BMR, vary from 13°C below body temperature in the prion Pachyptila, about 7°C in other petrels, 8-12°C in most ducks, 8-11°C in megapodes, 4-5°C in galliforms, and 4-10°C in most charadriiforms.

Thermogenesis

The last component of the heat budget of the chick is the capacity for heat production. Owing to difficulties of obtaining cold temperatures in field conditions, this has been measured in relatively few species (Table 1). As in the case of conductance, variation in capacity for heat production is related to the thermal independence of the chick,

Table 1. Thermal characteristics of young chicks of selected species of precocial birds.

Family and species	Age (d)	W (g)	BMR	(%)	Cond.	(%)	Max BMR	Tlc (°C)	Tmin (°C)	Ref
Procellariidae										
Pachyptila desolata	1	26.7	39.7	118	3.0	107	2.6	13.2	34.4	13
	2-3	31.7	42.3	132	3.0	119	2.5	14.1	35.3	
Puffinus puffinus	1	35	34.4	110	4.7	197	>2.7	7.3	19.4	4
Hydrobatidae										
Oceanodroma leucorhoa	1-5	6.7-13.7	50.0	115	7.0	141	2.8	7.1	20.1	14
Pelecanoididae										
Pelecanoides georgicus	1	16.7	29.9	79	4.2	114	2.2	7.1	15.7	13
P. urinatrix	1	15.5	36.1	93	5.4	141	1.2	6.7	8.0	13
	2	17.5	40.0	106	5.3	148	1.3	7.5	9.8	
Anatidae										
Anas crecca	1	16.8	36.4	96	6.0	164		6.1		12
A. penelope	1	26.4	36.8	109	4.1	146		9.0		
A. platyrhynchos	1	28.8	24.3	74	4.6	172		5.3		
Aythya ferina	1	40.1	28.9	96	3.6	164		8.0		
A. fuligula	1	34.1	28.0	89	3.6	148		7.8		
Bucephala clangula	1	32.4	36.0	113	3.7	148		9.7		
Melanitta fusca	1	54.7	31.0	112	3.0	163		10.3		
Mergus merganser	1	46.2	26.4	91	3.1	153		8.5		
M. serrator	1	44.2	29.7	101	3.1	149		9.6		
Somateria mollissima	1	61.4	25.5	95	2.2	128		11.6		
Megapodidae										
Alectura lathami	1	114.5	20.0	88	1.8	151	>2.8	11.1	>31.1	7
Leipoa ocellata	1	114	16.3	71	1.72	144	4.0	9.5	37.9	6
	1	114	16.3	71	2.2	184	4.0	7.4	29.6	
Tetraonidae										
Lagopus lagopus	1-3	17.5	33.6	89	6.7	188	1.7	5.0	8.5	2
	5-7	17			7.5	206				
	5-7	27			6.3	225				
	5-7	29			6.7	252				
Tetrao urogallus	1-3	32.5	39.7	125	5.94	239	2.8	6.7	18.7	11
	4-7	42.0	28.8	96	4.64	217	3.0	6.2	18.6	
	11-14	87.1			2.92	208	3.4			
	18-24	181.8			3.01	330	3.7			
Phasianidae										
Colinus virginianus	1	6.3	48.2	98	12.2	188	1.2	4.0	4.7	15
	3	7.3	53.2	112	13.9	233	>1.3	3.8	>5.0	
	6	9.5	73.5	166	16.4	321	1.2	4.5	5.4	
	10	11.5	57.1	136	10.1	222	>1.5	5.6	>8.4	
	14	16.4	52.2	137	7.2	193	>1.5	7.3	>10.9	
	18	21	48.6	135	5.64	176	>1.9	8.6	>16.4	

(Continued)

Table 1, continued

	Age									Ref
Laridae										
Larus atricilla	1	28.4	40.0	121	4.03	150	1.5	9.9	14.9	8
L. delawarensis	1	34.1	32.4	103	3.56	147	2.1	9.1	19.1	
L. occidentalis livens	1	65.4	27.3	103	3.02	182	1.8	9.0	16.3	
L. o. wymani	1	58.0	29.7	109	3.38	190	1.9	8.8	16.7	
Rissa tridactyla	1	33.3	26.8	85	4.97	202	1.7	5.4	9.2	5
Alcidae										
Cepphus grylle	1	35.0	29.7	95	2.38	100		12.5		9
Fratercula arctica	1	41.9	26.3	88	4.39	204	>2.1	6.0	>12.6	3
Synthliboramphus hypoleucus	1	24.8	36.2	106	9.9	340	3.5	3.7	12.8	10
	5+	c35			4.6	193				
Uria lomvia	1	74.1	25.3	99	4.36	283	1.6	5.8	9.3	1

Notes: Age 1 = day of hatching; percentages of basal metabolic rate (BMR; $J\ g^{-1}\ h^{-1}$) for adult, resting-phase nonpasserines of the same body weight (Aschoff and Pohl 1970), and of conductance ($J\ g^{-1}\ h^{-1}\ °C^{-1}$) for adult, resting-phase nonpasserines of the same body weight (Aschoff 1981); max = maximum metabolic rate under cold stress; Tlc = lower critical temperature gradient (body - ambient); Tmin = maximum sustainable temperature gradient; references (1) Aarvik and Vongraven in Bech et al. (1987), (2) Aulie (1976), Aulie and Moen (1975), (3) Bech et al. (1987), (4) Bech et al. (1982), (5) Bech et al. (1984), (6) Booth (1984), (7) Booth (1985), (8) Dawson and Bennett (1981), (9) Drent (1965), (10) Eppley (1984), (11) Hissa et al. (1983), (12) Koskimies and Lahti (1964), (13) Ricklefs and Roby (1983), (14) Ricklefs et al. (1980), (15) Spiers et al. (1985), weights from Stoddard (1931).

and strong thermogenesis by young chicks is not a prerequisite for breeding in cold environments. In most of the petrels, the thermogenic capacities of neonates vary between 2 and 3 times BMR; it is only 1.2 X BMR in the common diving petrel, which is brooded for up to two weeks. High thermogenic capacities also characterize megapodes, grouse (capercaillie Tetroa urogallus), Xantus's murrelet (Synthliboramphus hypoleucus), and probably many ducks (Koskimies and Lahti 1964). Ptarmigan (Tetraonidae), quail (Phasianidae), and most Laridae, Alcidae, and, undoubtedly, waders (Scolopacidae) (as well as altricial species) have low thermogenic capacities (generally < 2 X BMR). Conductance and thermogenic capacity together allow us to calculate the maximum temperature gradient that can be maintained between the body and the surroundings. This varies from 35-40°C in the prion and megapodes, 15-20°C in most of the petrels, the capercaillie, and most gulls, and less than 10°C in the common diving petrel, ptarmigan, quail, kittiwake, and murre.

Chicks vary in both conductance and thermogenic capacity, and there seems to be relatively little correlation between the two. Among species with low thermogenic capacity, one finds species with high conductance (kittiwake), which presumably allows active transfer of heat from the parent, and low conductance (common diving petrel Pelecanoides urinatrix), in which the parent's role must be primarily insulative. Young diving petrels are brooded continuously and presumably are never allowed to cool (Ricklefs and Roby 1983). Young kittiwake chicks may have to be rewarmed occasionally. Among species with high thermogenic capacity, megapodes and prions (Pachyptila) have low conductances, as one would expect; Xantus's murrelet and capercaillie have rather high conductances, again suggesting that active foraging in cool environments by these chicks may occasionally lead to body cooling and the need for active heat transfer from the adult to the chick.

CONCLUSIONS

The relationship between physiological characteristics of the young chick and its thermal relationship to the environment and the parent has not been thoroughly investigated, particularly for precocial species. Brooding conveys costs in terms of time, energy, and risk of predation. These may be reduced by increasing the thermogenic capacity of the chick. But such precocious development of thermogenic function also may convey costs in terms of reduced growth rate (Ricklefs 1979b). The optimum balance between the two seems to be determined primarily by the nature of the food supply. Where chicks can gather their own food, thus releasing the parent from the burden of food provisioning, high mobility and a certain level of thermal independence should develop. Differences in thermal dependence between most galliform chicks (high) and ducklings (low) suggest, however, that the compromise required for independent foraging can be achieved within broad limits. The great foraging distances of the petrels require either that the parent feed at sea or brood the chick for long periods. Because brooding halves the total foraging rate per pair, petrel chicks have apparently have been selected for very early thermal independence.

Low temperature clearly influences breeding and parent-offspring relationships among species that reproduce at high latitudes. However, the thermal characteristics of the young chick appear to be related more closely to the nature of the parent-chick interaction, often determined by the character and dispersion of the food supply, rather than by environmental temperature. Certainly cold does not preclude breeding by altricial species whose chicks depend totally on their parents for maintaining body temperature. But it is also evident that avifaunas of high latitudes contain a higher proportion of precocial species (especially Procellariiformes, Anseriformes, and Charadriiformes) than do those in hotter climates. While this selectivity may reflect food supply to some extent, it is hard to imagine that the consequences of low temperature for early chick development have not also played a role. Endothermy in young chicks varies from virtually none to near adult levels. It would be informative at this point to relate physiological capacity more closely to the thermal environment, chick activity, and the parent-offspring interaction.

ACKNOWLEDGMENTS

This work has been supported in part by grants from the National Science Foundation and the National Geographic Society. I am grateful to NATO for subvention of travel and accomodation for this symposium.

REFERENCES

Aulie, A., 1976, The pectoral muscles and the development of thermoregulation in chicks of willow ptarmigan (Lagopus lagopus), Comp. Biochem. Physiol., 53A:343.

Aulie, A., and Moen P., 1975, Metabolic thermoregulatory responses in eggs and chicks of willow ptarmigan (Lagopus lagopus), Comp. Biochem. Physiol., 51A:605.

Austin, G. T., 1974, Nesting success of the cactus wren in relation to nest orientation, Condor, 76:32.

Bakken, G. S., 1976, A heat transfer analysis of animals: unifying concepts and the application of metabolism chamber data to field ecology, J. Theor. Biol., 60:337.

Bakken, G. S., 1980, The use of standard operative temperature in the study of the thermal energetics of birds, Physiol. Zool., 53:108.

Barnett, L. B., 1970, Seasonal changes in temperature acclimatization of the house sparrow, Comp. Biochem. Physiol., 33:559.

Bartholomew, G. A., Howell, T. R., and Cade, T. J., 1957, Torpidity in the white-throated swift, anna hummingbird, and the poorwill, Condor, 59:145.

Bartholomew, G. A., Vleck, C. M., and Bucher, T. L., 1983, Energy metabolism and nocturnal hypothermia in two tropical passerine frugivores, Manacus vitellinus and Pipra mentalis, Physiol. Zool., 56:370.

Bech, C., Aarvik, F. J., and Vongraven, D., 1987, Temperature regulation in hatchling puffins (Fratercula arctica), J. Ornithol., 128:163.

Bech, C., Brent, R., Pederson, P. F., Rasmussen, J. G., and Johansen, K., 1982, Temperature regulation in chicks of the manx shearwater Puffinus puffinus, Ornis Scand., 13:206.

Bech, C., Martin, S., Brent, R., and Rasmussen, J., 1984, Thermoregulation in newly hatched black-legged kittiwakes, Condor, 86:339.

Blem, C. R., 1973, Geographic variation in the bioenergetics of the house sparrow, Ornithol. Monogr., 14:96.

Boersma, P. D., 1986, Body temperature, torpor, and growth in chicks of fork-tailed storm-petrels (Oceanodroma furcata), Physiol. Zool., 57:10.

Booth, D. T., 1984, Thermoregulation in neonate mallee fowl Leipoa ocellata, Physiol. Zool., 57:251.

Booth, D. T., 1985, Thermoregulation in neonate brush turkeys (Alectura lathami), Physiol. Zool., 58:374.

Bucher, T. L., and Worthington, A., 1982, Nocturnal hypothermia and oxygen consumption in manakins, Condor, 84:327.

Calder, W. A., 1971, Temperature relationships and nesting of calliope hummingbird, Condor, 73:314.

Calder, W. A., 1973, Microhabitat selection during nesting of hummingbirds in the Rocky Mountains, Ecology, 54:127.

Chappell, M. A., 1980, Thermal energetics of chicks of arctic-breeding shorebirds, Comp. Biochem. Physiol., 65A:311.

Clark, L., 1982, The development of effective homeothermy and endothermy by nestling starlings, Comp. Biochem. Physiol., 73A:253.

Clark, L., and Balda, R. P., 1981, The development of effective endothermy and homeothermy by nestling pinon jays, Auk, 98:615.

Dawson, W. R., and Bennett, A. F., 1981, Field and laboratory studies of the thermal relations of hatchling western gulls, Physiol. Zool., 54:155.

Dawson, W. R., and Carey, C., 1976, Seasonal acclimitization to temperature in cardueline finches. I. Insulative and metabolic adjustments, J. Comp. Physiol., 112:317.

Dawson, W. R., Marsh, R. L., Buttemer, W. A., and Carey C., 1983, Seasonal and geographic variation of cold resistance in house finches, Physiol. Zool., 56:353.

Drent, R. H., 1965, Breeding biology of the pigeon guillemot, Cepphus grylle, Ardea, 53:99.

Dunn, E. H., 1976, The relationship between brood size and age of effective homeothermy in nestling house wrens, Wilson Bull., 88:478.

Ellis, H. I., 1984, Energetics of free-ranging seabirds, in: "Seabird Energetics," G. C. Whittow and H. Rahn, eds., Plenum Press, New York.

Eppley, Z. A., 1984, Development of thermoregulatory abilities in Xantus' murrelet chicks *Synthliboramphus hypoleucus*, Physiol. Zool., 57:307.

Gates, D. M., 1980, "Biophysical Ecology," Springer-Verlag, New York.

Haftorn, S., 1972, Hypothermia of arctic tits in winter, Ornis Scand., 3:153.

Hainsworth, F. R., and Wolf, L. L., 1970, Regulation of oxygen consumption and body temperature during torpor in a hummingbird, *Eulampis jugularis*, Science, 168:368.

Hainsworth, F. R., and Wolf, L. L., 1978, The economics of temperature and torpor in nonmammalian organisms, in: "Strategies in Cold," L. Wang and J. W. Hudson, eds., Academic Press, New York.

Hart, J. S., 1962, Seasonal acclimatization in four species of small wild birds, Physiol. Zool., 35:224.

Hissa, R., Saarela, S., Rintamaki, H., Linden, H., and Hohtola, E., 1983, Energetics and development of temperature regulation in capercaillie *Tetrao urogallus*, Physiol. Zool., 56:142.

Horvath, O., 1964, Seasonal differences in rufous hummingbird nest height and their relation to nest climate, Ecology, 45:235.

Jorgensen, E., and Blix, A. S., 1985, Effects of climate and nutrition on growth and survival of willow ptarmigan chicks, Ornis Scand., 16:99.

Kendeigh, S. C., 1961, Energy of birds conserved by roosting in cavities, Wilson Bull., 73:140.

Knorr, O. A., 1957, Communal roosting of the pygmy nuthatch, Condor, 59:398.

Koskimies, J., 1948, On temperature regulation and metabolism in the swift, *Micropus a. apus* L., during fasting, Experientia, 4:274.

Koskimies, J., and Lahti, L., 1964, Cold-hardiness of the newly hatched young in relation to ecology and distribution in ten species of European ducks, Auk, 81:281.

Le Maho, Y., 1977, The emperor penguin: a strategy to live and breed in the cold, Amer. Sci., 65:680.

Mackenzie, J. M. D., 1959, Roosting of treecreepers, Bird Study, 6:8.

Mayer, L., Lustick, S., and Battersby, B., 1982, The importance of cavity roosting and hypothermia to energy balance of the winter acclimatized Carolina chickadee, Int. J. Biometeorol., 26:231.

Mertens, J. A. L., 1969, The influence of brood size on the energy metabolism and water loss of nestling great tits *Parus major major*, Ibis, 111:11.

Myhre, K., and Steen, J. B., 1979, Body temperature and aspects of behavioural temperature regulation in some neonate subarctic and arctic birds, Ornis Scand., 10:1.

Nice, M. M., 1962, Development of behavior in precocial birds, Trans. Linn. Soc. N. Y., 8:1.

Nir, I., Shapira, N., Nitsan, A., and Dror, Y., 1974, Force-feeding effects on growth, carcass and blood composition in the young chick, Br. J. Nutr., 32:229.

Nir, I., Nitsan, Z., Dror, Y., and Shapira, N., 1978, Influence of overfeeding on growth, obesity and intestinal tract in young chicks of light and heavy breeds, Br. J. Nutr., 39:27.

O'Connor, R. J., 1975, The influence of brood size upon metabolic rate and body temperature in nestling blue tits *Parus caeruleus* and house sparrows *Passer domesticus*, J. Zool., 175:391.

Osbaldiston, G. W., 1966, The response of the immature chicken to ambient temperature, in: "Physiology of the Domestic Fowl," C. Horton-Smith and E. C. Amoroso, eds., Oliver & Boyd, Edinburgh and London.

Remmert, H., 1980, "Arctic Animal Ecology," Springer-Verlag, Berlin.

Ricklefs, R. E., 1979a, Patterns of growth in birds. V. A comparative study of development in the starling, common tern, and japanese quail, Auk, 96:10.

Ricklefs, R. E., 1979b, Adaptation, constraint, and compromise in avian postnatal development, Biol. Rev., 54:269.

Ricklefs, R. E., 1982, Development of homeothermy in antarctic seabirds, *Antarctic J. 1982 Rev.*:177.
Ricklefs, R. E., 1983, Avian postnatal development, *Avian Biol.*, 7:1.
Ricklefs, R. E., and Hainsworth, F. R., 1969, Temperature regulation in nestling cactus wrens: the nest environment, *Condor*, 71:32.
Ricklefs, R. E., and Roby, D. D., 1983, Development of homeothermy in the diving petrels *Pelecanoides urinatrix exsul* and *P. georgicus*, and the antarctic prion *Pachyptila desolata*, *Comp. Biochem. Physiol.*, 75A:307.
Ricklefs, R. E., White, S. C., and Cullen, J., 1980, Energetics of postnatal growth in Leach's storm-petrel, *Auk*, 97:566.
Scholander, P. F., Walters, V., Hock, R., and Irving, L., 1950, Body insulation of some arctic and tropical mammals and birds, *Biol. Bull.*, 99:225.
Skowron, C., and Kern, M., 1980, The insulation in nests of selected North American songbirds, *Auk*, 97:816.
Spiers, D. E., Adams, T., and Ringer, R. K., 1985, Homeothermic development in the bobwhite (*Colinus virginianus*), *Comp. Biochem. Physiol.*, 81A:921.
Stoddard, H. L., 1931, "The Bobwhite Quail," Charles Scribner's Sons, New York.
Taylor, J. R. E., 1986, Thermal insulation of the down and feathers of pygoscelid penguin chicks and the unique properties of penguin feathers, *Auk*, 103:160.
Tiainen, J., Hanski, I. K., and Mehtala, J., 1983, Insulation of nests and the northern limits of three *Phylloscopus* warblers in Finland. *Ornis Scand.*, 14:149.
Walsberg, G. E., and King, J. R., 1978, The energetic consequences of incubation for two passerine species, *Auk*, 95:644.
Weathers, W. W., 1979, Climate adaptation in avian standard metabolic rate, *Oecologia*, 42:81.
Webb, D. R., and King, J. R., 1983, An analysis of the heat budgets of the eggs and nest of the white-crowned sparrow, *Zonotrichia leucophrys*, in relation to parental attentiveness, *Physiol. Zool.*, 56:493.
Webb, D. R., and King, J. R., 1983, Heat transfer relations of avian nestlings, *J. Thermal Biol.*, 8:301.
West, G. C., 1965, Shivering and heat production in wild birds, *Physiol. Zool.*, 38:111.
West, G. C., 1972, Seasonal differences in resting metabolic rate of Alaskan ptarmigan, *Comp. Biochem. Physiol.*, 42A:867.
West, G. C., and Norton, D. W., 1975, Metabolic adaptations of tundra birds, *in*: "Physiological Adaptations to the Environment," F. J. Vernberg, ed., Intext Educ. Publ., New York.
Yarbrough, C. G., 1970, The development of endothermy in nestling gray-crowned rosy finches, *Leucosticte tephrocotis griseonucha*, *Comp. Biochem. Physiol.*, 34:917.

ENERGY PARTITIONING IN ARCTIC TERN CHICKS (Sterna paradisaea)
AND POSSIBLE METABOLIC ADAPTATIONS IN HIGH LATITUDE CHICKS

Marcel Klaassen[1,2], Claus Bech[2], Dirkjan Masman[1,3] and Guri Slagsvold[2]

[1]Zoological Laboratory, University of Groningen
P.O. Box 14, 9750 AA Haren, The Netherlands
[2]Department of Zoology, University of Trondheim
N-7055 Dragvoll, Norway
[3]Laboratory of Isotope Physics
University of Groningen
Westersingel 34, 9718 CM Groningen, The Netherlands

INTRODUCTION

In order to survive and grow, neonates need to remain homeothermic. However, in the arctic, which is mostly regarded as a harsh environment with prevailing low ambient temperatures, achievement of homeothermy for chicks might cause problems. The small neonates have, besides an unfavorable volume area ratio, a less well developed plumage than adult birds. Therefore one might hypothesize that if no special cold adaptations have evolved, total energy expenditure of free living chicks is dominated by the costs of thermoregulation. Many physiologists working in extreme environments have focused on the ability of chicks to cope with low environmental temperatures (e.g. Maher ,1964; Norton, 1973; Aulie and Steen, 1976; Boggs et al., 1977; Pedersen and Steen, 1979; Bech et al., 1984; Jørgensen and Blix, 1985; Taylor, 1985; Boersma, 1986). However, to evaluate the importance of any adaptation to cold one needs to measure the actual contribution of thermoregulatory expenses to the total requirements of free living chicks. So far precise quantifications of the thermoregulatory costs in free living chicks in polar environments are only available for arctic tern chicks (Sterna paradisaea), studied on Spitsbergen (79 °N, 12 °W; Klaassen et al., 1989a,b), which are summarized here, after an analysis of possible metabolic adaptations in chicks to climatic conditions in general.

METABOLIC ADAPTATIONS TO COLD

Many adaptations to cold, behavioral, morphological and physiological, are possible (for overview see: Lustick, 1984), of which many are discussed in these proceedings. Most cold adaptations involve enlarged size and/or function of certain tissues. For instance increased muscle function capacity is necessary for a high thermogenetic shivering capacity which is according to current thinking the main heat generating mechanism in birds (Calder and King, 1974). Tissues such as liver and kidneys are indispensable in the support of muscle activity and it is likely that an increase of muscle function is associated with an increase in the supporting tissues. These hypothesized physiological changes, with associated morphological adjustments, in cold acclimatized birds, must go hand in hand with predictable shifts in the basal metabolic rate (BMR) of the animal. Recent work substantiates the parallelism between organ mass changes and BMR in experiments on the european kestrel (<u>Falco tinnunculus</u>; Daan et al., 1989). In our view, measuring the basal metabolic rate in chicks (BMR_c) is a useful first step towards unravelling cold adaptation mechanisms in chicks, since basal metabolic rate can be taken to reflect thermoregulatory capacity. Support for the hypothesized relation between thermoregulatory capacity and BMR_c, is given by the findings that BMR in adult birds seems to be geared to daily energy expenditure (King, 1974; Drent et al., 1978) and maximum sustained working level (Drent and Daan, 1980).

For the decision whether a measured basal metabolism is relatively high or low, the possibility of comparison is required. For adult birds allometric equations are available to calculate the expected BMR which can be used in comparative analyses (e.g. Lasiewski and Dawson, 1967; Aschoff and Pohl, 1970). However, for chicks such allometric equations are not available. Using allometric equations compiled on adult birds is a doubtful procedure to follow, as neonates are totally different in composition from adult birds. The composition of chicks is continuously changing from hatching to adulthood and so their metabolism. Thus each developmental stage needs its own allometric equation for the comparison proposed. Here we restrict ourselves to one clearly distinct phase in the development from zygote to adult, and analyze the basal metabolism of hatchlings only (BMR_h). The residuals from the allometric equation for BMR_h (Fig. 1) are not associated with developmental state of the chicks at hatching according to Nice (1962; Klaassen, in prep.). Thus the lumping of all chick data from the literature, irrespective of their developmental stage at hatching, seems to be a valid procedure to follow.

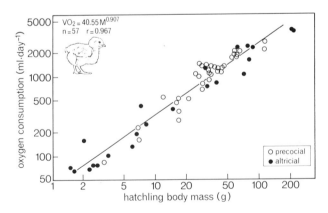

Fig. 1. Hatchling basal metabolism in relation with fresh body mass of the hatchlings.

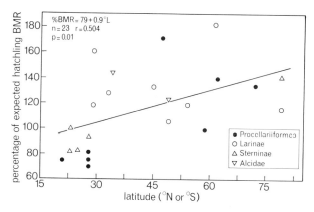

Fig. 2. Residual analysis of the relation presented in Fig. 6. with latitude as the dependent variable.

In line with the analysis of Ellis (1984) for adult seabirds, we made a similar residual analysis with latitude for seabird chicks. The plot of BMR_h residuals with latitude (Fig. 2) has a significant positive correlation. Thus chicks from high latitudes have a BMR_h elevated over those from low latitudes for which adaptations to cold might be hold responsible.

Also in adult birds there has been found a positive correlation between latitude and BMR (Weathers, 1979; Hails, 1983; Ellis, 1984). This correlation has been explained by regarding a reduced endogenous heat production in the tropics (Weathers, 1979; Ellis, 1984), and an elevated power output capacity at high latitudes (Ellis, 1984; Kersten and Piersma, 1987) as advantageous. The reduced endogenous heat production is in this view regarded as an adaptation towards a reduced chance on heat stress. In the arctic where weather conditions can be harsh and the breeding seasons are short, a possible strategy is to have the capability of a high energy expenditure capacity. The basal metabolism, which is a summation of the specific metabolism of all tissues in the animal, is accordingly high in birds which have a high maximum work output (Ellis, 1984; Kersten and Piersma, 1987).

Nevertheless, the variation found in the residual analysis of BMR_c with latitude (Fig. 2) is considerable and encourages the addition that we must not oversimplify the problem and should be wary of single-cause explanations. Many other ecological constraints than cold can act upon the physiology, morphology and thus metabolism of animals (e.g. Drent and Klaassen, McNab, this issue).

THERMOREGULATORY COSTS FOR ARCTIC TERN CHICKS

One easily focuses on ambient temperature when forming ideas about the impact of weather conditions on thermoregulatory costs. Ambient temperature on Spitsbergen is quite stable around 5 °C during summer. However, the operative temperature (Te), i.e. the temperature actually experienced by an animal, can deviate considerably under influence of humidity, wind, solar radiation, and precipitation. To measure operative temperature encountered by the tern chicks we measured Te using a tin cast covered with the skin of a one day old arctic tern chick. This model was placed in a typical habitat of arctic terns and Te was recorded continuously during a period of 6 weeks in the chick rearing period. Dividing the Te measurements in two groups using cloud cover as the discriminant, reveals that on clear days Te can be as much as 20 °C higher than ambient temperature (Fig. 3). This result fits in observations of panting birds at low ambient temperatures under intense solar radiation (e.g. Stonehouse, 1967; Lustick et al., 1978). Although it is clear that solar radiation can contribute to the heat budget of tern chicks and the sun indeed does not disappear beyond the horizon during the breeding season, high Te levels are only reached occasionally. First of all the summer climate of Spitsbergen is not characterized by many clear days. Furthermore the angle between sun and horizon determines the amount of shade and is very important for the possible increase of Te due to solar radiation. Nevertheless the data show that operative temperatue is consistently elevated over ambient temperature.

Measurements of thermal conductance and basal metabolic rate in arctic tern chicks during the course of development (Klaassen et al., 1989b) and the observed operative temperatures were used to calculate the costs for thermoregulation. The estimated thermoregulatory costs as a function of age are presented in Fig. 4. Mass specific thermoregulatory costs are high just after hatching and rapidly decrease afterwards, reaching a fairly stable level at about 10 days of age. Thermoregulatory costs expressed as multiples of BMR_c are 3.3 at hatch and between 0.4 and 0.7 from day 7 onwards. However, so far we did not take parental brooding into account. Busse (1983) reports for arctic terns at temperate latitudes that they nearly continuously brood their chicks during the first week after hatching and may still brood or shelter them the two subsequent weeks. The thermoregulatory costs after correction for the savings due to brooding,

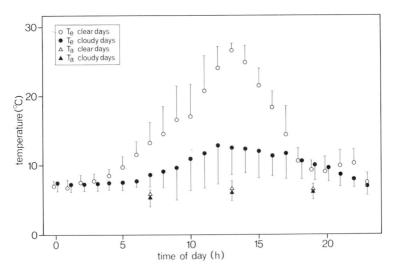

Fig. 3. Daily course of ambient and operative temperature ($\pm$ SD) for clear (cloud cover $\leq$ 4/8) and cloudy (cloud cover $\geq$ 4/8) days, in an arctic tern colony on Spitsbergen during the breeding season.

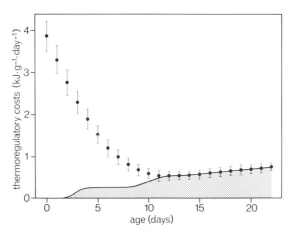

Fig. 4. Thermoregulatory costs in arctic tern chicks as a function of age, calculated from metabolic chamber and operative temperature estimates. Indicated variation ($= \pm$ SD) expresses variation in measured operative temperatures. Shaded area represents the estimated actual thermoregulatory costs after accounting for parental brooding (see text).

of which the derivation will be explained below, are indicated by the shaded area in Fig. 4.

We probably overestimated the thermoregulatory costs, also when taking parental brooding into account. The taxidermic model was immobile whereas we might expect real chicks to be continuously altering their position, when not brooded, to optimize not only the protection for possible predators but also their heat balance.

CONTRIBUTION OF THERMOREGULATORY COSTS TO TOTAL ENERGY EXPENDITURE

We used the doubly labeled water method (Lifson and McClintock, 1966; Nagy, 1980) to measure total energy expenditure (Edlw) in free living arctic tern chicks. The summation of total energy expenditure and energy accumulated in tissue (Etis), which was estimated by carcass analyses, resulted in an estimate of the total energy requirements (Ereq). Total energy expenditure was subdivided in costs for basal metabolism, tissue synthesis (Esyn), activity (Eact) and thermoregulation (Eth). Esyn was calculated assuming a synthesis efficiency of 75 % (Ricklefs, 1974). The estimated thermoregulatory costs, based on Te and laboratory measurements, were added to BMR_c and Esyn. This summation yielded values exceeding Edlw during the first 11 days of age, and we assumed this excess to represent the energy saved by parental brooding (Fig. 4). After the 11th day the summation of BMR_c, Esyn and Eth was lower than Edlw, and we assumed the remainder in energy expenditure to represent energy allocated to activity. It is clear that any overestimate of Eth thus results in an underestimate of Eact. Nevertheless, aside from this

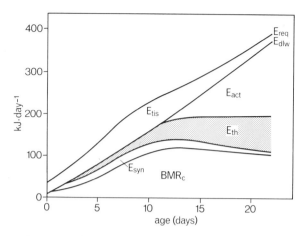

Fig. 5. Total energy requirements (Ereq) of free living arctic tern chicks on Spitsbergen, from hatching until fledging. Total energy expenditure (Edlw) is partitioned in BMR_c, tissue synthesis (Esyn), thermoregulation (Eth), and costs for activity (Eact). The accumulation of body tissue (Etis) completes the total energy requirements.

anomaly a precise energy budget for arctic tern chicks on Spitsbergen could be compiled (Fig. 5).

From the energy budget presented it was concluded that thermoregulatory costs never exceeded 21 % of the total energy requirements, the mean amounting to 16 % of Ereq and, as explained above, these values should still be regarded as a maximum. The low costs for thermoregulation are probably mainly due to the subsidy of parental brooding. The energy saved over the first 11 days of life due to brooding were estimated as 26 % of Ereq during the same 11-day period. Evaluation of the thermoregulatory costs for free-living arctic tern chicks should ideally be done by comparison of the energy budgets compiled on the same species both at arctic and temperate latitudes. Alas data on arctic tern chicks at temperate latitudes are not available. We therefore compared our data with data of common tern chicks (<u>Sterna hirundo</u>) studied by Ricklefs and White (1981) on Great Gull Island, New York (41 °N, 74 °W). During the chick rearing period on Great Gull Island, ambient temperatures were approximately 17 and 25 °C for night and day respectively. Unfortunately the activity costs were not estimated for the common tern, allowing only a partial comparison of the energy budgets of both tern species. Nevertheless, the strikingly small differences between these energy budgets suggests on average a minimal effect of low ambient temperatures, on the energy requirements of arctic tern chicks (Fig. 6).

Now the question arises whether the strong resemblance in both energy budgets under these different climatic conditions is brought about by parental behaviour (brooding: Fig. 2), chick behaviour or any physical constitution in the arctic tern chicks, as adaptations to the colder climate. Unfortunately for the moment it is impossible to throw more light on this dilemma as measurements of for instance thermal conductance are only

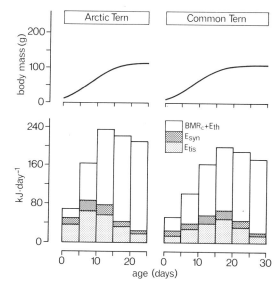

Fig. 6. Comparison of the growth curves and energy budgets of arctic and common tern chicks. The energy budgets are compiled out of costs for tissue accumulation (E_{tis}), tissue synthesis (E_{syn}), and basal metabolism plus thermoregulation ($BMR_c + E_{th}$).

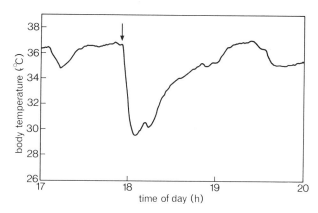

Fig. 7. Body temperature registration of a 3 day old arctic tern chick measured with a 0.85 g temperature transmitter implanted in the abdominal cavity. The swallowing of an approximately 10 cm long polar cod (Boreogadus saida) resulted in a body temperature decrease of 8 °C.

available for arctic terns (Klaassen et al., 1989b). As a further consequence the suggested metabolic adaptation of a high basal metabolic rate in chicks compensating for low temperatures can not yet directly be supported by a comparative analysis.

One should keep in mind that, although the average daily costs of thermoregulation can be low, the chicks can incidentally be exposed to extreme cold. Such is the case for instance when the parents desert the chick to chase a potential predator or when chicks swallow large prey items. All prey is caught in polar waters, having a water temperature of approximately 2 °C and transported at ambient temperatures of only 5 °C. Hence prey ingestion can lead to extreme cooling by up to a decrease of 8 °C in body temperature of the chick (Slagsvold et al., in prep.; Fig. 7). Thus, although total daily thermoregulatory costs are low, possibly due mainly to behavioural adaptations in the parents, this finding does not exclude the adaptive value of physiological cold adaptations.

ACKNOWLEDGMENTS

This study was supported by grants from the Dutch Foundation for Arctic Research in the Natural Sciences, the Nansen Foundation (75/87) and the Norwegian Polar research Institute (30/86). Rudi Drent kindly commented on the text.

LITERATURE CITED

Aschoff, J., and Pohl, H., 1970, Der Ruheumsatz von Vögeln als funktion der Tageszeit und der Körpergrösse, J. Orn. 111:38.
Aulie, A., and Steen, J. B., 1976, Thermoregulation and muscular development in cold exposed willow ptarmigan chicks (Lagopus lagopus L.), Comp. Biochem. Physiol., 55A:291.
Bech, C., Martini, S., Brent, R., and Rasmussen, J., 1984, Thermoregulation in newly hatched black-legged kittiwakes, Condor, 86:339.
Boersma, P. D., 1986, Body temperature, torpor, and growth in chicks of fork-tailed storm-petrels (Oceanodroma furcata), Physiol. Zool., 59:10.
Busse, K., 1983, Untersuchungen zum Ehe- und Sozialleben der Küstenseeschwalbe (Sterna paradisaea PONT.) mit besonderer berüksichtigung des lang zeitlichen Wandels der individuellen Beziehungen, Ecol. Birds 5:73.
Calder, W. A., Jr., and King, J. R., 1974, Thermal and caloric relations of Birds, in: "Avian Biology, Vol. IV," D. S. Farner, and J. R. King, eds., Academic Press, New York and London.
Daan, S., Masman, D., Strijkstra, A., and Verhulst, S., 1989, Intraspecific allometry of basal metabolic rate: relations with body size, temperature, composition and circadian phase in the kestrel, Falco tinnunculus:, J. Biol. Rhythms., (in press).
Drent, R. H., Ebbinge, B., and Weijand, B., 1978, Balancing the energy budgets of arctic-breeding geese throughout the annual cycle: a progress report, Verh. orn. Ges. Bayern 23:239.
Drent, R. H., and Daan, S., 1980, The prudent parent: energetic adjustments in avian breeding, Ardea, 68:225.
Ellis, H. I., 1984, Energetics of free ranging seabirds, in:

"Seabird Energetics," G. C. Whittow, and H. Rahn, eds., Plenum Press, New York and London.

Hails, C. J., 1983, The metabolic rate of tropical birds, Condor 85:61.

Jørgensen, E., and Blix, A. S., 1985, Is the rate of body cooling in cold exposed neonatal willow ptarmigan chicks a regulated process ?, Acta Physiol. Scand. 124:404.

Kersten, M., and Piersma, T., 1987, High levels of energy expenditure in shorebirds; metabolic adaptations to an energetically expensive way of life, Ardea, 75:175.

King, J. R., 1974, Seasonal allocation of time and energy resources in birds, in: "Avian Energetics," R. A. Paynter, Jr., ed., Cambridge, Massachusetts, Nuttall Ornithol. Club.

Klaassen, M., Bech, C., Masman, D., and Slagsvold, G., 1989a, Growth and energetics of arctic tern chicks (Sterna paradisaea), Auk, 106:(in press).

Klaassen, M., Bech, C., and Slagsvold, G., 1989b, Basal metabolic rate and thermal conductance in arctic tern chicks and the effect of heat increment of feeding on thermoregulatory expenses, (in prep.).

Lasiewski, R. C., and Dawson, W. R., 1967, A reexamination of the relation between standard metabolic rate and body weight in birds, Condor 69:13.

Lifson, N., and McClintock, R., 1966, Theory of use of the turnover rates of body water for measuring energy and material balance, J. Theor. Biol., 12:46.

Lustick, S., Battersby, B., and Kelty, M., 1978, Behavioral thermoregulation: orientation toward the sun in herring gulls, Science, 200:81.

Lustick, S., 1984, Thermoregulation in adult seabirds, in: "Seabird Energetics," G. C. Whittow, and H. Rahn, eds., Plenum Press, New York and London.

Maher, W. J., 1964, Growth rate and development of endothermy in the snow bunting (Plectrophenax nivalis) and lapland longspur (Calcarius lapponicus) at Barrow Alaska, Ecology, 45:520.

Nagy, K. A., 1980, CO_2 production in animals: analysis of potential errors in the doubly labeled water method, Amer. J. Physiol., 238:R466.

Nice, M. M. 1962, Development of behavior in precocial birds, Trans. Linn. Soc. New York 8:1.

Norton, D. W., 1973, Ecological energetics of calidridine sandpipers breeding in Northern Alaska, Diss. Univ. Alaska, Fairbanks Alaska.

Pedersen, H. C., and Steen, J. B., 1979, Behavioural thermoregulation in willow ptarmigan chicks (Lagopus lagopus), Ornis Scand., 10:17.

Ricklefs, R. E., 1974, Energetics of reproduction in birds, in: "Avian Energetics," R. A. Paynter, Jr., ed., Cambridge, Massachusetts, Nuttall Ornithol. Club.

Ricklefs, R. E., and White, S. C., 1981, Growth and energetics of chicks of the sooty tern (Sterna fuscata) and common tern (S. hirundo), Auk, 98:361.

Stonehouse, B., 1967, The general biology and thermal balance of penguins, in: "Advances in Ecological Research," J. B. Cragg, ed., Vol. 4, Academic Press, London.

Taylor, J. R. E., 1985, Ontogeny of thermoregulation and energy metabolism in pygoscelid penguin chicks, J. Comp. Physiol., 155B:615.

Weathers, W. W., 1979, Climatic adaptations in avian standard metabolic rate, Oecologia (Berlin), 42:81.

ENERGETICS OF AVIAN GROWTH:

THE CAUSAL LINK WITH BMR AND METABOLIC SCOPE*

Rudolf Drent[1] and Marcel Klaassen[1,2]

[1]Zoological Laboratory, University of Groningen
P.O. Box 14, 9750 AA Haren, The Netherlands
[2]Research Institute for Nature Management (RIN)
P.O. Box 9201, 6800 HB Arnhem

INTRODUCTION

Our point of departure is Lack's (1968) viewpoint that avian growth rate is one of the parameters adjusted in the course of evolution to help match the needs of the brood to the foraging ability of the parents in nidicolous species. A second selective pressure, valid especially for nidifugous species, is to minimize the period of heightened predation risk when the chicks are small and cannot yet fly. Furthermore, Lack reasoned that the seasonal timing of growth required synchrony, between the period of maximal demand of the growing young and the period of greatest food abundance. A further selective pressure for rapid growth rate might be expected in strongly seasonal environments where it is imperative to complete development before the onset of unfavourable conditions. These considerations are based on the premise that changes in growth rate bring about large changes in the daily ration required to raise the chick, i.e. the energetic consequences of alteration of growth rate loom large in the daily energy budget.

There is unfortunately little direct information on the relationship between growth rate and energy budget with which to test these ideas. In his review Ricklefs (1983) concluded, perhaps surprisingly, that the energy savings of slow growth may not in fact be substantial. Ricklefs' crown witness is his comparative study (Ricklefs and White 1981) on two related tern species, the temperate common tern, Sterna hirundo, with a brood of two to three chicks experiencing a relatively rapid growth rate, as contrasted to the tropical sooty tern, Sterna fuscata, whose single chick shows a markedly retarded growth rate (the fledging periods are 30 and 60 days respectively). Although the energy budgets of growing chicks of these species differed in a number of points, preliminary calculations suggested that allowing the sooty tern a more rapid growth rate would have only a minor effect on the daily food requirements (a doubling of the growth rate would increase maximal demand by only 20 %, and the total energy input by only 5 % over the entire nestling period).

---

*Dedicated in friendship to Robert Ricklefs for his stimulation.

Evidence from Ricklefs and White's (1981) growth studies of terns thus argue against an energy limitation in setting growth rates. Their calculations on the repercussions of growth rate on the energy requirement of the sooty tern chick depend however on the assumption that the maintenance cost is unaffected (indeed for simplicity a mass-independent maintenance cost was assumed). As described in the contribution by Klaassen et al. (this volume) these preliminary budgets disregarded the cost of activity. Maintenance as employed by Ricklefs and White (1981) embraces basal metabolism and cost of thermoregulation. Manipulation of growth rate in their treatment thus involves only the other two components of the budget namely energy deposited in the body tissues plus cost of synthesis (an efficiency of 75 % was assumed). The main problem to be tackled here is to investigate whether basal metabolism is indeed independent of growth rate.

GROWTH STUDIES IN GULLS AND TERNS

There are at present too few studies describing basal metabolism throughout nestling life (BMR_c) to enable sophisticated approaches to this problem, but there are sufficient measurements of hatchling BMR_c to enable a first step. There is a difficulty in defining BMR_c in an actively growing animal that may rarely if at all be in a truly post-absorptive state (Klaassen et al. 1987), but in keeping with common usage by BMR_c we mean minimal values of oxygen consumption obtained when chicks are at rest. An overall relationship relating BMR_c to hatchling body mass has been described by Klaassen et al. (this volume) based on 57 species and this enables us to evaluate metabolic rate of a given species in relation to expectation.

For growth rates we relied on the pioneering analyses of Ricklefs (1968, 1973) who demonstrated the tight relationship

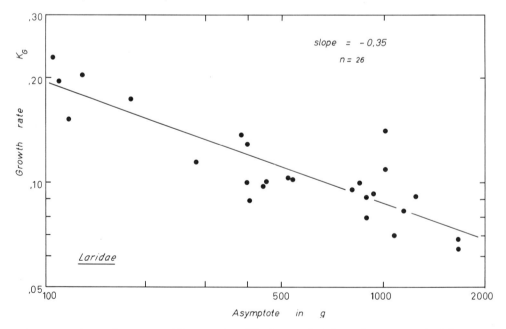

Fig. 1. Growth rate (Gompertz K) in relation to asymptotic weight in gulls and terns (double log plot). Species with broodsize one omitted.

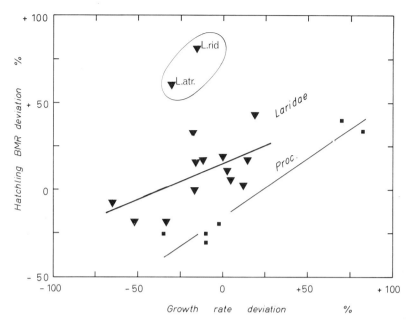

Fig. 2. Residual analysis of hatchling BMR$_c$ (deviation from allometric equation for 57 species) in relation to growth rate (deviation with respect to group specific expectation as in Fig. 1) for gulls and terns (<u>Laridae</u>, triangles) and <u>Procellariiformes</u> (squares). Data sources: Appendix.

between adult body mass (the growth asymptote) and growth rate, larger birds growing more slowly. Again, these interspecies relations allow us to put the individual species in context, and define retarded, normal, or enhanced growth rate in relation to eventual adult size. For example, Ricklefs (1973) analysed growth rate (employing the Gompertz growth constant) in gulls and terns and defined the normal pattern for the group on the basis of data on 16 species with a brood of two or more, and found that three additional species with reduced (singleton) broods indeed showed strongly retarded growth rates. When we repeated this analysis on the basis of the 26 species for which data are now at hand, we found (Fig. 1) the Gompertz growth constant to depend on asymptotic weight according to the formula:

$$K_G = 10^{-0.024} A^{-0.345}$$

(A = asymptote in g, r=-0.87, p<0.01), virtually indistinguishable from Ricklefs' original result (he found a slope of -0.42).

For 15 members of the <u>Laridae</u> (gulls and terns) hatchling BMR$_c$ has been measured and residual analysis of deviation from expected BMR$_c$ (the 57 species line taken as the standard) as against deviation in growth rate (the 26 species line for gulls and terns presented above) is given in Fig. 2. A correlation is evident, more rapid growth going hand in hand with an increased BMR$_c$. There are two embarrassing points however (encircled): <u>Larus</u> <u>ridibundus</u> and <u>L. atricilla</u> both have a much higher BMR$_c$ than related species with the same relative growth rates. This aberration was earlier commented upon by Dawson et al. (1976) in their compilation of hatchling BMR_c in relation to (hatchling) body mass. One ecological correlate comes to mind: both species traditionally nest in unstable habitats and the young are mobile at a relatively early age. It would be worth examining whether the leg muscles are relatively larger than in other gulls at that page. Provisionally, we conclude that with the exception of these two hooded gulls the other 13 species tend to support the contention that growth rate and BMR_c covary (Y=14.00+0.44X, r=0.63, p=0.02).

GROWTH STUDIES IN OTHER BIRDS

A second group for which a reasonable sample of hatchling BMR_c is available is the Procellariiformes. All of the birds in this group lay but a single egg, and we have assembled growth data on eight species (from Ricklefs 1973) ranging from Oceanodroma (75 g asymptote) to Diomedea (3200 g) to define the relation between growth rate (Gompertz K) and the asymptote ($K_G=10^{-0.419}A^{-0.332}$, r=-0.90, p<0.01). The six hatchling BMR_c data points show (Fig. 2) a significant relation to growth rate in the residual analyses plot (Y=-9.02+0.72X, r=0.97, p<0.01).

Of a special interest is a third group, Charadriiformes (waders or shorebirds), where in some species (Haematopus, the Oystercatchers) the young are fed by the parents but in the majority the chicks must find all of their food themselves, although guided and brooded by the parents (Scolopacidae and Charadriidae). Analysis of the Gompertz growth constant in relation to the asymptotic weight has recently been carried out for this group by Visser and Beintema (1989). On the basis of 16 species a predictive formula for shorebirds has been assembled for the first time (Ricklefs could only draw on seven species for his 1973 compilation). For six species hatchling BMR_c has been measured and these data (Fig. 3) fit the pattern for the other groups (Y=2.80+0.80X, r=0.85, p=0.03).

These correlations imply a link between basal metabolic rate and growth rate. Ricklefs (1968) commented on the similarity of the exponent (0.72) linking absolute rate of growth (g/day) to the asymptote (drawing on data from temperate passerines and raptors) to that relating basal metabolism to body mass in the adult bird. This correspondence in the exponent led O'Connor (1984) to suggest that "the ability of the nestlings to gain further weight is proportional to their species-specific metabolism" and this conclusion is supported by our present analyses.

Although we have restricted ourselves to growth rates of the chick, a close correlation between growth in the egg and the post-hatching period has been presupposed in the older literature (see Drent 1975). This implies that the fit could equally have been demonstrated taking hatchling BMR_c and looking back to growth rate in the egg. More recently, Ricklefs (1987) has cast doubt on the validity of the assumption that post-natal growth rate is closely linked to embryonic growth rate. However, providing one restricts the comparison to related forms, the correlation between growth rates in the two phases remains fairly tight. Ricklefs (1987) refers to Nelson's (1978) compilation on the genus Sula as a possible exception to this generality, but extracting the relevant data from Nelson's treatise confirms that chick growth (estimated as asymptote divided by time needed to reach this weight) is closely correlated to embryonic growth (egg mass divided by incubation period, r=0.94, p<0.01, n=6). This

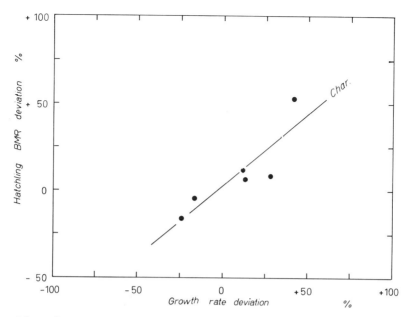

Fig. 3. Residual analysis of hatchling BMR_c (deviation from allometric equation for 57 species) in relation to growth rate deviation for waders (see text). BMR_c values from Visser and Beintema (1989) and Klaassen (in prep.).

point seems to have escaped notice on account of the large differences in egg mass within the genus, precluding a simple comparison on the basis of incubation period alone.

GROWTH AND CHICK METABOLISM BEYOND HATCH

Taking the oystercatcher, Haematopus ostralegus, as an example (Fig. 4), we see that gross energy intake (GEI) rises steeply during growth, averaging 2 x BMR_c and sometimes reaching 3 x BMR_c. Kirkwood (1983) pointed out that maximal ME values of birds and mammals during rapid growth, refeeding after fasting, or during extreme cold exposure followed the following expression:

$$ME_{max} = 1712 \, Kg^{0.72} \quad kJ/day.$$

For the average non-passerine this is equivalent to 4 x BMR.

BMR when expressed relative to the expected value offers a means of scaling the metabolic potential of the animal concerned, reflecting as it does such functions as protein turnover, gluconeogenesis, and as we believe potential rates of tissue synthesis. A case can be made for considering relative BMR as indicative of metabolic scope (the simplest formulation commensurate with the data is to express the values as multiples of BMR rather than absolute values). Support for this view comes from the examination of changes in starving animals. The fall in BMR of kestrels, Falco tinnunculus, kept on short rations, is matched by a decrease in the relative size of kidney and liver (which both contribute importantly to BMR). Reduction of these

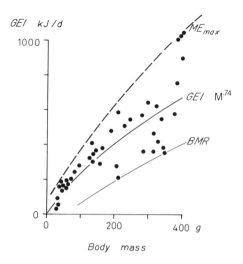

Fig. 4. Chick metabolism in the oystercatcher: gross energy intake (GEI) in relation to body mass. The points refer to individual measurements of GEI over a 24 hour period. For comparison the empirical level of BMR_c is shown, as well as the expected maximal GEI according to Kirkwood (1983, broken line). From Klaassen in prep.

organs under conditions of restricted food reflect a rescaling of the metabolic machinery of the body and of course entail a considerable savings in daily energy expenditure (Daan et al. 1989). This problem is further touched upon by Dawson (this volume) but if we extend our horizons to include the mammals the resolution of an old paradox can now be offered. For at least 80 years it has been known that marsupial tend to show lower metabolic rates than eutherian mammals. The most satisfactory explanation and one which puts the marsupials and eutherians on one line is to relate BMR to protein turnover as measured in liver slices. As discussed fully by Hume (1982), the lower BMR of marsupials affords economy in daily cost of free existence at the expense of less rapid tissue synthesis and rates of energy storage.

For three tern species (<u>Sterna</u> <u>sandvicensis</u>, <u>S. hirundo</u>, and <u>S. paradisaea</u>) extensive measurements of BMR_c as well as total metabolisable energy (ME) have been collected during the past summer spanning the entire growth period. When these data are displayed in relation to body mass (Fig. 5) it will be noted that the cloud of points describing individual determinations of ME are enclosed by the limits of 1 x BMR_c (the few points below this line must represent periods of stagnation of growth) and at the upper end 4 to 5 times BMR_c (the 4 x BMR_c line expected from the Kirkwood, 1983, equation is included for reference). ME most closely approaches this upper level in terms of BMR_c in the region of the inflection point of the growth curve (at body mass of 80, 55, and 45 g, resp.) when absolute gain (g/day) peaks.

The contribution of BMR_c to total ME throughout nestling growth can be computed for the four charadriiform species considered, to which can be added the dunlin, <u>Calidris</u> <u>alpina</u>,

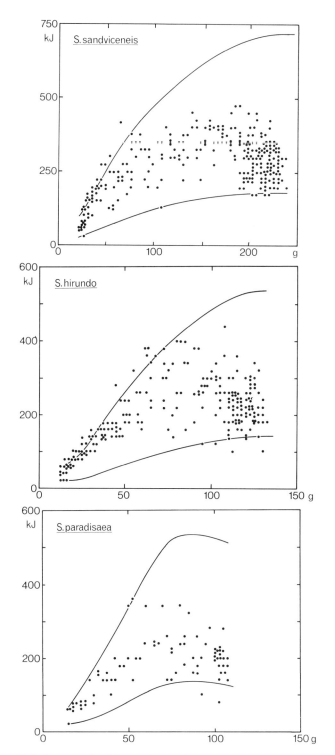

Fig. 5. Chick metabolism for three tern species: ME (kJ/day) for captive chicks in relation to body mass (g). Empirical level of BMR_c (34, 29, 20 readings resp.) shown, and four times this level. From Klaassen and Zwaan in prep.

studied by Norton (1973). In all cases approximately half of the total energy metabolised by the chick throughout development is expended in basal metabolism (Table 1). Any change in BMR_c will thus have considerable repercussions on the total energy budget. If we now return to the problem posed by the sooty tern studied by Ricklefs and White (1981), we recall that considering growth around the point of inflection (20 days) these authors concluded that a doubling of the growth rate would entail a daily energy increment of 42 kJ, $1/3$ above the normal value at that age (and 20 % above the peak which falls at the end of the nestling period). With our new information we will give a reconstruction. Since at the point of inflection the Gompertz growth constant (K_G) is proportional to absolute daily growth gain (Ricklefs, 1968, 1973) we can take the relative change in BMR_c associated with a 100 % change in the K_G in relation to expectation as a first approximation of the alteration in BMR_c associated with a doubling in growth rate. From Fig. 2, we learn that this involves a 44 % increase in BMR_c for gulls and terns. From the data of Ricklefs and White (1981) the BMR_c of chicks at 20 days can be computed as 79 kJ/day hence in our reasoning BMR_c would be raised to (1.44 x 79 =) 114 kJ/day. The total expected increment in daily energy cost (ME) is thus in our view 41.6 kJ (tissue added + cost of synthesis, Ricklefs and White's figure) plus 34.9 kJ for the adjustment of BMR_c for a total 76½ kJ. This represents an increase of 63 % over the observed value, and exceeds the maximal ME by 46 %. Under normal conditions the sooty terns studied by Ricklefs and White experienced a growth rate of 4 g/day at this age (in keeping with growth under favourable conditions in Ashmole's 1963 study) hence a doubling of growth rate entails a daily increment of 4 g. The additional cost of these 4 g according to our calculations was 76 kJ, or 19 kJ/g, close to the value expected for growth at inflection on the basis of an inter-species comparison (24 kJ/g, Drent and Daan, 1980). The reality of BMR_c adjustment coupled with growth rate should preferably be examined by analysing individual differences in performance and we intend to pursue this line with chicks raised in captivity. Comparing populations of the same species under widely differing conditions offers another promising avenue.

Table 1. Energy budgets of charadriiform chicks.

Species	Fledging age (days)	Asymptote (g)	ME (kJ)	BMR_c (%)	E_{tis} (%)
Sterna sandvicensis	25	225	6017	53	32
Sterna hirundo	21	130	3572	48	31
Sterna paradisaea	21	110	3097	61	23
Haematopus ostralegus	31	466	10197	50	19
Calidris alpina[1]	21	44	1357	54	19

ME = metabolisable energy over entire nestling period
BMR_c = basal metabolic rate over entire nestling period (%ME)
E_{tis} = energy content of tissue deposited throughout growth (%ME)
[1] calculated from Norton, 1973, Ecological energetics of calidridine sandpipers breeding in Northern Alaska, Unpubl. diss., Univ. Alaska (Fairbanks).

Considerable interest has been generated recently by considering metabolic rate in birds in relation to latitude, in both adults (Weathers, 1979; Hails, 1983; Ellis, 1984; Gabrielsen et al., 1988) and hatchlings (Ricklefs, 1976; Bryant and Hails, 1983; Taylor, 1985; Klaassen et al., this volume). We do not claim that these differences are solely due to adaptive change of growth rates, but suspect that such changes provide a partial explanation. Indeed, growth rate deviations in relation to latitude reveal a statistically significant relationship for gulls and terns ($r=0.66$, $p<0.01$, $n=15$) and procellariiforms ($r=0.96$, $p<0.01$, $n=6$). One of the adaptations for brooding at high latitudes is thus an increase in growth rate as commented upon by Mehlum et al. (1987) with regard to procellariiform birds. We have argued that an increased growth rate is only possible with concomitant increase in metabolic rate, reflecting a relative increase in the capacity of the organism to absorb and transform nutrients. The enhanced metabolic rate of high latitude breeders also extends their metabolic scope, which may be vital in surviving periods of climatic vicissitudes.

ACKNOWLEDGMENTS

Financial assistance was received from the Dutch Foundation for Arctic Research in the Natural Sciences, and we thank the University of Trondheim and the Norwegian Polar Research Institute for many courtesies.

LITERATURE CITED

[1]Ackerman, R. A., Whittow, G. C., Paganelli, C. V., and Pettit, T. N., 1980, Oxygen consumption, gas exchange and growth of embryonic wedge-tailed shearwaters (Puffinus pacificus chlororhynchus), Physiol. Zool., 53:210-221.

Ashmole, N. P., 1963, The biology of the wideawake or sooty tern Sterna fuscata on Ascension Island, Ibis, 103b:297-364.

[2]Bech, C., Brent, R., Pederson, P. F., Rasmussen, J. G., and Johansen, K., 1982, Temperature regulation in chicks of the manx shearwater(Puffinus puffinus),Ornis Scand., 13:206-210.

[3]Bech, C., Martini, S., Brent, R., and Rasmussen, J., 1984, Thermoregulation in newly hatched black-legged kittiwakes, Condor, 86:339-341.

[4]Bech, C., Mehlum, F., and Haftorn, S., 1988, Development of chicks during extreme cold conditions: the antarctic petrel (Thallassoica antarctica), Proc. Ornithol. Congr. 1986.

Bryant, D. M., and Hails, C. J., 1983, Energetics and growth patterns of three tropical bird species, Auk, 100:425-439.

Daan, S., Masman, D., Strijkstra, A., and Verhulst, S., 1989, Intraspecific allometry of basal metabolic rate: relations with body size, temperature, composition and circadian phase in the kestrel, Falco tinnunculus:, J. Biol. Rhythms., (in press).

[5]Dawson, W. R., Bennett, A. F., and Hudson, W. J., 1976, Metabolism and thermoregulation in hatchling ring-billed gulls, Condor, 78:49-60.

[6]Dawson, W. R., and Bennet, A. F., 1980, metabolism and thermoregulation in hatchling western gulls (Larus occidentalis livens), Condor, 82:103-105.

[7]Dawson, W. R., Hudson, J. W., and Hill, R. W., 1972, Temperature

regulation in newly hatched laughing gulls (Larus atricilla), Condor, 74:177-184.

Dorward, D. F., 1963, The fairy tern Gygis alba on Ascension Island, Ibis, 103b:365-378.

[8]Drent, R. H., 1970, Functional aspects of incubation in the herring gull (Larus argentatus Pont.), Behaviour suppl., 17:1-132.

Drent, R. H., 1975, Incubation, Pp. 333-420, in: "Avian Biology," Vol. V, D. S. Farner, and J. R. King, eds., Academic Press, New York.

Drent, R. H., and Daan, S., 1980, The prudent parent: energetic adjustments in avian breeding, Ardea, 68:225-252.

Ellis, H. I., 1984, Energetics of free ranging seabirds, Pp 203-234, in: "Seabird Energetics," G. C. Whittow, and H. Rahn, eds., Plenum Press, New York and London.

Gabrielsen, G. W., Mehlum, F., and Karlsen, H. E., 1988, Thermoregulation in four species of arctic seabirds, J. Comp. Physiol. B, 157:703-708.

Glutz von Blotzheim, U. N., and Bauer, K. M., 1982, Handbuch der Vögel Mitteleuropas, 8:I:1-699, Akad. Verlag, Wiesbaden.

Hails, C. J., 1983, The metabolic rate of tropical birds, Condor 85:61-65.

Hume, I. D., 1982, Digestive Physiology and Nutrition of Marsupials, Cambridge Univ. Press.

Kirkwood, J. K., 1983, A limit to metabolisable energy intake in mammals and birds, Comp. Biochem. Physiol., 75A:1-3.

[9]Klaassen, M., Slagsvold, G., and Bech, C., 1987, Metabolic rate and thermostability in relation to availability of yolk in hatchlings of black-legged kittiwake and domestic chicken, Auk, 104:787-789.

Lack, D., 1968, Ecological Adaptations for Breeding in Birds, Methuen, London.

Langham, N. P., 1983, growth strategies in marine terns, Stud. Avian Biol., 8:73-83.

Mehlum, F., Bech, C., and Haftorn, S., 1987, Breeding ecology of the antarctic petrel Thalassoica antarctica in Mühlighofmannfjella, Dronning Maud Land, Proc. NIPR Symp. Polar Biol., 1:161-165.

Nelson, J. B., 1978, The Sulidae, Gannets and Boobies, Oxford Univ. Press.

O'Connor, R. J., 1984, The Growth and Development of Birds, John Wiley, New York.

[10]Palokangas, R., and Hissa, R., 1971, Thermoregulation in young black-headed gulls (Larus ridibundus L.), Comp. Biochem. Physiol., 38A:743-750.

[11]Pettit, T. N., Grant, G. S., Whittow, G. C., Rahn, H., and Paganelli, C. V., 1981, Respiratory gas exchange and growth of white tern embryos, Condor, 83:355-361.

[12]Pettit, T. N., Grant, G. S., Whittow, G. C., Rahn, H., and Paganelli, C. V., 1982, Embryonic oxygen consumption and growth of the laysan and black-footed albatross, Am. Journ. Physiol. 242:R121-R128.

[13]Pettit, T. N., Grant, G. S., Whittow, G. C., Rahn, H., and Paganelli, C. V., 1982, Respiratory gas exchange and growth of bonin petrel embryos, Physiol. Zool., 55:162-170.

[14]Pettit, T. N., and Whittow, G. C., 1983, Embryonic respiration and growth in two species of noddy terns, Physiol. Zool., 56:455-464.

Pettit, T. N., Whittow, G. C., and Ellis, H. I., 1984, Food and energy requirements of seabirds at french frigate shoals, Hawaii, Pp 265-282, in: Proc. 2nd Symp. Res. Inv., R. W. Grigg, and K. Y. Tanoue, eds., Univ. Hawaii, Honolulu.

Ricklefs, R. E., 1968, Patterns of growth in birds, Ibis, 110:419-451.
Ricklefs, R. E., 1973, Patterns of growth in birds II. Growth rate and mode of development, Ibis, 115:177-201.
Ricklefs, R. E., 1983, Avian Post-Natal Development, Pp 1-83, in: "Avian Biology," Vol. VII, D. S. Farner, and J. R. King, eds., Academic Press, New York.
Ricklefs, R. E., 1987, Comparative analysis of avian embryonic growth, J. Exp. Zool. Suppl., 1:309-323.
[15]Ricklefs, R. E., and White, S. C., 1981, Growth and energetics of chicks of the sooty tern (Sterna fuscata) and common tern (S. hirundo), Auk, 98:361-378.
Schreiber, E. A., and Schreiber, R. W., 1980, Breeding biology of laughing gulls in Florida, J. Field Orn. 51:340-355.
Spaans, A. L., 1971, On the feeding ecology of the herring gull in the northern part of the Netherlands, Ardea, 59:73-188.
Taylor, J. R. E., 1985, Ontogeny of thermoregulation and energy metabolism in pygoscelid penguin chicks, J. Comp. Physiol., 155B:615-627.
[16]Visser and Beintema, 1989, Energetics of growth in waders, Ardea, 77:(in press).
Weathers, W. W., 1979, Climatic adaptations in avian standard metabolic rate, Oecologia (Berlin) 42:81-89.

APPENDIX

	lat. °N,°S	K_G	A g	BMR_G ml/day	mass g	authors
Puffinus puffinus	62	.069	650	1280	31.0	R, 2
Pterodroma hypoleuca	28	.054	250	757	31.8	P, 13
Puffinus pacificus	21	.043	390	840	39.1	P, 1
Thalassoica antarctica	72	.071	735	2343	63.2	4
Diomedea immutabilis	28	.016	3200	3820	208.0	R, 12
Diomedea nigripes	28	.022	3200	3680	215.0	R, 12
Philomachus pugnax	53	.092	125	480	13.4	R, 16
Tringa totanus	53	.070	137	441	14.2	16
Vanellus vanellus	53	.054	236	449	17.1	16
Limosa limosa	53	.085	273	873	27.8	16
Haematopus ostralegus	54	.081	466	1392	30.9	K
Numenius arquata	53	.051	990	1603	55.2	16
Sterna paradisaea	79	.230	105	547	12.0	R, K
Sterna paradisaea	54	.196	110	457	13.0	R, K
Sterna hirundo	54	.204	130	537	14.9	R, K
Gygis alba	28	.064	115	470	16.1	D, 11
Anous minutus	23	.152	117	524	16.8	L, 14
Sterna fuscata	25	.073	205	529	21.0	R, 15
Sterna sandvicensis	54	.175	180	763	25.3	R, K
Larus ridibundus	61	.115	281	1460	26.8	G, 10
Larus atricilla	29	.088	350	1356	28.4	S, 7
Anous stolidus	23	.103	190	690	28.5	R, 11
Rissa tridactyla	79	.097	449	1123	33.1	K, 3,9
Larus delawarensis	45	.099	403	1345	34.6	G, 5
Larus argentatus	54	.083	1150	1901	57.4	Sp, 8
Larus glaucescens	49	.094	927	1766	60.3	W, 8
Larus occidentalis livens	29	.080	900	2119	65.4	R, 6

Numbers (BMR_G, mlO_2/day): consult literature cited. Growth data: D = Dorward 1963, G = Glutz 1982, K = Klaassen (et al.) in prep., L = Langham 1983, P=Pettit 1984, R = Ricklefs 1973, S = Schreiber 1980, Sp = Spaans 1971, W = Ward pers. comm.

STRATEGIES OF HOMEOTHERMY IN EIDER DUCKLINGS (SOMATERIA MOLLISSIMA)

Johan B. Steen, Hans Grav*, Berit Borch-Iohnsen* and Geir W. Gabrielsen**

Dept. of Gen. Physiol., Univ of Oslo, Box 1051, Blindern 0316 Oslo 3, Norway
* Dep. Nutrition, Univ. of Oslo, Box 1046, 0316 Oslo 3 Norway
** Norwegian Polar Research Inst. 1330 Oslo Lufthavn, Norway

INTRODUCTION

While most nest-dwelling birds depend on parental brooding to maintain normal body temperature(T_B) during the first days of life (Riclefs, 1974), nest-fleers appear to rely on endogenous heat (Koskimies and Lahti,1964). We have studied eider ducklings breeding near the Research Station of The Norwegian Polar Institute at Ny Ålesund, Svalbard ($79°55'$). The young eiders leave the nest once they have dried at the age of 6-8 hours and from then on spend most of their time on the ice-strewn water.

In this study we have described temperature regulation of 1 day old ducklings in terms of their resting V_{O_2}, lower critical temperature (LCT) and the maximum V_{O_2}. The development of homeothermy was studied by recording V_{O_2} and T_B in 0 to 1 day old chicks exposed to a standard cold stress. In addition, we have studied the fine structure and measured the oxydative capacity of muscle and liver tissue from chicks up to one day old.

MATERIAL AND METHODS

This study was carried out during the first half of July in 1985 and 1986. Mean ambient temperature (T_A) was $6°C$. There were occasional sunny days (Tmax=$15°C$), but also spells of high winds and snow. Pipped eggs were collected from nearby islands and hatched in an incubator at $36.5°C$. The process of hatching usually takes about 12 hours. The first visual sign of the upcoming event is the pipping when the embryo pecks the first hole in the eggshell. Usually it proceeds to peck a circular slit around the blunt end of the egg before it finally pushes the lid open. Once freed from the shell, it is no longer called an embryo, but instead a chick, or, in the case of a duck's offspring, a duckling. As they hatched, the ducklings were marked and weighed and kept without food and water in the incubator for at most 48 hours. Remaining chicks were released and immediately adopted by wild eider females.

Oxygen consumption of ducklings placed in a TV-surveyed plexiglass box was measured by analysing the gas before and after it was sucked through the system. T_A was measured by a thermocouple inside the box and T_B by a thermocouple carefully positioned 2 cm into the cloacum prior to the experiment, or 2-3 cm down the oesophagus at the end of the experiment. V_{O_2} at STPD, based on values for gas flow and composition obtained when the chick was physically passive, was calculated according to conventional procedures.

Oxygen consumption of pipped eggs was measured by a manometric method (Scholander 1949). In this case T_B was obtained via a thermocouple threaded through the hole in the shell and down between the neck and belly of the embryo.

EMG was recorded using a 3-prong electrode fastened with tape to the down-plucked thigh muscle.

Further details on the experimental procedures are found in Steen and Gabrielsen 1986 and 1988.

The oxydative capacity of tissues was obtained by measuring cytochrome c oxidase activity according to Aulie and Grav (1979). Samples of thigh muscles and liver were removed from freshly killed birds and immediately frozen in liquid N_2. Samples for EM studies and for histochemistry were taken from m. vastus lateralis in the thigh. Details of these procedures are described in Grav et. al.(submitted).

Temperature regulation in 1 day old ducklings

Ducklings 1 to 2 days of age were exposed to T_A between 0 and $35^0 C$ for at least 1 hour while we measured V_{O_2} and T_B (Steen and Gabrielsen 1986). We also tested them while they were floating on ice water. Figure 1 shows that they maintained normal body temperature at all T_A tested. V_{O_2} was constant down to $23^0 C$. With further lowering of the temperature, it increased linearly. Extrapolation to $V_{O_2} 0$ gave an x-value of $41^0 C$ which is very close to the T_B actually measured. The thermal conductance was calculated to be 0.45 and 0.66 cal. g^{-10}, hr^-1, $^0 C^-1$, for ducklings in air and with feet in water, respectively. We conclude that eider ducklings have a well developed defence system against cold already at the age of one day by a combination of a well insulating down pelt and the ability to increase their heat production by a factor of 4.

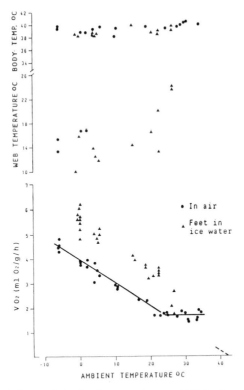

Fig. 1. Oxygen uptake and body temperature of 1-2 day old eider ducklings at varying ambient temperatures with and without feet in water (From: Steen and Gabrielsen 1986).

Where is heat produced?

The body composition of the ducklings is a guide to where the main sources of cold induced heat might be. Three 1 day old eider ducklings with average live weight = 66.8 gram were dissected (Table 1).

Tab.1. Body composition of 3 day-old eider ducklings. Average live body weight = 66.8 gram.

Organ	Weight of organ (gram)	Organ wt in % of total
Pelt w. skin	9.5	14.3
Wings	1.3	2.0
Feet w. tarsi	5.4	8.1
Head, neck, skeleton	22.0	33.0
Liver	2.4	3.6
Heart	0.5	0.8
Dissectable fat	2.8	4.1
Pectoralis	0.6	0.9
Yolk	5.9	8.9
Fluids	4.6	6.8
Leg muscles	6.7	9.8

From: Steen and Gabrielsen (1986)

The fat was white and showed none of the features typical for brown adipose tissue. The two most likely sources of heat production are the liver and the leg muscles.

To investigate this assumption more clcosely, one day old ducklings were exposed to $-20°C$ for 20 min. Immediately after being decapitated a thermocouple mounted in a hypodermic needle was inserted into the desired organs. Thigh temperatures averaged $37.7°C$, liver temperatures $35.7°C$. But the area between the yolk sac and the spine consistently read the highest temperatures. Similar results were reported by Hohtala at the Loen workshop (see his paper in this volume). The explanation appears to be associated with the fact that birds have a portal circulation to the kidneys (Akester 1971). Thereby heated blood from the legs is brought to the kidneys where it is further heated by the normal kidney metabolism.

To check if ducklings had in fact developed shivering at hatching we recorded EMG in 1 hr old ducklings exposed on 12 and $27°C$. In all cases the mV reading at $12°C$ was 2-7 times higher than at $27°C$. There did not appear to be any major differences in response pattern between 1 hour and 24 hour old birds.

If the leg muscles are the main source of cold induced thermogenesis, then the ducklings should be unable to maintain normal T_B during cold if the legs were put out of function. Pilot experiments to this end were conducted by temporarily tying off both legs as close to the body as possible by a tourniquet while we measured T_B. Such ducklings were unable to maintain normal T_B both when exposed to $2°C$ (Fig.2) and $15°C$. Thus temperature induced heat production in the legs seems indispensable for these downy ducklings.

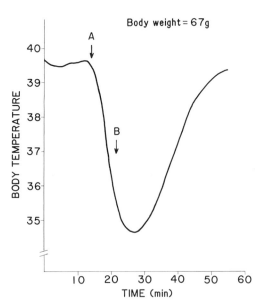

Fig. 2. Body temperature of a one day old eider duckling exposed to $2°C$ before, during and after a tourniquet was applied around both thighs.

The time course of the developement of homeothermy

To describe the time course of development of homeothermy we did measurements of VO_2 and T_B at varying ambient temperatures on eggs at the stage of pipping and on ducklings up to the age of 1 day.

Embryoes at pipping rapidly loose heat when exposed to cold. All our measurements of oxygen uptake were therfore obtained from hypothermic embryos. Still, the values were considerably higher - up to 2X - than values obtained at thermoneutrality (Fig.2). This demonstrates that already at this age is the duckling-to-be capable of considerable temperature induced thermogenesis.

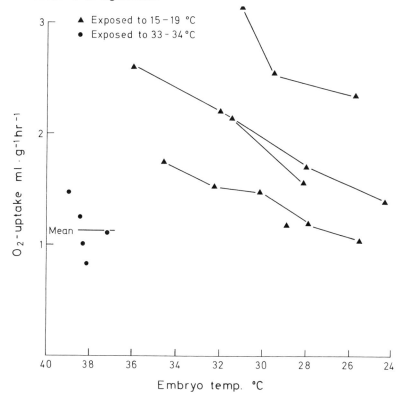

Fig.3. Oxygen consumption at 33-34°C (circles) and 15°C (triangles) of eider embryoes at the stage of pipping. Each curve describes results from one egg. The absissa gives the body temperature at each measurement. (From: Steen and Gabrielsen 1988).

At hatching the duckling is wet. When exposed to cold it rapidly looses body heat which makes it difficult to evaluate its temperature induced thermogenetic capacity. To overcome this, we dried them by means of a toothbrush and a hair drier. This procedure took 5 min after which the chick stayed in the incubator for at least 30 min to regain the heat lost by evaporation.

We also studied 24-hr old ducklings, wetted with detergent and water to match the weight loss of about 2.5 g measured during drying.

Figure 4 shows data obtained on ducklings from hatching to the age of one day. During the first post-hatching "wet" hours, the ducklings were unable to maintain normal T_B, despite efforts to increase the oxygen uptake. However, if artificially dried so that the down pelt becomes fluffy and offers improved insulation against heat loss, they maintain normal T_B for at least one hour at $0°C$. It is remarkable that at $T_A = 0°C$ the VO_2 of the 1-3 hour old ducklings was only half as big as for the 24 hr old ones. Still they are both normothermic at the same cold stress. This can only be so if the conductance is lower in newly hatched than in day old birds. And since the pelt most likely is much the same at both ages, we are led to the conclusion that the newly hatched and artifi-

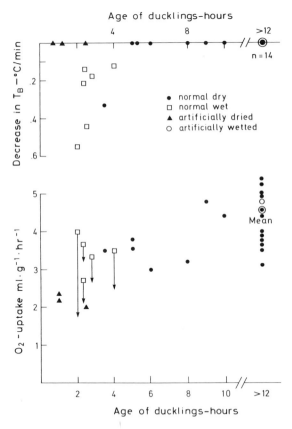

Fig.4. Thermoregulatory response in eider ducklings from hatching to 1 day of age when exposed to ambient temperature of $2°C$. Arrows indicate the decrease in in VO_2 over a 30 min period. Each point represents measurements on one animal. Filled circles-normally dried. Filled triangles-artificially dried. Open squares-normal wet. Open circles-artificially wetted (From: Steen and Gabrielsen 1988).

cially dried maintain normothermy by a combination of a moderate increase in heat production and a decrease in heat loss due to developement of a cold periphery.

The VO_2 of 1 hr old ducklings at thermonutrality was 1.2 ml $O_2 \cdot hr^{-1} \cdot g^{-1}$ as compared to 1.7 in day old ducklings. VO_2 of 1 hour old and of 1 day old ducklings is compared in fig.4.

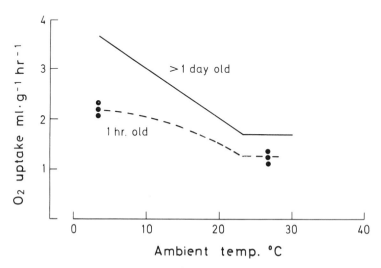

Fig.5. Metabolic respons to $T_A = 2°C$ of eider ducklings 1 day old and naturally dried (upper curve=fig 1), compared to that of 1 hour old and artificially dried. The dotted line is curved to indicate that the 1 hour olds presumably have the same down-insulation (= same lower critical temperature as 1 day olds), but develop a progressively thicker cold periphery with increasing cold stress (From: Steen and Gabrielsen 1988).

The development of oxydative capacity of liver and leg muscle

EM pictures of longitudinal sections of m. vastus lateralis revealed a dramatic developement from the embryonic stage to the duckling was 1 day old (Grav et al submitted). Embryonic fibers had few mitichondria in both sarcolemmar and intrafibrillar locations. At this stage the mitochondria consistently contained electron lucid spaces and had few cristae. Just after hatching, the mitochondria had become more numerous at both locations. The mitochondrial profiles were more elongated and the packing of cristae membranes was more dense. One day old ducklings showed a further increase in denseness of cristae membrane packing, probably indicative of an enlargement of the total cristal area per mitochondrion. Numerous lipid droplets were evident at all stages consistent with the predominantly aerobic fiber type.

These results suggest an increase in metabolic capacity with age of the leg muscles. The dramatic increase in the density of intrafibrillar as well as subsarcolemmar mitochondria also suggest that both shivering and non-shivering-thermogensis may be operating in these muscles during cold stress.

The developement of oxydative capacity in these muscles were studied histochemically by incubating sections of muscles from ducklings at different stages for NADH-tetrazolium reductase. Enzyme activity showed the most pronounced increase from the embryonic to the hatchling stage followed by a slower, further enhancement in activity as the duckling approached 2 days of age.

The oxydative capacity of liver and leg muscles was also investigated biochemically by an amperimetrical method (Aulie and Grav 1979) (Tab.2).

Table 2. Tissue oxidative capacity of liver and thigh muscle in the eider duckling during transition from embryonal to the hatchling stage. N = number of individuals studied.

Develop.stage	Oxyd.capacity ($mlO_2 \cdot h^{-1} \cdot g$ wet tissue^{-1})			
	Liver	N	Tigh muscle	N
Embryo	43	6	18	4
Just hatched	46	4	67	8
12 hr old	54	6	57	8
24 hr old	54	4	69	7
48 hr old	53	2	79	6
Adult	47	4	14	12

(From: Grav et al. 1989)

Conslusions

The striking finding in the studies reported here, is the coincident development of whole body thermogenesis, thigh muscle respiratory chain enzyme activity, numbers of thigh muscle mitochondria and total surface area of cristae membranes, all events occurring within a few days in connection with hatching. Once the duckling is dry from embryonic fluids, it is able to maintain normal adult body temperature while swimming in ice-water at an ambient temperature near zero. This is achieved by a combination of a well-insulating down pelt (lower critical temperature = $23°C$) and the potential to increase heat production by a factor of 4. Similar mechanisms have been demonstrated on the long-tailed duck from Svalbard (Steen and Gabrielsen 1986).

The leg muscles and the liver make out 9.8 and 3.6% respectively of the live weigt of eider hatchlings. Together they make out about one half of the metabolising tissue of the animal (Tab.1). The oxydative capacity of these organs can be obtained by combining the organ weight with their oxydative capacity. These come out to be 460 $mlO_2.hr^-1$ for the leg muscles and 120 for the liver of a 60 gr bird. A comparison to the maximum measured VO_2 of 360 $mlO_2.hr^-1$ shows that these tissues alone have more than enough capacity to cover the demands of the entire animal.

The values obtained for oxydativ capacity of the leg muscles of eider ducklings correspond closely to the values found for the liver of cold acclimated bantam chicks (Aulie and Grav 1979) and on muscovy ducklings (Barre et al 1987). However, they are only 1/3 of the values found for brown adipose tissue of cold acclimated rats (Barre et al 1987).

The newly hatched eider ducklings obviously shiver during cold exposure (Steen and Gabrielsen 1988). Cold-acclimated muscovy ducklings combat cold by a combination of shivering and non-shivering thermogenesis (Barre et. al 1985); the latter possibly operating by free fatty acid loose-coupling as proposed for mice (Skulachev 1963), seal pups (Grav and Blix 1979) and muscovy ducklings (Barre et al 1986). The high concentration of subsarcolemmar mitochondria (Grav et al summitted) in addition to the intracellular ones, may be taken as suggestive evidence for the presence of non-shivering thermogenes in eider ducklings.

References

Akester, A. R. 1971. The Blood vascular system. In: "Physiology and biochemistry of the domestic fowl". O. J. Bell and B. M. Freman, eds. Acad. Press N. Y. , London

Aulie, A. and Grav H. J. 1979. Effect of cold acclimation on oxidative capacity of skeletal muscles and liver in young bantam chicks. Comp. Biochem. Physiol.62A: 335-338.

Barre, H., Geloen, A., Chatonnet, J., Dittmar, A. ans Rouanet, J. L. 1985. Potentiated muscular thermogenesis in cold-acclimated muscovy ducklings. Am. J. Physiol. 249: (Reg. Int. Comp. Physiol. 18:) R533-R538.

Barre, H., Nedergaard, J. and Cannon, B. 1986. Increased respiration in skeletal muscle mitochondria from cold-acclimated muscovy ducklings: Uncoupling effects of free fatty acids. Comp. Biochem. Physiol. 85B: 343-348.

Barre, H., Bailly, L. and Rouanet, J. L. 1987. Increased oxidative capacity in skeletal muscles from cold-acclimated ducklings: Comparison with rats. Comp. Biochem. Physiol. 88B: 519-522.

Grav, H. J. and Blix, A. S. 1979. A source of nonshivering thermogenesis in fur seal skeletal muscle. Science 204: 87-89.

Grav, H. J., Borch-Ionsen, B., Dahl, H. A., Gabrielsen, G. W. and Steen, J. B. (submitted J. Comp. Physiol. B).

Koskimes, J. and Lathi, L. 1964. Cold-hardiness of the newly hatched young in relation to ecology and distribution in ten species of European ducks. Auck. 81: 281-307.

Ricklefs, R. E. 1974. Energetics of reproduction in birds. In: "Avian energetics". R. A. Paynter, ed. Nuthall Orith. Club. Publ. 15.

Scholander, P. F. 1949. Volumetric respirometer for aquatic animals. Rev. Scient. Instr. 20: 885-887.

Skulachev, V. P., Maslov, S. P., Sivkova, V. G., Kalinichenko, L. P. and Maslova, G. M. 1963. Uncoupling of oxidation from phosphorylation in muscles of cold-adapted white mice. Biochemistry (Biokhimya) 28: 54-60.

Steen, J. B. and Gabrielsen, G. W. 1986. Thermogenesis in newly hatched eider (Somateria mollissima) and long-tailed (Clangula hyemalis) ducklings and barnacle (Branta leucopsis) goslings. Polar Res. 4: 181-188.

Steen, J. B. and Gabrielsen, G. W. 1988. The development of homeothermy in common eider ducklings (Somateria mollissima), Acta Physiol. Scand. 132: 557-561.

BODY TEMPERATURES UNDER NATURAL CONDITIONS

OF TWO SPECIES OF ALCID CHICKS

Dag Vongraven, Frode J. Aarvik and Claus Bech

Department of Zoology
University of Trondheim - AVH
N-7055 Dragvoll, Norway

INTRODUCTION

Body temperatures of birds have been recorded using a number of different techniques. Most often these have been types of "grab and jab"-methods that cause stress and disturb the birds to an extent inconsistent with the objectives of unbiased biological monitoring procedures. The use of telemetric methods has made it possible to monitor physiological parametres with minimal interference to the birds being studied. This method has now been used to record the natural variations in the body temperature of chicks of two species of alcids, the puffin Fratercula arctica and the Brünnichs guillemot Uria lomvia. They are both arctic, semi-precocial species, which could be expected to show some thermoregulatory adaptations to life in a cold climate.

MATERIAL AND METHODS

The puffin chicks were studied on the island of Sklinna (65°12'N, 11°00'E), off the coast of central Norway, those of Brünnichs guillemot in the vicinity of Ny-Ålesund (79°15'N, 12°00'W), on Spitsbergen, in July 1984 and in July 1985, respectively. The radio transmitters (type L-M and T-M, Mini-Mitter Co., Oregon) were implanted in the intraperitoneal cavity under general anaesthesia and left implanted for 3 to 5 days. The radio signals were received by a walkie-talkie and recorded manually using a stopwatch, either immediately or later on after being initially recorded on tape. A total of 7 puffin chicks and 6 Brünnichs guillemot chicks were monitored during these studies. The ages of the chicks were estimated using published growth curves (Ashcroft, 1979; Nettleship and Birkhead, 1985).

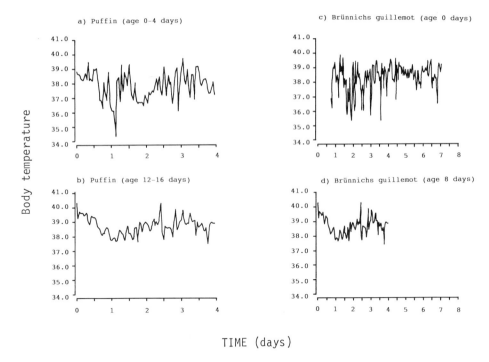

Fig. 1 Normal body temperatures of free-living alcid chicks (the ages shown are the estimated ages at the start of the recording period).

RESULTS

Fig. 1 shows the variation in body temperatures recorded under natural conditions of chicks of different ages of the two alcid species studied.

Chicks of both species had subadult body temperatures. Guillemot chicks achieve the ability to maintain a high and stable body temperature at an age of 5-6 days (own unpublished results). The highest recorded stable body temperature for chicks of both the puffin and Brünnichs guillemot was ca. 38.5°C. The oldest chicks studied were about 20 and 14 days old, respectively. The body temperatures of adults of these two species have been found to be respectively 40.1°C (for incubating individuals; Barrett, 1984) and 39.9°C (one individual in the laboratory; own results).

DISCUSSION

The most striking feature of both species is the great range in the body temperatures of all the individual chicks studied. Even chicks at an age they could have been expected to have a fully-developed capacity for thermoregulation, such wide fluctuations were quite regularly recorded. Changes in body temperature of 2-3°C within an hour were regularly recorded throughout the day for both species and changes in body temperature of a magnitude of 9°C in 12 hours and of 7°C in 2 hours were also quite frequently noted. Marked falls in body temperature were associated with feeding periods, bad weather conditions and the physical condition of the chicks (Vongraven et al., 1987; Vongraven and Aarvik, in prep.).

CONCLUSION

The chicks of both these arctic bird species spend much of their time in a hypothermic state, even after homeothermy is fully developed. This could be regarded as being in part a passive cooling effect of feeding. However, it can also be considered as a regulated condition. Such an adaptation would serve to reduce the chicks' energy demands, a very useful feature considering their thermal environment and the changes experienced in the availability and the energy content of their food supply.

REFERENCES

Ashcroft, R. E., 1979, Survival rates and breeding biology of Puffins on Skomer Island, Wales, Ornis Scand., 10:100-110.

Barrett, R. T., 1984, Adult body temperatures and the development of endothermy in the Puffin Fratercula arctica, Razorbill Alca torda and Guillemot Uria aalge, Fauna norv. Ser. C, Cinclus, 7:119-123.

Nettleship, D. N., and Birkhead, T. R., (eds.), 1985, The Atlantic Alcidae, Academic Press, London, 574 pp.

Vongraven, D., Aarvik, F. J., and Bech, C., 1987, Body temperature of Puffin Fratercula arctica chicks. Ornis Scand., 18:163-167.

Vongraven, D., and Aarvik, F. J., In prep., Body temperature of free-living chicks of Brünnichs guillemot Uria lomvia.

PARTICIPANTS

Arnfinn Aulie
Veterinary College of Norway
Department of Physiology
Box 8146, Dep. 0033-Oslo 1
Norway

Martha K. Bakkevig
Department of Zoology
University of Trondheim
N-7055 Dragvoll
Norway

Hervé Barré
Univ. Claude Bernard (CNRS)
Lab. Therm. Metab. Energ.
  (L.A. 181)
8, Avenue Rockefeller
F-69373 Lyon Cedex 08
France

Claus Bech
Department of Zoology
University of Trondheim
N-7055 Dragvoll
Norway

Martin Berger
Westfalisches Museum für
  Naturkunde
Sentruper Str. 285
D-4400 Münster, F.R.G.

Ralph J. Berger
Department of Biology
University of California
Santa Cruz,
CA 95064, U.S.A.

Marvin H. Bernstein
Department of Biology
New Mexico State University
Box 3AF/Las Cruces
New Mexico 88003-0032
U.S.A.

Theresa L. Bucher
Department of Biology
University of California
405 Hilgard Avenue
Los Angeles, CA 90024
U.S.A.

William A. Buttemer
Department of Zoology
University of Washington
Seattle, Washington 98195
U.S.A.

*Barbara Cannon
The Wenner-gren Institute
Dept. of Metabolic Research
University of Stockholm
S-10691 Stockholm
Sweden

Cynthia Carey
Department of EPO Biology
University of Colorado
Boulder, CO 80309
U.S.A.

*William R. Dawson
Div. of Biological Sciences
Museum of Zoology
University of Michigan
Ann Arbor, Michigan 48109-1079
U.S.A.

R. H. Drent
Department of Zoology
University of Groningen
Postbus 14,
NL-9750 Haren AA
The Netherlands

Claude Duchamp
Univ. Claude Bernard (CNRS)
Lab. Thermoreg. Met. energ.
  (L.A. 181)
8, Avenue Rockefeller
F-69373 Lyon Cedex 08
France

Morten Ekker
Department of Zoology
University of Trondheim
N-7055 Dragvoll
Norway

Geir W. Gabrielsen
Norw. Polar Research Institute
P.O. Box 158
N-1330 Oslo Lufthavn
Norway

Rudolf Graf
MPI für Neurologische
Forschung
Ostmerheimer Str. 200
D-5000 Köln 91 (Merheim)
F.R.G.

*H. Graig Heller
Dept. of Biological Sciences
Stanford University
California 94305-2493
U.S.A.

Esa Hohtola
Department of Zoology
University of Oulu
SF-90100 Oulu
Finland

Bjørn M. Jenssen
Department of Zoology
University of Trondheim
N-7055 Dragvoll
Norway

Hege Johannesen
Department of Zoology
University of Trondheim
N-7055 Dragvoll
Norway

Marcel Klaassen
Department of Zoology
University of Groningen
Postbus 14
NL-9750 Haren AA
The Netherlands

John O. Krog
Dept. of General Physiology
University of Oslo
Box 1051, Blindern
N-0316 Oslo 3
Norway

*Yvon Le Maho
Lab. d'Etude des Régulations
  Physiologiques, associé à
  l'Université Louis Pasteur
CNRS, 23 rue Becquerel
F-67087 Strasbourg
France

Richard L. Marsh
Department of Biology
360 Huntington Ave.
Northeastern University
boston, MA 02115
U.S.A.

Svein Martini
Department of Biomedical
  Engineering
University of Trondheim
Eiriks Jarls gt. 10
N-7030 Trondheim
Norway

Dirkjan Masman
Department of Zoology
University of Groningen
Postbus 14
NL-9750 Haren AA
The Netherlands

Brian K. McNab
Department of Zoology
University of Florida
Gainesville
FL 32611, U.S.A.

James B. Mercer
Div. of Arctic Biology
University of Tromsø
P.O. Box 635
N-9001 Tromsø, Norway

*Uffe Midtgård
Institute of Cell Biology
  and Anatomy
University of Copenhagen
Universitetsparken 15
DK-2100 Copenhagen Ø
Denmark

Joachim Ostheim
Fakultät für Biologie
A.G. Temperaturregulation
Ruhr-Universität Bochum
D-4630 Bochum 1
F.R.G.

*Johannes Piiper
MPI für Experimentelle Medizin
Abteilung physiologie
Herman-rein-Strasse 3
D-3400 Göttingen
F.R.G.

Michael E. Rashotte
Department of Psychology
Florida State University
Tallahassee,
FL 32306-1051, U.S.A.

Werner Rautenberg
Fakultät für Biologie
A.G. Temperaturregulation
Ruhr-Universität Bochum
D-4630 Bochum 1
F.R.G.

Randi E. Reinertsen
Department of Zoology
University of Trondheim
N-7055 Dragvoll, Norway

*Robert E. Ricklefs
Department of Biology
University of Pennsylvania
Philadelphia, PA 19104-6018
U.S.A.

Seppo Saarela
Department of Zoology
University of Oulu
SF-90570 Oulu
Finland

*Eckhart Simon
MPI für Physiologische und
 Klinische Forschung
W.G. Kerckhoff-Institut
D-6350 Bad Nauheim
F.R.G.

Guri Slagsvold
Department of Zoology
University of Trondheim
N-7055 Dragvoll
Norway

Sherrie L. Souza
Department of Biology
University of California
Riverside, CA 92521
U.S.A.

Johan B. Steen
Dept. of General Physiology
University of Oslo
Box 1051, Blindern
N-0316 Oslo 3
Norway

Øivind Tøien
Dept. of General Physiology
University of Oslo
Box 1051, Blindern
N-0316 Oslo 3
Norway

Jan E. Østnes
Department of Zoology
University of Trondheim
N-7055 Dragvoll
Norway

* Giving introductory lecture

INDEX

Abdominal air sac, oxygen content, 168-169, 176-177
Acanthis, 289
Acclimatization
  to cold, 85, 86, 141, 151
  enzyme activity, 97-98, 108-109
  metabolic, 83, 87, 89, 91, 118
  seasonal 87, 88, 90, 91, 105, 110, 116, 121, 131-132
Aegolius funereus, 298
Albatross, laysan, 294
Alcidae, 334
Alectoris chukar, 171
Alectura lathami, 333
Alle alle, 137, 139
Altitude (see high altitude)
Anas crecca, 333
  penelope, 333
  platyrhynchos, 88, 163-165, 173, 333
Anastomoses, 215
Anatidae, 333
Anorexia, 296
Anous minutus, 359
  stolidus, 359
Antidiuretic hormone, 5
Apnea, during torpor, 183
Aptenodytes foresteri, 214, 294, 315
  patagonicus, 59, 85
Arginine vasotocin, 5
Arterial-venous P_{O2}-difference, 165
Arteriovenous heat exchange, 211
Assimilated energy, 129
Auk, little, 137, 139
Aythia ferina, 333
  fuligula, 333

Bantam cock, 77
Bantam hen, 77, 217, 296, 305

Basal metabolic rate (BMR), 37, 83-85, 118,
  in chicks, 333-334, 359
  and food habits, 284
  and use of torpor, 286
Beta-hydroxyacyl-CoA-dehydrogenase (HOAD), 96, 97, 98, 109
Beta-oxidation, 90-91, 111-112
Body composition, 363
Body temperature
  of arctic chicks, 371-373
  during incubation, 320
  effect of meal, 345
Bolborynchus lineola, 164
Brain cooling, 215
Brain temperature, 202, 208-210
Branta canadensis, 297
  nigricans, 84
Brant, black, 84
Breeding season, 295
Brood patch
  heat transfer 309-312
  temperature, 217
  vasomotor activity, 311-312
Buchephala clangula, 333
Bunting, snow, 330
  yellow, 84

Cairina muschata, 39, 49
Calidris alpina, 354, 356
Canary, 294
Capercaillie, 115, 334
Caprimulgiformes, 286
Carbohydrate metabolism, 110
Cardinal, 84
Cardinalis, 84
Cardiorespiratory system, and thermoregulation, 228
Cardueline finches, 105
  carduelis, 84
  chloris, 84, 115
  flammea, 86
  pinus, 86

Cardueline finches (continued)
  spinus, 84, 115
  tristis, 84, 95, 106, 121
Carpodacus mexicanus, 84, 95, 107, 121
Catbird, 100
Cepphus grylle, 137, 139, 334
Charadriiformes, 332, 352
Chickadee, black-capped, 289
Chicken, leghorn, 38
Chukar, 171
Circadian cycles of body temperature, 234, 251, 256, 267, 269
  ambient temperature, 247-248
  food deprivation, 250-252
  light intensity, 249-250
Circadian cycles of light regime, 268, 272
Circulatory adjustments
  to cold, 215-220
  to web immersion, 216
Citrate, 109
Citrate synthase, 96, 97, 98, 100, 102
Clangula hyemalis, 147-152
CO_2-receptors, interpulmonary, 157
CO_2-transfer in bird lungs, 159-160
Coccothraustes vespertinus, 84
Cold exposure, and peripheral heterothermia, 287-288
Cold tolerance, 288
Colibri coruscans, 179
Coliidae, 286
Colinus virginianus, 333
Columba livia, 28, 84, 207, 223, 235, 237, 255-263, 266
Columbidae, 286
Conductance, 120, 190-192 (see also insulation)
  in brood patch, 311-312
  in chicks, 332-334
  in Rufus hummingbird, 189
  in seabirds, 139, 148, 150
Cooling (see temperature stimulation)
Coot, european, 164
Corvus caurinus, 84
  cryptoleucus, 197
Cost of activity, 298, 301
Cost of maintenance, 296-297, 300
Cost of thermoregulation
  in chicks, 341-343
  in kestrels, 128-132
Crane, sandhill, 288

Cross current model of gas exchange, 157, 159
Crow, nortwest, 84
Cygnus olor, 84

Daily energy expenditure
  chicks, 344, 356
  emus, 322
  seabirds 137-143
Development of homeothermy, in eiders, 365
Diomedea, 352
  immutabilis, 294, 359
  nigripes, 359
Domestic fowl, 38, 84, 296
Dove, collared turtle, 84
Drepanidae, 286
Dromaius novaehollandiae, 315
Duck, pekin, 9, 17, 25, 27, 163, 164
  muscovy, 39, 49
  long-tailed, 147-152
Dumatella carolinensis, 100
Dunlin, 354

Ectothermy, in chicks, 330
Eider, common, 17, 147, 325-327, 332, 361
Electromyography (EMG), 3, 70, 73, 248, 364
  in m. iliotibialis, 78-80
  in m. pectoralis, 78-80
Emberiza citrinella, 84
Emu, 9, 315
Energy budget, of chicks, 356
Energy conservation, in pigeons, 273, 280
Energy intake, 125
Energy requirement in the cold, 294
Energy substrates
  carbohydrates, 90, 108-109
  fatty acids, 89-90
  ketone bodies, 111
Eudypta c. chrysolophus, 87
Eugenes fulgens, 191
Eulampis jugularis, 191
Evaporation, respiratory, 200

Falco sparverius, 204, 294
  tinnunculus, 123-133, 294, 340, 353
Fasting, 299-300, 353
  effect on nocturnal body temperature, 250-251
  in pigeons, 237, 271
  in kestrels, 130
Fat storage, 299
Feeding strategies
  during cold, 260

Feeding strategies (continued)
  during food scarcity, 259
Fever, 251
Field metabolic rate, 140
Flight, and metabolic rate, 198
Food intake, reduced, 296
Food scarcity, 294-295 (see also fasting)
  and thermoregulation, 256-260
  in pigeon, 280, 276
Fowl, 296
Fratercula arctica, 334, 371
Fulica atra, 164
Fulmar, 139
  giant, 216
  southern giant, 330
Fulmarus glacialis, 137, 139

Galliformes, 332
Gallinago gallinago, 331
Gallus gallus, 77, 84, 217, 296, 305
Gas exchange, in bird lungs, 157-159
Geese, canada, 297
Glucose infusion, 268-271
Glucose turnover, 90
Glycogen, 299
Glycolysis, 90
Goldfinch, american, 84, 95, 106, 121
  european, 84
Greenfinch, 84, 115
Grosbeak, pine, 289
  evening, 84, 121
Grouse, black, 49, 115
Growth, 330, 335, 349-357
  and latitude 357
  and metabolic rate, 351-356
  in tern chicks 345
Grus canadensis, 288
Guillemot, black, 137, 139
  Brünnichs, 137, 139, 371-372
Gull, glaucous, 137, 139
Gull, Ivory, 137, 139
Gygis alba, 359

Haematopus ostralegus, 352, 353, 356, 359
Haldane effect, 159-160
Heart rate
  during incubation, 321
  cold acclimation, 219
Heat increment of feeding
  annual variation, 131-132
  diurnal variation, 131-132
  in kestrels, 127
Heat loss, from birds feet, 213
Hesperiphona vespertina, 121

Hexokinase (HK), 96-102
Hibernation, 2, 232
High altitude, 199-204, 207-210
High latitude adaptation, in chicks, 339
Hirundinidae, 286
Homeothermia, development of, 365-368
House finch, 84, 86, 95, 107, 121
House sparrow 84, 121
Huddling, 332
Hummingbird, broad-tailed, 187 297
  rufous, 187, 188
  West Indian, 191
Hydrobatidae, 333
Hypometabolism, 231
Hypothalamus, 4
Hypothermia, 2, 231-242
  in chicks 331
  and food deprivation, 237, 240, 250
  nocturnal, 236-241
  in pigeons 247, 269, 280
  and season, 239-240
Hypoxia, and flying, 199

Iguana, 12
Incubation, 77, 294, 305-312, 319-323, 325
Insulation (see also conductance)
  in air and water, 150
  in chicks, 332
  vasomotor control, 2
Interclavicular air sac, oxygen content, 168-169, 176-177

Junco, dark-eyed, 88
Junco hyemalis, 88

Kestrel, 123, 204, 294, 340
Ketone bodies, 108, 111-112
Kittiwake, 137, 139, 164, 332
Kiwi, 323

Lagopus lagopus, 296, 332, 333
  mutus, 289
Lanius excubitor, 85
Laridae 334, 351
Larus argentatus, 359
  atricilla, 334, 351, 359
  delawarensis, 334, 359
  glaucescens, 359
  hyperboreus, 137, 139
  occidentalis, 334, 359
  ridibundus 351, 359

Latitude (see also high latitude)
  and growth, 357
  and metabolic rate in chicks, 341
Leg muscle, 96, 97, 98, 99
Leipoa ocellata, 333
Limosa limosa, 359
Liver temperature, 364
Lungs of birds, 153-155
Lyruus tetrix, 115

Macronectes giganteus, 216, 330
Mallard, 88, 173-177
Maximum heat production, 118, 119
  cold exposed chicks, 332-334
Megapodes, 332, 333
Melanitta fusca, 333
Meliphagidae, 286
Mergus merganser, 333
  serrator, 333
Metabolic rate, 1
  in chicks, 333-334, 340, 351, 354, 355
  and clutch size, 308
  effect of meal, 129
  during incubation, 307, 326
  in kestrels, 126-127
  in long-tailed ducks, 149
  maximum in chicks, 333-334
  in seabirds, 139
Microenvironment of chicks, 330
Migration, 289
Mitochondria, 367-368
Molt, 293
Murrelet, xantus's, 334

Nervous control, 1
Neuronal discharge
  and Ca^{++}, 11
  and Mg^{++}, 11
  and spinal temperature, 12
Nitrogen retention, 301
Nonshivering thermogenesis, 2-3, 37, 369
  acclimation to cold, 37-38, 49-52
  in birds, 40, 52, 61
  brown adipose tissue, 2-3, 44, 53
  epinephrine, 63
  glucagon, 41, 52, 62-64
  muscle tissue, 2, 43, 54, 55, 65-66
  norepinephrine, 38, 40, 41, 63-64
  in rats, 41-42
  thermogenin, 45
  visceral tissus, 76

Numenius arquata, 359
  phaeopus, 332
Nyctea scandiaca, 289

Oceanodroma, 352
  leucorhoa, 333
Operative temperature, 132, 342
Ostrich, 323
Owl, barn, 298
  snowy, 289
  Tengmalm's, 298
Oxidative capacity of muscles, 367-369
Oxygen diffusing capacity, 169, 171, 179,
Oxygen extraction
  of bird lungs, 164-171
  during flight, 201
  during torpor, 183, 193
  in mallard 174-175
Oystercatcher, 352

Pachyptila, 332, 335
  desolata, 333
Pagophila eburnea, 137, 139
Panterpe insignis, 191
Panting, 157
Parakeet, linneated, 164
Paridae, 286
Partridge, 70
Parus atricapillus, 289
  caeruleus, 297
  major, 84
  montanus, 239
  palustris, 294
Passer domesticus, 84, 121
Pectoralis muscle, 96, 98, 99
Pelecanoides georgicus, 333
  urinatrix, 333, 334
Pelecanoididae, 333
Penguin, 9
  adelie, 27, 330
  emperor, 214, 294, 315
  king, 41, 59, 85
  macaroni, 87
Perdix perdix, 70
Petrel, 332
  bonin, 294
  common diving, 334
Phalacrocorax atriceps, 330
Phalaenoptilus nuttallii, 86, 285
Phalarope, northern, 332
Phalaropus labatus, 332
Phasianidae, 333
Phasianus colchicus, 70
Pheasant, 69, 70
Philomachus pugnax, 359
Phosphofructokinase (PFK), 96, 98, 108

Photoperiod, effect on T_b in pigeons, 271-272
Picoides pubescens, 289
Pigeon, 9, 28, 12, 40, 84, 207, 223, 235, 237, 255-263, 265-274
Piloerection, 2
Pineal, and photoperiodic input 234
Pinicola enucleator, 289
Pipridae, 286
Plasma osmolality, 5
Plectrophenax, 289
Plectrophenax nivalis, 330
Poorwill, 86, 235, 285
Pre-optic and anterior hypothalamic nuclei (POAH), 232
Prion 332, 335
Procellariiformes, 332, 333, 351, 352
Protein utilization, 300-301, 327
Psophia crepitans, 288
Ptarmigan, 9, 332
  rock, 289
  svalbard, 296
Pterodroma hypoleuca, 294, 359
Puffin, 371-372
Puffinus pacificus, 359
  puffinus 331, 333, 359
Pygoscelis adeliae, 330

Q_{10}, 5-8
  in emu eggs, 318-319
  synaptic, 6-7

Raven, white-necked, 197
Rat, 41
Redpoll, common, 86
Respiratory responses
  during torpor, 181-186
  to cold, 163
  to external cooling, 174
  to flying, 197
  to oesophageal cooling, 175
  to spinal cooling, 224-225
Resting metabolic rate (RMR), 84-85, 138, 141, 325, 326
Rete mirabile, 213-214
Rissa tridactyla, 137, 139, 164, 332, 334, 359
RQ, 269-270, 327

Saline infusion, 268-271
Scolopacidae, 352
Selasphorus platycercus, 187, 297
  rufus, 187, 188

Sericornis frontalis, 85
Serin, 84
Serinus canarius, 84, 294
Shag, blue-eyed, 330
Shearwater, manx, 331
Shivering thermogenesis, 3, 70, 73
  m. iliotibialis, 79
  m. pectoralis, 79
  skeletal muscle, 106
  and ventilation, 224-227
Shrike, Grey, 85
Siskin, 84, 115
  pine, 86
Skua, south polar, 330
Sleep, 231, 233, 296
  in pigeons, 268-270
  Slow wawe sleep (SWS), 236, 268-270
  Rapid eye movement sleep (REM), 236, 268-270
Snipe, 331
Somateria mollissima, 33, 147, 325, 332, 361
Sparrow, white-crowned, 300
Spermophilus lateralis, 296
Spinal cord temperature
  daily cycles in pigeons, 248
  during fasting in pigeons, 238
  during stimulation, 31
Squirrel, ground, 296
Starling, 84, 126
Starvation, incubating eiders, 327
Stercorarius maccrormicki, 330
Sterna fuscata, 349, 359
  hirundo, 349, 354-356, 359
  paradisaea, 335, 339, 354-356, 359
  sandvicensis, 354-356, 359
Streptopelia decaocto, 84
Sturnus vulgaris, 84, 121
Suprachiasmatic nuclei, 233
Surface area, effect on heat balance in chicks, 331
Swan, mute, 84
Synthliboramphus hypoleucus, 334

Telemetry, 345, 371
  use in greenfinch, 116
  use in pigeon, 266, 276
Temperature regulation
  behaviour, 276-280
  in chicks, 330-335
  in ducklings, 362
  instrumental, 2, 3, 275-276
  posture, 2

Temperature stimulation
  body, 17, 224
  hypothalamic, 4, 5, 9, 19
  oesophageal, 18, 25, 175
  spinal cord, 12, 28, 29, 224, 225, 234, 238, 251
Tern, arctic, 339
  common, 344, 349
  sooty, 349
Tetrao urogallus, 115, 333, 334
Tetraonidae, 333
Thalassoica antarctica, 359
Thermal sensitivity (see thermosensitivity)
Thermogenesis
  cold-induced, 297
  energy substrates, 107-112
  skeletal muscle, 106
Thermogenic capacity, 85-86, 118-119
  in chicks, 332-334
Thermogenic endurance, 87, 88, 99, 105, 107, 113
Thermogenin, 45
Thermoneutral zone (TNZ), 1
Thermoregulatory set-point, and entrence into torpor, 232
Thermoregulatory system, interaction with cardio-respiration, 228
Thermosensitivity
  deep body tissues, 2
  hypothalamic, 232-233
  lower brain stem, 2
  oesophagal, 25
  skin, 2, 11
  spinal cord 10, 27, 34, 235, 247, 272
  total body, 17, 24-26
Thermosensors, spinal cord, 2
Thyroxine, 327
Tit, blue, 297
  great, 84
  marsh, 294
  willow, 239
Torpor, 179, 187, 235-239, 289, 297
  and basal metabolic rate, 286
  and body mass, 285-287
  in mammals 232-233
Triiodothyronine, 327
Tringa totanus, 359
Trochilidae, 286
Trumpeter, gray-winged, 288
Tyto alba, 298

Uria lomvia, 137, 139, 334, 371-372

Vanellus vanellus, 359
Vascular system
  legs, 212
  head, 214-215
  wings, 213-214
Vasoactive intestinal polypeptide (VIP), 218, 312
Vasodilatation, cold induced, 218
Vasomotor activity, in brood patch, 311-312
Vasomotor control, 2,
  in brood patch, 217
Venae comitantes system, 214
Ventilation
  of bird lungs, 155-157
  during cold exposure, 166
  during torpor, 182-184, 192
Violeteat, sparkling, 179
Vulture, turkey, 214

Water turnover rate, 327
Whimbrel, 332
Winter fattening, 88-89
Woodpecker, downy, 289
Wren, white-browed scrub, 85

Zonotrichia leucophrys, 300